T0200895

# REINFORCED CONCRETE DESIGN WITH FRP COMPOSITES

# REINFORCED CONCRETE DESIGN WITH FRP COMPOSITES

## Hota V. S. GangaRao
## Narendra Taly
## P. V. Vijay

CRC Press
Taylor & Francis Group
Boca Raton London New York

CRC Press is an imprint of the
Taylor & Francis Group, an **informa** business

CRC Press
Taylor & Francis Group
6000 Broken Sound Parkway NW, Suite 300
Boca Raton, FL 33487-2742

© 2007 by Taylor & Francis Group, LLC
CRC Press is an imprint of Taylor & Francis Group, an Informa business

No claim to original U.S. Government works

ISBN-13: 978-0-8247-5829-5 (hbk)
ISBN-13: 978-0-367-38984-0 (pbk)

This book contains information obtained from authentic and highly regarded sources. Reasonable efforts have been made to publish reliable data and information, but the author and publisher cannot assume responsibility for the validity of all materials or the consequences of their use. The authors and publishers have attempted to trace the copyright holders of all material reproduced in this publication and apologize to copyright holders if permission to publish in this form has not been obtained. If any copyright material has not been acknowledged please write and let us know so we may rectify in any future reprint.

Except as permitted under U.S. Copyright Law, no part of this book may be reprinted, reproduced, transmitted, or utilized in any form by any electronic, mechanical, or other means, now known or hereafter invented, including photocopying, microfilming, and recording, or in any information storage or retrieval system, without written permission from the publishers.

For permission to photocopy or use material electronically from this work, please access www.copyright.com (http://www.copyright.com/) or contact the Copyright Clearance Center, Inc. (CCC), 222 Rosewood Drive, Danvers, MA 01923, 978-750-8400. CCC is a not-for-profit organization that provides licenses and registration for a variety of users. For organizations that have been granted a photocopy license by the CCC, a separate system of payment has been arranged.

**Trademark Notice:** Product or corporate names may be trademarks or registered trademarks, and are used only for identification and explanation without intent to infringe.

| Library of Congress Cataloging-in-Publication Data |
| --- |

GangaRao, Hota V. S.
    Reinforced concrete design with FRP composites / by Hota V.S. GangaRao, Narendra Taly, P.V. Vijay.
        p. cm.
    Includes bibliographical references and index.
    ISBN 0-8247-5829-3 (alk. paper)
    1. Fiber reinforced plastics. 2. Reinforced concrete construction. I. Taly, Narendra. II. Vijay, P. V. III. Title.

TA455.P55G36 2006
624.1'8341--dc22                                                        2006045236

**Visit the Taylor & Francis Web site at**
**http://www.taylorandfrancis.com**

**and the CRC Press Web site at**
**http://www.crcpress.com**

# Dedication

Advancement of Knowledge — whose illumine recedes all (other) actions to embers

## ज्ञानाग्निः सर्वकर्माणि भस्मसात्कुरुते तथा

**Bhagawad Gita 4.37**

ENGLISH PRONUNCIATION

GYaanaagniH sarvakarmaaNi bhasmasaatkurute tathaa

# Acknowledgment

Author P.V. Vijay wishes to thank his parents, B.S. VenkataRayappa and N. Anjana for their constant encouragement, and his wife Sucharitha and children Aishwarya and Arjun for their patience, support, and sacrifice during the writing of this book. Sincere thanks are also due to several professionals in this field for their valuable suggestions, and to my M.S./Ph.D. graduate students at West Virginia University, where many of the topics in this book have been taught as course materials for the last six years.

# Preface

Fiber reinforced polymer (FRP) composites (the combination of two or more materials) have emerged as an evolutionary link in the development of new materials from conventional materials. Used more often in the defense and aerospace industries, advanced composites are beginning to play the role of conventional materials (commodities) used for load-bearing structural components for infrastructure applications. These unique materials are now being used worldwide for building new structures as well as for rehabilitating in-service structures. Application of composites in infrastructural systems on a high-volume basis has come about as a result of the many desirable characteristics of composites that are superior to those of conventional materials such as steel, concrete, and wood.

The increased use of composites in thousands of applications — domestic, industrial, commercial, medical, defense, and construction — has created a need for knowledgeable professionals as well as specific literature dedicated to advancing the theory and design of composites to provide a compendium of engineering principles for structural applications in general and concrete structures in particular.

A rich body of literature — texts and handbooks — exists on composites, such as the manufacturing of composites and the analysis and design of composite lamina based on laminated plate theory. In spite of considerable work that has been carried out over the past 50 years, notably in the U.S., Canada, Japan, and several European countries involving concrete and composites, literature in the form of comprehensive texts that can be practically used for composites in conjunction with concrete construction is sparse.

This book, *Reinforced Concrete Design with FRP Composites*, presents readers with specific information needed for designing concrete structures with FRP reinforcement as a substitute for steel reinforcement and for using FRP fabrics to strengthen concrete members. Separate chapters have been provided that discuss both of these topics exhaustively, supplemented with many practical examples and fundamental theories of concrete member behavior under different loading conditions.

This book is self-contained in that it presents information needed for using FRP composites along with concrete as a building material. It has been written as a design-oriented text and presents in a simple manner the analysis, design, durability, and serviceability of concrete members reinforced with FRP. Mechanics of composites and associated analysis involving differential equations have been intentionally omitted from this book to keep it simple and easy to follow. An extensive glossary of terms has been provided following Chapter 8 for the readers' quick reference.

The idea of writing this book evolved in 1996 while all three authors were attending the Second International Conference on Advanced Composite Materials for Bridges and Structures in Montreal, Canada. Since then, the authors have focused on preparing a state-of-the-art book on analysis and design of concrete members with fiber reinforced polymer reinforcement (bars and fabrics). The material presented in this book is extensively referenced, and based on the authors' extensive research experience of many years and the knowledge-base developed at the Constructed Facilities Center of West Virginia University, Morgantown, West Virginia. In addition to the theory of design of concrete members with FRP bars and fabrics, this book presents an in-depth discussion of the analysis and design approaches recommended by the ACI Committee 440 guide documents.

This book is intended to serve as a text for adoption in colleges and universities teaching a course in concrete design using FRP reinforcement, both as internal reinforcement (as a substitute for steel reinforcement) and as external reinforcement for strengthening concrete members. The authors hope that it will serve as a good resource for practical design, construction, and as a rehabilitation guide for practicing engineers.

Great care has been exercised in organizing and presenting the material in this book; however, readers may inevitably find some controversial ideas and even a few errors. The authors would be grateful to readers for communicating with them any controversies or errors that they might find, and welcome any suggestions or comments offered to enhance the usefulness of this book.

**Hota V.S. GangaRao**
**Narendra Taly**
**P.V. Vijay**

# Contents

# 1 Frequently Asked Questions about Composite Materials

This chapter is intended to present basic concepts of composite materials in a simple, direct, reader-friendly question-and-answer format. Although self-generated, the questions represent the mindset of the uninitiated in that they are frequently asked questions about composite materials. Many topics discussed in the following presentation are discussed in detail in subsequent chapters of this book. Readers will also encounter many new terms related to composites, which are defined in the **Glossary** provided at the end of this book.

## What are composite materials?

*Composite materials* (often referred to as *composites*) are man-made or natural materials that consist of at least two different constituent materials, the resulting composite material being different from the constituent materials. The term composite material is a generic term used to describe a judicious combination of two or more materials to yield a product that is more efficient from its constituents. One constituent is called the *reinforcing* or *fiber phase* (one that provides strength); the other in which the fibers are embedded is called the *matrix phase*. The matrix, such as a cured resin-like epoxy, acts as a binder and holds the fibers in the intended position, giving the composite material its structural integrity by providing shear transfer capability.

The definitions of composite materials vary widely in technical literature. According to Rosato [1982], "a composite is a combined material created by the synthetics assembly of two or more components — a selected filler or reinforcing agent and a compatible matrix binder (i.e., a resin) — in order to obtain specific characteristics and properties…The components of a composite do not dissolve or otherwise merge into each other, but nevertheless do act in concert…The properties of the composite cannot be achieved by any of the components acting alone." Chawla [1987] provides the operational definition of a composite material as one that satisfies the following conditions:

1. It is a manufactured material (not a naturally occurring material, such as wood).

2. It consists of two or more physically or chemically distinct, suitably arranged phases with an interface separating them.
3. It has characteristics that are not depicted by any of the constituents in isolation.

The *EUROCOMP Design Code* [1996] provides a rather technical definition of a composite or composite material as "a combination of high modulus, high strength and high aspect ratio reinforcing material encapsulated by and acting in concert with a polymeric material." In a practical sense, the term "composite" is thus limited to materials that are obtained by combining two or more different phases together in a controlled production process and in which the content of the dispersed phase in the matrix is substantially large.

Portland cement concretes, and asphalt concretes are examples of man-made composite materials with which civil engineers are familiar. These clearly heterogeneous composites consist of a binding phase (Portland cement phase or asphalt) in which aggregates up to about 1 inch in size are dispersed. The composites used for construction include fiber-reinforced composites, in which fibers are randomly dispersed in cement or polymer matrix, and laminated composites made of a layered structure.

### You mentioned wood as being a naturally occurring composite material. Can you explain?

Wood is a composite material that occurs in nature; it is not a manufactured material (hence not generally discussed as a composite). However, it is a composite material in that it consists of two distinct phases: cellulose fibers, and lignin that acts as a binder (matrix) of fibers.

### What is meant by plastic?

According to the *EUROCOMP Design Code*, plastic is a material that contains one or more organic polymers of large molecular weight, is solid in its finished state, and can be shaped by flow at some state in its manufacturing, or processing into finished articles.

From a chemistry standpoint, *plastics* are a class of materials formed from large molecules (called *polymers*), which are composed of a large number of repeating units (called *monomers*). The monomers react chemically with each other to form extended molecular chains containing several hundred to several thousand monomer units. Most monomers are organic compounds, and a typical polymer is characterized by a carbon chain backbone, which can be linear or branched. The molecular structure of the unit that makes up very large molecules controls the properties of the resulting material, the polymer or plastic. The rigidity of the chains, density and the regularity of packing (i.e., crystallinity) within the solids, and interaction between the molecular chains can be altered and thus change the bulk properties of the plastic.

### Do other composite materials occur naturally?

Bone that supports the weight of body is an example of a naturally occurring composite material. Weight-bearing bone is composed of short and soft collagen fibers that are embedded in a mineral matrix called *apatite*. The human body (Figure 1.1) is an excellent example of a living structure made from composites.

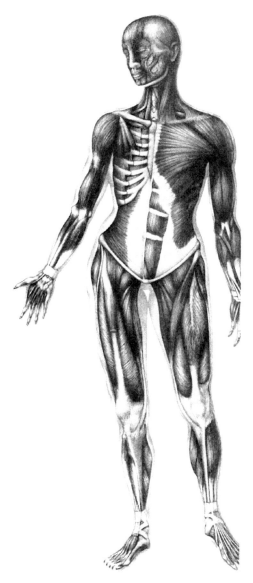

**FIGURE 1.1** The human body — an example of a perfect composite structure. (Image by Dorling Kindersley.)

### What are some of the most commonly used fiber types?

A variety of fibers are used in commercial and structural applications. Some common types are glass, carbon, aramid, boron, alumina, and silicon carbide (SiC).

### How are fibers arranged within a composite?

The arrangement of fibers in a composite is governed by the structural requirement and the process used to fabricate the part.

**What determines the mechanical and thermal properties of a composite?**
The mechanical and thermal properties of a composite depend on the properties of the fibers, the properties of the matrix, the amount and the orientation of fibers.

**What are some common types of matrix materials?**
A number of matrix materials are used by the industry (and their number is growing), the more frequently used matrix materials are:

1.  Thermoplastic polymers: polyethylene, nylon, polypropylene, polystyrene, polyamids
2.  Thermosetting polymers: polyesters, epoxy, phenolic, polymide
3.  Ceramic and glass
4.  Carbon
5.  Metals: aluminum, magnesium, titanium

**What are some of the resins used in composites?**
Various polymers include epoxy, phenol, polyester, vinyl ester, silicone, alkyd, fluorocarbon, acrylic, ABS (acrylonitrile-butadiene-styrene) copolymer, polypropylene, urethane, polyamide, and polystyrene [Rosato 1982].

**Are resins the same as matrix materials?**
Yes, but not all matrix materials are resins, as stated earlier. A matrix is a *cured phase* of resin. Fabrics made of fibers such as nylon, polyester, polypropylene, and so on are simple forms of cured resins that have different chemical structures.

**Do resins, epoxies, and polymers mean the same thing?**
A *resin* is a semisolid or pseudosolid organic material that has often high molecular weight, exhibits a tendency to flow when subjected to stress, and usually has a softening or melting range. In reinforced plastics, the resin is the material used to bind together the fibers. Generally speaking, the polymer with additives is called *resin system* during processing and *matrix* after the polymer has cured (solidified). *Epoxies* are a class of resins (or polymers) that are most commonly used.

**How are these resins classified?**
Resins (or polymers or plastics) can be classified in several different ways. The earliest distinction between types of polymers was made long before developing any in-depth understanding of their molecular structure; it was based on their reaction to heating and cooling.

On this basis, resins or polymers are classified as *thermoset* (or thermosetting) and *thermoplastic*. From a chemical molecular chain standpoint, the main difference between the two is the nature of bonds between the molecular chains: secondary van der Waals in the thermoplastics and chemical crosslinks in thermosets [Young et al. 1998].

A thermoset resin (or polymer) is characterized by its ability to change into a substantially infusible and insoluble material when cured by the application of heat or by chemical means. Although upon heating, these polymers soften and can be made to flow under stress once, they will not do so reversibly — i.e., heating causes them to undergo a "curing" reaction. Further heating of these polymers leads only to degradation, but it will not soften them. Bakelite is a good example of a thermosetting plastic (polymer or resin). Because Bakelite is a strong material and also a poor conductor of heat and electricity, it is used to make handles for toasters, pots, and pans; for molding common electrical goods, such as wall outlets and adapters; and for such diverse items as buttons and billiard balls. Resins similar to Bakelite are used in fiberboard and plywood.

A thermoplastic resin (or polymer) is characterized by its ability to soften and harden repeatedly by increases and decreases (respectively) in temperature, with minimal change in properties or chemical composition [Mott 2002]. These polymers soften upon heating and can then be made to flow under pressure. Upon cooling, they will regain their solid or rubbery nature. Thermoplastics mimic fats in their response to heat; thermosets are more like eggs (boiled and hardened eggs cannot be transformed back into egg yolks; the change caused by heat is irreversible). Note that thermoplastics lose their properties dramatically after four or five cycles of heating and cooling.

Plastics or polymers can also be classified on the basis of their molecular chains. When molecules are strung together like a string of paper clips, in one or three dimensions, the resulting compound is called a *polymer* or *macromolecule*. One class of polymers consists of linear chains, i.e., the chains extend only in one dimension. These polymers are *linear polymers* and are referred to as thermoplastic. They gradually soften with increasing temperature and finally melt because the molecular chains can move independently. An example is polyethylene, which softens at 85°C. The polymers in the other group have cross-links between chains, so that the material is really one three-dimensional giant molecule — these are called *crosslinked polymers* and referred to as thermosetting [Selinger 1998].

### Can you give some examples of thermoset and thermoplastic polymers?

Examples of thermoplastics include ABS, acetals, acrylics, cellulose acetate, nylon, polyethylene, polypropylene, polystyrene, and vinyls. Examples of thermosetting polymers include alkyds, allyls, aminos, epoxies, phenolics, polyesters, silicones, and some types of urethanes.

### In addition to acting as binders, do resins serve any other functions?

As mentioned earlier, resin systems used in composites act not only as binders of fibers but also add structural integrity and protection from environmental hazards such as moisture, corrosive agents in the environment, freeze-thaw cycles, and ultraviolet (UV) radiation.

### What is meant by resin system?

Typically, resins used in composites are combined with several additives or modifiers (e.g., fillers, catalysts, and hardeners); the term *resin system* rather than resin is used as an all-inclusive term for a binder ready for use at the time of process or manufacturing of a composite. The purpose of these additives is to modify the properties of the resin to provide protection to fibers from moisture ingress and ultraviolet radiation, add color, enhance or reduce translucence, modify surface tension or wettability during the low-viscosity period before curing, and so on.

### Why are so many resin systems used in composites?

Different resin systems offer different advantages and also have their own drawbacks. For example, polyester has a low cost and an ability to be made translucent. Its drawbacks include service temperatures below 77°C, brittleness, and a high shrink-age factor of as much as 8% during curing. Phenolics are low-cost resins that provide high mechanical strength but have the drawback of a high void content. Epoxies provide high mechanical strength and good adherence to metals and glasses, but high costs and processing difficulties are their drawbacks. Thus, each resin system offers some advantages but also has some limitations. The use of a particular resin system depends on the application.

**Are many kinds of composites used? How are they classified?**
Composites are classified generally in one of the two ways [Kaw 1997]: (1) by the geometry of reinforcement — particulate, flake, and fibers; or (2) by the type of matrix — polymer, metal, ceramic, and carbon. *Particulate* composites consist of particles immersed in matrices such as alloys and ceramics. *Flake* composites consist of flat fiber reinforcement of matrices. Typical flake materials are glass, mica, aluminum, and silver. These composites provide a high out-of-plane modulus, higher strength, and lower cost. Fiber composites consist of a matrix reinforced by short (discontinuous) or long (continuous) fibers and even fabrics. Fibers are generally anisotropic, such as carbon and aramid.

**What is meant by reinforced plastic?**
*Reinforced plastic* is simply a plastic (or polymer) with a strength greatly superior to those of the base resin as a result of the reinforcement embedded in the compo-sition [Lubin 1982].

**What are FRPs?**
"FRP" is an acronym for *fiber reinforced polymers*, which some also call fiber rein-forced plastics, so called because of the fiber content in a polyester, vinyl ester, or other matrix. Three FRPs are commonly used (among others): composites containing glass fibers are called *glass fiber reinforced polymers* (GFRP); those containing carbon fibers are called *carbon fiber reinforced polymers* (CFRP); and those reinforced with aramid fibers are referred to as *aramid fiber reinforced polymers* (AFRP).

**You referred to FRP as fiber reinforced polymer. How is this different from fiber reinforced plastic?**
Actually, there is no difference. *Plastics* are composed of long chain-like molecules called *polymers*. The word "polymer" is a chemistry term (meaning a high molecular weight organic compound, natural or synthetic, containing repeating units) for which the word "plastic" is used as a common descriptive. The word "polymer" rather than "plastic" is preferred in the technical literature. However, the term "fiber reinforced plastics" continues to enjoy common usage because of the physical resemblance of the FRPs to commonly used plastics. The term "FRP" is most often used to denote glass fiber reinforced polymers (or plastics). The term "advanced composites" is usually used to denote high-performance carbon or aramid fiber-reinforced polymers (or plastics).

**Are FRPs different from regular plastics that are used for making common products such as plastic bottles, jugs, spoons, forks, and bags (i.e., grocery and trash bags)? What is the difference?**
FRPs are very different from ordinary plastics; the key phrase in FRP is "fiber reinforced." Generally, plastic is a material that is capable of being shaped into any

**FIGURE 1.2** Common household plastic products.

form, a property that has made it a household name. Figure 1.2 shows a variety of unreinforced plastic products, such as a milk jug, coffee maker, bottles to contain medicine and liquids, compact disc, plastic bags, and other items found in common households. Literally hundreds of thousands of unreinforced plastic products are in use all over the world. These include our packing and wrapping materials, many of our containers and bottles, textiles, plumbing and building materials, furniture, flooring, paints, glues and adhesives, electrical insulation, automobile parts and bodies, television, stereo, and computer cabinets, medical equipment, cellular phones, compact discs, and personal items such as pens, razors, toothbrushes, hairsprays, and plastic bags of all kinds. In fact, we can say that we live in the Plastic Age (similar to the prehistoric Stone Age).

**Can you describe some of these plastics and their commercial uses?**
Table 1.1 lists several commercial and industrial applications of plastics along with desirable properties pertinent to those applications and suitable plastics. Some of these plastics are categorized as hard and tough — they have high tensile strength and stretch considerably before breaking. Because of these superior properties, they are relatively expensive and have specialized applications. For example, consider *polyacetals*, which have high abrasion resistance and resist organic solvents and water. Therefore, they are used in plumbing to replace brass or zinc in showerheads, valves, and so on. Furniture castors, cigarette lighters, shavers, and pens are also often made from polyacetals as they give a nonstain as well as satin finish.

**TABLE 1.1**
**Major Types of Commercial Plastics and Applications**

| Type of Plastic | Characteristics | Typical Commercial Applications |
|---|---|---|
| Low-density polyethylene (LDPE) | Low melting; very flexible; soft; low density | Bags for trash and consumer products; squeeze bottles; food wrappers; coatings for electrical wires and cables |
| Poly(vinyl chloride), also known as PVC | Tough; resistant to oils | Garden hoses; inexpensive wallets, purses, keyholders; bottles for shampoos and foods; blister packs for various consumer products; plumbing, pipes, and other construction fixtures |
| High-density polyethylene (HDPE) | Higher melting, more rigid, stronger, and less flexible than low-density polyethylene | Sturdy bottles and jugs, especially for milk, water, liquid detergents, engine oil, antifreeze; shipping drums; gasoline tanks; half to two-thirds of all plastic bottles and jugs are made of this plastic |
| Polypropylene | Retains shape at temperatures well above room temperature | Automobile trim; battery cases; food bottles and caps; carpet filaments and backing; toys |
| Polystyrene | Lightweight; can be converted to plastic foam | Insulation; packing materials including "plastic peanuts"; clear drinking glasses; thermal cups for coffee, tea, and cold drinks; inexpensive tableware and furniture; appliances; cabinets |
| Poly(ethylene terephthalate), also known as PET | Easily drawn into strong thin filaments; forms an effective barrier to gases | Synthetic fabrics; food packages; backing for magnetic tapes; soft-drink bottles |
| Phenol-formaldehyde resins | Strongly adhesive | Plywood; fiberboard; insulating materials |
| Nylon, phenolics, tetrafluoroethylene (TFE)-filled acetals | Easily drawn into strong thin filaments; resistant to wear, high tensile and impact strength, stability at high temperatures, machinable | Synthetic fabrics; fishing lines; gears and other machine parts |
| Tetrafluoroethylene (TFE) fluorocarbons, nylons, acetals | Low coefficient of friction; resistance to abrasion, heat, corrosion | High-strength components, gears, cams, rollers |
| Acrylics, polystyrene, cellulose, acetate, vinyls | Good light transmission in transparent and translucent colors, formability, shatter resistance | Light-transmission components |

*Source*: Adapted from Selinger, B., *Chemistry in the Market Place*, 5th ed., Marrickville, NSW, Australia: Harcourt Brace & Co. Australia, 1998; Snyder, C.H., *The Extraordinary Chemistry of Ordinary Things*, New York: John Wiley & Sons, 2003.

*Polycarbonate* is an example of a plastic that finds use in a wide variety of commercial applications. Polycarbonates are often used instead of glass because they are transparent, dimensionally stable, and impact resistant, even when subjected to a wide range of temperatures. Babies' bottles, bus-shelter windows, plastic sheeting for roofing, and telephones are examples of polycarbonates. In sporting equipment, they are used in helmets for team players, motorcyclists, and snowmobilers. Because of their superior fire resistance, they are used in firemen's masks, interior moldings of aircraft, and in electronic equipment [Selinger 1998].

*Nylon* is yet another example of plastic that has excellent mechanical properties and resists solvents. As such, it is an ideal material for gears and bearings that cannot be lubricated. About 50% of molded nylon fittings go into cars in the form of small gears (for wipers), timing sprockets, and all sorts of clips and brackets.

**What are advanced composites and high-performance composites?**
Fiber reinforced composites are composed of fibers embedded in a matrix. The fibers can be short or long, continuous or discontinuous, and can be oriented in one or multiple directions. By changing the arrangements of fibers (which act as reinforcement), properties of a composite can be engineered to meet specific design or performance requirements. A wide variety of fibers, such as glass, carbon, graphite, aramid, and so on, are available for use in composites and their number continues to grow.

An important design criterion for composites is the performance requirement. In low-performance composites, the reinforcements — usually in the form of short or chopped fibers — provide some stiffness but very little strength; the load is carried mainly by the matrix. Two parameters that are used to evaluate the high-performance qualities of fibers are *specific strength* and *specific stiffness*. Fibers having high specific strength and high specific stiffness are called *advanced* or *high-performance fibers* and were developed in the late 1950s for structural applications. Composites fabricated from advanced fibers of thin diameters, which are embedded in a matrix material such as epoxy and aluminum, are called *advanced* or *high-performance composites*. In these composites, continuous fibers provide the desirable stiffness and strength, whereas the matrix provides protection and support for the fibers and also helps redistribute the load from broken to adjacent intact fibers. These composites have been traditionally used in aerospace industries but are now being used in infrastructure and commercial applications. The advanced composites are distinguished from *basic composites*, which are used in high-volume applications such as automotive products, sporting goods, housewares, and many other commercial applications where the strength (from a structural standpoint) might not be a primary requirement.

**What is meant by specific strength and specific stiffness? What is their significance?**
*Specific strength* is defined as the ratio of the tensile strength of a material to its unit weight. *Specific stiffness* (also called *specific modulus*) is defined as the ratio of the modulus of a material to its unit weight. These properties are often cited as indicators of the structural efficiency of a material; they form very important and often critical design considerations for the many products for which composites offer unique advantages.

**TABLE 1.2**
**Properties of Selected Composites and Steel**

| Material | Tensile Strength (ksi) | Modulus of Elasticity ($\times10^6$ psi) | Specific Weight, $\gamma$ (lb/in$^3$) | Specific Strength ($\times10^6$ in) | Specific Modulus ($\times10^8$ in) |
|---|---|---|---|---|---|
| Steel | | | | | |
| AISI 1020 HR | 55 | 30.0 | 0.283 | 0.194 | 1.06 |
| AISI 5160 OQT 700 | 263 | 30.0 | 0.283 | 0.929 | 1.06 |
| Aluminum | | | | | |
| 6061-T6 | 45 | 10.0 | 0.098 | 0.459 | 1.02 |
| 7075-T6 | 83 | 10.0 | 0.101 | 0.822 | 0.99 |
| Titanium | | | | | |
| Ti-6A1-4V quenched and aged at 1000°F | 160 | 16.5 | 0.160 | 1.00 | 1.03 |
| Glass/epoxy composite | | | | | |
| 60% fiber content | 114 | 4.0 | 0.061 | 1.87 | 0.66 |
| Boron/epoxy composite | | | | | |
| 60% fiber content | 270 | 30.0 | 0.075 | 3.60 | 4.00 |
| Graphite/epoxy composite | | | | | |
| 62% fiber content | 278 | 19.7 | 0.057 | 4.86 | 3.45 |
| Graphite/epoxy composite | | | | | |
| Ultrahigh modulus | 160 | 48.0 | 0.058 | 2.76 | 8.28 |
| Aramid/epoxy composite | | | | | |
| 60% fiber content | 200 | 11.0 | 0.050 | 4.00 | 2.20 |

*Source*: Adapted from Mott, R.L. *Applied Strength of Materials,* Upper Saddle River, NJ: Prentice Hall, 2002.

**Can you list a few important mechanical properties of composites and metals such as steel and aluminum to make a valid comparison and explain why composites are considered superior structural materials?**
Table 1.2 lists several key properties (tensile strength, modulus of elasticity, specific weight, specific strength, and specific modulus) of metals such as steel, aluminum, titanium alloys, and selected composites [Mott 2002]. It also lists the two important parameters — specific strength and specific stiffness — that are often cited as indicators of structural efficiency of a material.

**What does this information mean to a designer?**
The information provided in Table 1.2 is very important to a designer while selecting a material type for complex structural systems. Figure 1.3 gives a comparison of the specific strength and specific stiffness of selected composite materials [Mott 2002].

For example, consider a boron/epoxy composite having a specific weight of 0.075 and steel (AISI 5160 OQT 700) having a specific weight of 0.283; their strengths are comparable — 270 ksi and 263 ksi, respectively. However, the specific strength of the boron/epoxy composite (3.60) is almost four times that of steel

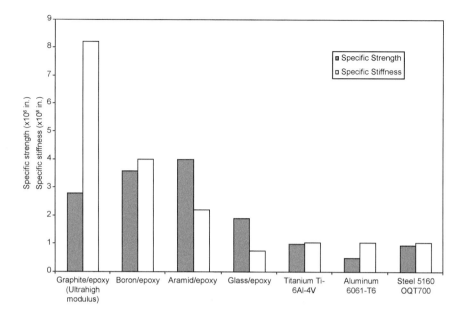

**FIGURE 1.3** Comparison of specific strength and specific stiffness of selected materials. (Adapted from Mott, R.L. *Applied Strength of Materials,* Upper Saddle River, NJ: Prentice Hall, 2002.)

(0.929). Now consider the simple case of a rod designed to carry an axial tensile force. The cross-section of the boron/epoxy composite rod need only be one-fourth that of a steel rod. This reduction in cross-sectional area translates into reduced space requirements and also reduced material and energy costs. Figure 1.4 shows a plot of these data with specific strength on the vertical axis and specific modulus on the horizontal axis [Mott 2002]. Note that when weight is critical, the ideal material would be found in the upper right part of Figure 1.4.

**Can you describe the basic anatomy of composites?**
As stated earlier, the two main constituents of composites are fibers and resins. Additionally, they contain quantities of other substances known as *fillers* and *additives*, ranging from 15 to 20% of the total weight (and hence the term *resin system*). Fibers are the backbone of a composite. Their diameters are very thin — much thinner than a human hair. For reasons explained earlier, their small diameter is the reason for their extreme strength. The tensile strength of a single glass filament is approximately 500 ksi. However, the thin diameter of a fiber is also a major disadvantage in that the compressive strength of a thin fiber is very small due to its vulnerability to buckling.

To harness the strength of fibers, they are encased in a tough polymer matrix, which gives the composite its bulk. The matrix serves to hold fibers together in a structural unit and spread the imposed loads to many fibers within the composite, and to protect the fibers from environmental degradation attributed to moisture,

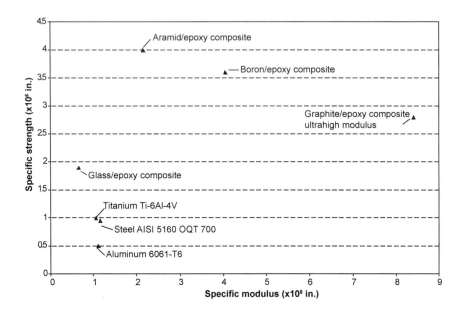

**FIGURE 1.4** Specific strength vs. specific modulus for selected metals and composites (Adapted from Mott, R.L. *Applied Strength of Materials,* Upper Saddle River, NJ: Prentice Hall, 2002.)

ultraviolet rays, corrosive chemicals, and to some extent, susceptibility to fire and from damage to fibers during handling. The function of the matrix is somewhat analogous to that of concrete in a conventional reinforced concrete member wherein the concrete surrounding reinforcing bars maintains the alignment and position of bars, spreads (or distributes) the imposed loads to all the reinforcing bars present in the member, and also provides environmental and fire protection to the bars. However, a major difference is noted. In a reinforced concrete member, concrete itself also shares the imposed loads (e.g., in beams and columns), whereas in a composite the matrix shares a negligible amount of load but helps transfer the load to fibers through interlaminar and in-plane shear; the entire imposed load is taken practically by the fibers alone.

The third constituent — fillers — are particulate materials whose major function is not to improve the mechanical properties of the composite but rather to improve aspects such as extending the polymer and reducing the cost of the plastic compound (fillers are much less expensive than the matrix resin). One of the earliest examples of filler is wood flour (fine sawdust), long used in phenolics and other thermosets. Calcium carbonate is used in a variety of plastics. Polypropylene is often filled with talc [Rosen 1993]. Hollow glass spheres are used to reduce weight. Clay or mica particles are used to reduce cost. Carbon particles are used for protection against ultraviolet radiation.

Alumina trihydrate is used for flame and smoke suppression [Katz 1978]. Fillers can also be used to improve certain properties of plastics. They almost all reduce mold shrinkage and thermal expansion coefficients, and also reduce warpage in molded parts. Mica and asbestos increase heat resistance.

In addition to the above described three constituents, *coupling agents* are used to improve the fiber surface wettability with the matrix and create a strong bond at the fiber-matrix interface. For example, coupling agents are used with glass fibers to improve the fiber-matrix interfacial strength through physical and chemical bonds and to protect them from moisture and reactive fluids [Mallick 1993]. The most common coupling agents are silanes.

For maximum fiber efficiency, stress must be efficiently transferred from the polymer to the reinforcing agents. Most inorganics have *hydrophilic* surfaces (i.e., they have an affinity for absorbing, wetting smoothly with, tending to combine with, or capable of dissolving in water), while the polymers are *hydrophobic* (i.e., incapable of dissolving in water), which results in a poor interfacial adhesion. This problem is exacerbated by the tendency of many inorganic fibers — particularly glass — to absorb water, which further degrades adhesion [Rosen 1993]. Carbon fiber surfaces are chemically inactive and must be treated to develop good interfacial bonding with the matrix. Similarly, Kevlar 49 fibers also suffer from weak interfacial adhesion with most matrixes [Mallick 1993].

Other constituents are added to composites in minute quantities for various important reasons. Most polymers are susceptible to one or more forms of degradation, usually as a result of environmental exposure to oxygen or ultraviolet radiation, or to high temperatures during processing. *Stabilizers* are added to inhibit degradation of polymers.

Plastics are often colored by the addition of pigments, which are finely powdered solids. If the polymer is itself transparent, a pigment imparts opacity. A common pigment is titanium dioxide, which is used where a brilliant, opaque white is desired. Sometimes pigments perform other functions as well. An example is calcium carbonate, which acts both as a filler and a pigment in many plastics. Black carbon is another example that acts both as a stabilizer and a pigment.

Dyes are another constituent used in minute quantities in producing plastics. Dyes are colored organic chemicals that dissolve in the polymer to produce a transparent compound, assuming that the polymer is transparent to begin with [Rosen 1993].

Flammability of polymers is a serious concern when designing with composites. Being composed of carbon and hydrogen, most synthetic polymers are flammable. Flame-retardants are added to polymers to reduce their flammability. The most common flame-retarding additives for plastics contain large proportions of chlorine or bromine [Rosen 1993]; however nonhalogenated flame retardants (other than chlorine or bromine) are being researched and implemented actively.

### What are the major considerations when designing with reinforcing fibers?

Composites are engineered materials for which fibers and resins are selected based on their intended function. Selection of appropriate fibers and resins are two major engineering decisions to be made in designing composites.

Generally speaking, three considerations must be met when designing with fiber reinforcement:

1. Fiber type: glass, carbon, aramid, or others
2. Fiber form: roving, tow, mat, woven fabrics, or others
3. Fiber architecture, i.e., orientation of fibers

The fiber architecture or fiber orientation refers to the position of the fiber relative to the axes of the element. Fibers can be oriented along the longitudinal axis of the element (at 0° to the longitudinal axis), transverse to the longitudinal axis (at 90° to the longitudinal axis), or in any other direction at the designer's discretion to achieve optimum product efficiency. This customization flexibility is unique to the fabrication of composites, which gives them versatility in applications. Although fiber orientation in a composite can be so varied that the resulting product is virtually an isotropic material with equal strength in all directions, in most cases composite structural elements are designed with the greatest strength in the direction of the greatest load. For example, for composite reinforcing elements such as bars and tendons, fibers are oriented longitudinally (i.e., in the direction of the applied or anticipated tensile force).

Once the fiber type and orientation are determined, an appropriate resin and the fiber-resin volume ratio are selected. The strength of a composite depends on the fiber-resin volume ratio — the higher the ratio, the stronger and lighter the resulting composite. Of course, higher fiber content results in increased product cost, especially for composites containing carbon and aramid fibers, including process difficulties.

Production of composites is amenable to a variety of processes, which can be fully automated or manual. Automated processes involve production of composites completely in a factory. Manually, the fibers and resin can be combined and cured on site. *Pultruded products* (so called because they are produced through a mechanical process called *pultrusion*) such as various structural shapes (e.g., beams, channels, tubes, bars) are examples of composites that are produced in a factory in their entirety and the finished products shipped to sites for the end use. Other automated processes for producing composites for construction applications include filament winding and molding. Filament winding can take place at a plant facility or at a construction site. Molding processes (several kinds exist) are also used in a plant facility. Alternatively, for low-volume applications such as structural repair and retrofit, fibers and resins can be mixed and cured on site, a manual process referred to as *hand-* or *wet-lay up* systems. In all cases, the fiber reinforcing material must be completely saturated with resin, compacted to squeeze out excess resin and entrapped air bubbles, and fully cured prior to applying loads. A variation of wet-lay up system is "prepreg" (short for pre-impregnated), which consists of unidirectional fiber sheets or fabrics that are pre-impregnated (i.e., precoated) with a resin system and ready for application on site. Machine applied systems are also available but are not commonly used because of the complexities of field applications.

**You alluded to the term "pultrusion." What does this mean?**
The term *pultrusion* refers to a continuous, mechanical process (see Figure 1.5) for manufacturing composites that have uniform cross-sectional shapes — such as "I,"

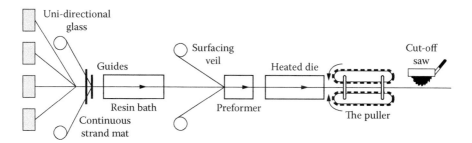

**FIGURE 1.5** Schematic of pultrusion process. (Courtesy of Strongwell Corporation, Bristol, VA.)

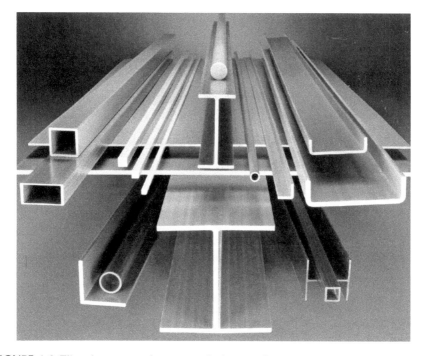

**FIGURE 1.6** Fiberglass composite structural elements formed by pultrusion. (Courtesy of Strongwell Corporation, Bristol, VA.)

"L," "T," rectangular, and circular sections — and hollow rectangular and circular tubes (similar to steel shapes) as shown in Figure 1.6 The process is automated; it involves pulling a fiber-reinforcing material through a resin bath and then through a heated (shaping) die where the resin is cured. Pultrusion is a cost-effective production process and is the dominant manufacturing process used for producing structural shapes, reinforcing bars, and prestressing tendons.

### What are the main types of fibers used for producing composites and on what basis are they selected?

Composites used for civil engineering applications are produced typically from three types of fibers: glass, carbon, and aramid. The selection of fiber type depends on the specific needs for a particular structural application. Various factors important for a composite design include the required strength, stiffness, corrosion resistance, durability, and cost. Cost is a major consideration and often plays a pivotal role in the selection process. Glass fibers are the least expensive. Carbon fibers are much more expensive than glass fibers, and aramid fibers are the most expensive. Typically, these fibers can cost several dollars per pound; some cost as high as $30 per pound in the year 2005. By comparison, structural steel in the year 2005 cost approximately $0.50 to $1.00 per pound. Considerable cost differences are found in terms of composite types, i.e., glass composite vs. carbon composite vs. aramid composite.

**Can you describe various types of fibers?**
Let us briefly discuss three types of commonly used fibers: glass, carbon, and aramid. The quality of these fibers in terms of greater strength and corrosion resistance continues to improve as new technologies evolve.

Glass fibers are produced from silica-based glass compounds that contain several metal oxides. A variety of glass fibers are produced to suit specific needs. The *E-glass* (E stands for electrical), so called because its chemical composition gives it excellent insulation properties, is one of the most commonly used glass fibers because it is the most economical. *S-glass* (S stands for structural) offers greater strength (typically 40% greater at room temperature) and also greater corrosion resistance than provided by E-glass. The corrosion-resistant *E-CR glass* provides even better resistance to corrosive materials such as acids and bases.

Carbon (and graphite, the two terms are often used interchangeably) fibers are produced from synthetic fibers through heating and stretching. Polyacrylonitrile (PAN), pitch (a by-product of the petroleum distillation process), and rayon are the three most common precursors (raw materials) used for producing carbon fibers. Some of these fibers have high strength-to-weight and modulus-to-weight ratios, high fatigue strength, and low coefficient of thermal expansion, and even negative coefficient of thermal expansion.

Aramid fiber is an aromatic polyamide that provides exceptional flexibility and high tensile strength. It is an excellent choice as a structural material for resisting high stresses and vibration.

**What is Kevlar?**
Kevlar® is the trademarked name for the aramid fiber produced by the DuPont Company.

**In what forms are various fibers commercially available?**
An individual fiber of indefinite length used in tows, yarns, or rovings is called *filament*. Because of their small diameters, filaments are extremely fragile, the primary reason for which they are sold in bundles. The industry uses different terminology for describing bundles of filaments that is based on the fiber type. For example, glass and aramid fibers are called *strands*, *rovings*, or *yarns*. The term *strands* refers to a collection of continuous glass or aramid filaments, whereas the term *rovings* refers to a collection of untwisted strands. Untwisted carbon strands are called *tows*. The term *yarns* refers to a collection of filaments or strands that are twisted together. Carbon fiber is commercially available as "tow," i.e., a bundle of untwisted fiber filaments. For example, 12K tow has 12,000 filaments and is com-

monly sold in a variety of modulus categories: standard or low (33 to 35 msi), intermediate (40 to 50 msi), high (50 to 70 msi), and ultra-high (70 to 140 msi).

Fibers for infrastructure applications are most commonly supplied in the form of rovings, tows, and fabrics. Both tows and rovings can be used to produce a wide variety of reinforcing materials such as mats, woven fabrics, braids, knitted fabrics, preforms, and hybrid fabrics. *Mats* are nonwoven fabrics that provide equal strength in all directions. They are available in two forms: chopped and continuous strand. Chopped-strand mats are characterized by randomly distributed fibers that are cut to lengths ranging from 1.5 to 2.5 in. Continuous-strand mats are formed by swirling continuous-strand fiber onto a moving belt and finished with a chemical binder that serves to hold fibers in place. Continuous-strand mat is stronger than the chopped-strand mat. Because of its higher strength, continuous-strand mat is used in molding and pultrusion processes. Chopped-strand mat is relatively weaker than the continuous-strand mat but is relatively cheaper.

Fabricated on looms, woven fabrics are available in a variety of weaves, widths, and weights. Bidirectional woven fabrics provide good strength in 0° and 90° directions and are suitable for fast fabrication. A disadvantage with woven fabrics is that fibers get crimped as they pass over and under each other during weaving, which results in a lower tensile strength of the fabric. Hybrid fabrics are manufactured with different fiber types.

Braided materials are produced by a complex manufacturing process. Although they are more efficient and stronger than woven fabrics, they cost more. Braided materials derive their higher strength from three or more yarns intertwined with one another without twisting any tow yarns around each other. Braids are continuously woven on the bias and have at least one axial yarn that is not crimped in the weaving process. Braided materials can be flat or tubular. Flat braids are used primarily for selective reinforcement (e.g., strengthening specific areas in pultruded parts), whereas tubular braids are used to produce hollow cross-sections for structural tubes.

Knitted fabrics permit placement of fibers exactly where they are needed. They are formed by stitching layers of yarn together, which permits greater flexibility in yarn alignment as the yarns can be oriented in any desired direction by putting them atop one another in practically any arrangement. A major advantage of knitted fabrics is the absence of crimping in the yarns as they lay over one another rather than crossing over and under one another (as in the case of woven fabrics). The absence of yarns' crimping results in utilization of their inherent strength, and helps create a fabric that is more pliable than woven fabrics.

**What is so special about composites? What is wrong with using the conventional materials such as steel, concrete, aluminum, and wood?**

Nothing is wrong with using the conventional building materials such as steel, concrete, aluminum, and wood. However, for certain applications, composite materials offer an attractive, and often the preferred, alternative because of the many properties that are superior to those of conventional building materials. Composites evolved as more efficient structural materials because of their many superior properties: ultra-high strength, corrosion resistance, lightweight, high fatigue resistance, nonmagnetic, high impact resistance, and durability. Because composites are man-made materials, they can be engineered (i.e., their shapes or profiles can be produced at designer's discretion) to meet the needs of specific applications. Structures built with composites also have low life-cycle costs.

## REFERENCES AND SELECTED BIBLIOGRAPHY

Ball, P., *Designing the Molecular World*, Princeton, NJ: Princeton University Press, 1994.

Benjamin, B.S., *Structural Design with Plastics*, New York: Van Nostrand Reinhold Co., 1982.

CFI, *Composites for Infrastructure*, Wheatridge, CO: Ray Publishing, 1998.

Chawla, K.K., *Composite Materials: Science and Engineering*, New York: Springer-Verlag, 1987.

Emsley, J., *Molecules at an Exhibition*, Oxford: Oxford University Press, 1998.

EUROCOMP, Structural design of polymer composites, in *EUROCOMP Design Code and Handbook*, J.L. Clarke (Ed.), London: E & FN Spon, 1996.

Gay, D., Hoa, S.V., and Tsai, S.W., *Composite Materials: Design and Applications*, Boca Raton, FL: CRC Press, 2003.

Harper, C.A. and Petrie, E.M., *Plastic Materials and Processes: A Concise Encyclopedia*, New York: Wiley Interscience, 2003.

Herakovich, C.T., *Mechanics of Fibrous Composites*, New York: John Wiley & Sons, 1998.

Hollaway, L.C. and Leeming, M.B., *Strengthening of Reinforced Concrete Structures Using Externally-Bonded FRP Composites in Structural and Civil Engineering*, Boca Raton, FL: CRC Press, 1999.

Katz, H.S. and Milewski, J.V. (Eds.), *Handbook of Fillers and Reinforcements for Plastics*, New York: Van Nostrand Reinhold Co., 1978.

Kaw, A.K., *Mechanics of Composites Materials*, Boca Raton, FL: CRC Press, 1997.

Lubin, G. (Ed.), *Handbook of Composites*, New York: Van Nostrand Reinhold Co., 1982.

Mallick, P.K., *Fiber Reinforced Composite Materials: Manufacturing and Design*, New York: Marcel Dekker, 1993.

Marshall, A., *Composite Basics Seven*, Walnut Creek, CA: Marshall Consulting Co., 2005.

McCrum, N.G., Buckley, C.P., and Bucknall, C.P., *Principles of Polymer Engineering*, New York: Oxford University Press, 1988.

Miller, T.E. (Ed.), *Introduction to Composites*, 4th ed., New York: Composites Institute of the Society of Plastics Industry, 1997.

Mott, R.L., *Applied Strength of Materials,* Upper Saddle River, NJ: Prentice Hall, 2002.

Potter, K., *An Introduction to Composite Products*, New York: Chapman & Hall, 1997.

Rosato, D.V., An overview of composites, in *Handbook of Composites*, G. Lubin (Ed.), New York: Van Nostrand Reinhold Co., 1982.

Rosen, S., *Fundamental Principles of Polymeric Materials*, New York: John Wiley & Sons, 1993.

Schwartz, M., *Composites Materials Handbook*, 2nd ed., New York: McGraw-Hill Co., 1992.

Selinger, B., *Chemistry in the Market Place*, 5th ed., Marrickville, NSW, Australia: Harcourt Brace & Co. Australia, 1998.

Snyder, C.H., *The Extraordinary Chemistry of Ordinary Things*, New York: John Wiley & Sons, 2003.

Young, J.F., Mindess, S., Gray, R.J., and Bentur, A., *The Science and Technology of Civil Engineering Materials*, Upper Saddle River, NJ: Prentice Hall, 1998.

# 2 Properties of Constituent Materials: Concrete, Steel, Polymers, and Fibers

## 2.1 INTRODUCTION

Development of composite materials represents a milestone in the history of our civilization. Along with conventional building materials such as steel, concrete, aluminum, and wood, composite materials offer an excellent alternative for a multitude of uses. Use of composite materials was pioneered by the aerospace industry beginning in the 1940s, primarily because of the material's high-performance and lightweight qualities. Today their potential is being harnessed for many uses. Advanced composite materials — so called because of their many desirable properties, such as high-performance, high strength-to-weight and high stiffness-to-weight ratios, high energy absorption, and outstanding corrosion and fatigue damage resistance — are now increasingly used for civil engineering infrastructure such as buildings and bridges.

Composite materials are manufactured from two or more distinctly dissimilar materials — physically or chemically — that are suitably arranged to create a new material whose characteristics are completely different from those of its constituents. Basically, composites consist of two main elements: the structural constituent, which functions as the load carrying element, and the body constituent called *matrix*, which encloses the composite. The structural constituent can be in the form of fibers, particles, laminae (or layers), flakes, and fillers. The matrix performs important dual functions: It acts as a binder to hold the fibrous phase in place (i.e., holds the fibers in a structural unit) protecting fibers from environmental attack, and under an applied load, it deforms and distributes the load to the high modulus fibers.

Composites discussed in this book are those fabricated from fibers such as glass, carbon, aramid, and boron. Unlike materials such as steel and aluminum alloys, fiber composites are anisotropic, hygrospic, and hygrothermally sensitive. The mechanical properties of a composite that influence its behavior and performance as structural material depend on the properties of its constituent — fibers and matrix — and the process used for its manufacture. Understanding these properties is very important for proper application of the composite. This chapter discusses the properties of various matrixes and fibers and methods of manufacturing carbon and glass fibers.

In addition to using composites for stand-alone load-carrying components (such as prefabricated bridge modules, beams, and girders), they can be also used in conjunction with concrete and steel as load-sharing elements. For example, glass fiber reinforced polymer (GFRP) bars can be used as reinforcement for concrete members in lieu of conventional steel reinforcing bars. Prestressing bars and strands made from carbon fibers can be used as tendons for prestressed concrete construction. This type of application is commonly referred to as *internal reinforcement*. Similarly, carbon fiber reinforced polymer (CFRP) strips can be used as *external reinforcement* for increasing the load-carrying capacity of conventional steel and reinforced concrete beams. In this type of application, the CFRP strips are bonded to the exterior tensile face of a beam to complement its flexural capacity. A brief discussion of properties of concrete and reinforcing steel is presented in this chapter. Although these properties are discussed in the many texts on design of reinforced concrete and steel structures, they are briefly reviewed in this chapter to preserve completeness of the subject matter.

## 2.2  INGREDIENTS OF CONCRETE

*Concrete* is a composite material consisting of fine and coarse aggregates and a binding material formed from a mixture of hydraulic cement and water. Admixtures or additional ingredients can also be used to alter the properties of concrete. ACI 318-02 defines concrete as "a mixture of portland cement or any other hydraulic cement, fine aggregate, coarse aggregate, and water with or without admixtures" [ACI 318-02]. The properties of concrete so formed are entirely different from any of its constituent materials, and that qualifies concrete as a composite material.

### 2.2.1  AGGREGATES

The term *aggregate* refers to the granular material, such as crushed stone, sand, or blast-furnace slag. Aggregates form the bulk of the finished concrete product. To broadly differentiate between the particle sizes of the aggregate, the latter is divided into two categories: *coarse aggregate* and *fine aggregate*. The particle sizes are referenced to the sieve number. The coarse aggregate (e.g., gravel) refers to the particle sizes larger than 4.75 mm (i.e., not passing through a No. 4 sieve). The fine aggregate (e.g., sand) refers to the particle size smaller than 4.75 mm but larger than 75 μm (i.e., not passing through a No. 200 sieve). In addition, blast-furnace slag — a by-product of the iron industry usually obtained by crushing blast furnace slag — is also used sometimes as coarse aggregate [Mehta 1986].

### 2.2.2  CEMENT

The binding material used to make concrete is formed from a pulverized material called *cement* and water. Cement by itself is not a binder; rather, it develops adhesive and cohesive properties as result of *hydration*, a chemical reaction between the cement minerals and water. Accordingly, such cements are called *hydraulic* cements.

Properly proportioned and mixed, cement and water together form a paste with stable adhesive characteristics that bind the constituent aggregates. The most commonly used cement is *portland cement*, which consists mainly of calcium silicates. ASTM C 150 defines portland cement as a hydraulic cement produced by pulverizing clinkers consisting essentially of hydraulic calcium and aluminum silicates. The addition of water to these minerals results in a paste that achieves stone-like strength upon hardening.

Cements can be hydraulic or nonhydraulic. Hydraulic cements are characterized by their ability to harden by reacting with water, and also to form a water-resistant product. Products formed from nonhydraulic cements are not resistant to water. Cements formed from calcination of gypsum or carbonates, such as limestone, are nonhydraulic.

Cements are classified on the basis of their rate of strength development, heat of hydration (low, moderate, or high), and resistance to sulfates (moderate or high). ASTM C 150 recognizes eight types of portland cements (Types I, IA, II, IIA, III, IIIA, IV, and V) that are commercially available to suit various needs. In addition, ASTM C 595-95 lists several categories of blended hydraulic cements.

### 2.2.3 ADMIXTURES

*Admixtures* are chemical agents that are added to constituents of concrete before or during the mixing of concrete. ASTM C 125 defines an admixture as a material other than water, aggregates, hydraulic cements, and fiber reinforcement, which is used as an ingredient of concrete mix and added to the batch immediately before or during mixing. The admixtures serve to modify the properties and performance of concrete to suit specific job requirements or for better economy. These include improved workability, increased strength, retarding or accelerating strength development, and increasing frost resistance. ASTM C 494-92 classifies admixtures as air-entraining admixtures, accelerating admixtures, water-reducing and set-controlling admixtures, admixtures for flowing concrete, and miscellaneous. ACI Committee 212 further classifies miscellaneous admixtures into 12 other types.

Conforming to ASTM C 260, *air-entraining admixtures* are used to increase resistance of concrete to freezing and thawing and provide better resistance to the deteriorating action of de-icing salts. The overall purpose is to increase concrete's durability. Accelerating admixtures are used to increase the rate of early strength development of concrete. They are also used — particularly in cold weather — to expedite the start of finishing operation and reduce the time required for curing and protection. Retarding admixtures are used to permit placement and finishing; to overcome damaging and accelerating effects of high temperatures; and to control setting time of large structural units to keep the concrete workable through the entire placing period [ACI Com. 212].

## 2.3  TYPES OF CONCRETE

Concrete is generally classified in two ways: by unit weight and by compressive strength.

### 2.3.1 CLASSIFICATION BASED ON UNIT WEIGHT

Based on unit weight, concrete is classified into three types:

1. *Normal weight concrete*: This weighs about 145 to 155 lb/ft$^3$ (145 lb/ft$^3$ is commonly used for calculating the modulus of elasticity of normal weight concrete, $E_c$, according to the ACI Code) and uses a maximum aggregate size of 3/4 in. This is the most commonly used type of concrete for structural applications.
2. *Lightweight concrete*: The unit weight of structural lightweight concrete varies between 90 and 120 lb/ft$^3$ (about 3000 lb/yd$^3$). ACI 318-99 defines structural lightweight concrete as "concrete containing lightweight aggregate that conforms to Section 3.3.1 of ACI 318-02 and has air-dry unit weight as determined by 'Test Method for Unit Weight of Structural Lightweight Concrete'" (ASTM C 567), not exceeding 115 lb/ft$^3$. Lightweight aggregate is defined as aggregate with a dry loose weight of 70 lb/ft$^3$.
3. *Heavyweight concrete*: Produced from high-density aggregates, the heavyweight concrete weighs between 200 and 270 lb/ft$^3$. It is a special-purpose concrete, used at times for radiation shielding in nuclear power plants when limitation of usable space requires a reduction in the thickness of the shield.

### 2.3.2 CLASSIFICATION BASED ON STRENGTH

This classification is based on the 28-day compressive strength of concrete ($f_c'$) [Mehta 1986]:

1. Low-strength concrete ($f_c'$ less than 3000 psi)
2. Medium-strength concrete ($f_c'$ between 3000 and 6000 psi)
3. High-strength concrete ($f_c'$ between 6000 and 10,000 psi)
4. Ultra high-strength concrete ($f_c'$ above 10,000 psi)

For common construction purposes, such as buildings and bridges, normal weight concrete in the strength range of 3000 to 4000 psi is usually used. For prestressed concrete, concrete in the strength range of 5000 to 6000 psi is used. Use of concrete over 6000 psi strength (high-strength concrete) is not very common. Use of high-strength concrete usually requires a special permit from jurisdictional building officials to ensure quality control and the production of concrete of consistent strength.

## 2.4  STRENGTH OF CONCRETE

As a building material, concrete is strong in compression but weak in tension. To overcome this deficiency, the tensile zone of a concrete member is reinforced with steel or *fiber reinforced polymer* (FRP) reinforcing bars. The compressive strength of concrete is the most important design parameter in reinforced concrete design. Other strength parameters — such as tensile strength, shear strength, bearing strength, and modulus of rupture — are assumed as some fractions of the compressive strength [ACI 318-02].

The compressive strength of concrete, determined through a standard ASTM procedure, depends on several factors as follows:

1. Proportions of ingredients including aggregate quality
2. Effect of specimen size
3. Water-cement ratio
4. Type of vibration (i.e., low to high degree of compaction)
5. Type of curing
6. Rate of loading
7. Age-strength relationship

Two batches of concrete produced under identical conditions can vary widely in their properties.

With the onset of *curing of concrete\** mixture, the compressive strength rapidly increases with age up to a certain time, after which the increase in strength is not much. The age-strength relationship for various curing periods and under various humidity conditions is shown in Figure 2.1 [Kosmatka and Panarese 1992]. The

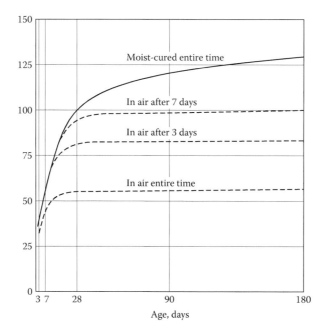

**FIGURE 2.1** Age-strength relationships for the curing period of concrete. (Modified from Kosmatka, S.H. and Panarese, W.C., *Design and Control of Concrete Mixtures*, 13th ed, Chicago: Portland Cement Association, 1992. With permission.)

---

\* The term "curing of concrete" refers to procedures devoted to promote cement hydration, such as control of time, temperature, and humidity conditions immediately after the placement of a concrete mixture into formwork.

compressive strength of concrete at 28 days can be seen to be about 1.5 times that at 7 days. After 180 days, for a given water-cement ratio, the strength of continuously moist-cured concrete is seen to be about three times the strength of continuously air-cured concrete. Humidity has a significant influence on the increase in strength of concrete with age. The curing age would have no effect on the strength of concrete if curing is carried in the absence of moisture, e.g., chemical coating on concrete surface to prevent moisture escape.

A minimum period of 7 days is generally recommended for moist-curing of concrete. The strength of concrete at 28 days can be calculated from its 7-day strength from the empirical relationship given by Equation 2.1 [Davis et al., 1964]:

$$f'_{c(28)} = f'_{c(7)} + 30\sqrt{f'_{c(7)}} \tag{2.1}$$

where $f'_{c(28)}$ and $f'_{c(7)}$ are, respectively, the concrete compressive strengths at 28 and 7 days.

## 2.5 STRENGTH CHARACTERISTICS OF CONCRETE

### 2.5.1 STRESS–STRAIN RELATIONSHIP

The stress-strain relationship for concrete is obtained from compression tests on concrete specimens, cylinders, or cubes at the age of 28 days. In the United States, the standard test specimen used to determine $f'_c$ is a 6 in $\times$ 12 in cylinder. During testing, the strains are recorded at various loading increments until the cylinder ruptures. Compression test results are sensitive to the rate of loading — the slower the rate of loading, the larger the strain (Figure 2.2). Larger strains at failure occur because of the phenomenon of creep (discussed later), a time-dependent character-istic of concrete that causes strains in concrete under constant load or stress to increase with time.

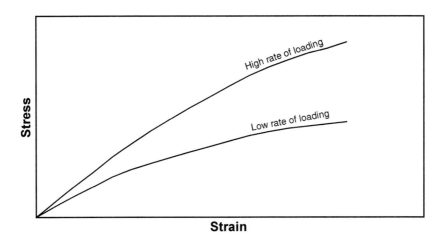

FIGURE 2.2 Stress-strain diagrams of concrete for various rates of compression loading.

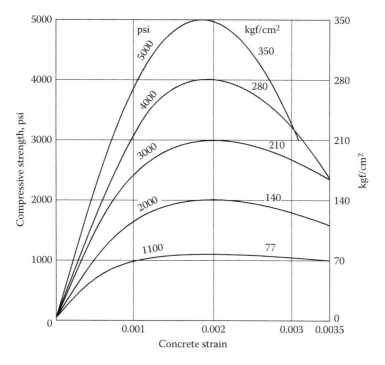

**FIGURE 2.3** Representative stress-strain curves for various concrete compression strengths. (From Wang, P.T. et al., *ACI Journal,* 75–62, 606, 1978. With permission.)

Figure 2.3 shows typical stress-strain curves for concretes having 28-day compression strengths in the range of 2000 to 10,000 psi. The following behavioral characteristics of concrete can be clearly observed from these curves:

1. The initial portion of all curves is practically straight, so that in this region concrete can be assumed to follow *Hooke's law.* The strain is proportional to stress and is reversible upon unloading of the test specimen; consequently, it is called *elastic strain.* The ratio between stress and this reversible strain is called the *modulus of elasticity* (or *Young's modulus*) of concrete.
2. The maximum compressive stress occurs at a strain of approximately 0.002.
3. Failure occurs when the strain is about 0.003. Although the strain at failure generally varies from 0.003 to 0.004, a strain level of 0.003 is conservatively assumed as the strain at which concrete fails, and this strain value is taken as a standard for design calculations. This is one of the fundamental assumptions in the *strength theory* of concrete.

## 2.5.2 MODULUS OF ELASTICITY OF CONCRETE

The static modulus of elasticity of concrete is calculated from the stress-strain curve obtained from the compression test under uniaxial loading. Methods of determining the Young's modulus of concrete are described in ASTM C 469-74. According to

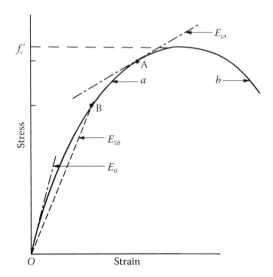

**FIGURE 2.4** Stress-strain curve and the modulus of elasticity of concrete. Legend: $f_c'$ = 28-day cylinder strength of concrete; $E_{ti}$ = initial tangent modulus; $E_{SB}$ = secant modulus at point B; $E_{tA}$ = tangent modulus at point A; $a$ = ascending branch of the stress-strain curve; $b$ = descending branch of the stress-strain curve; tends to be flatter than the ascending branch ($a$). For concretes with compressive strengths greater than about 10,000 psi, the descending branch can be nearly vertical.

Hooke's law, the modulus of elasticity is defined by the slope of the tangent to the stress-strain curve at a given point. However, because the stress-strain curve for concrete is nonlinear, three methods of computing the modulus are in use as shown in Figure 2.4 and described as follows:

1. The *initial tangent modulus* is given by the tangent to the stress-strain curve drawn at the origin. It corresponds to very small instantaneous strain and is referred to as the *dynamic modulus of elasticity*. Its value is 20%, 30%, and 40% higher than the static modulus of elasticity for high-, medium-, and low-strength concretes, respectively [Mehta 1986].
2. The *tangent modulus* is given by the slope of the line drawn tangent to the stress-strain curve at any point on the curve.
3. The *secant modulus* of elasticity is given by the slope of a line drawn from the origin to a point on the stress-strain curve corresponding to $0.5 f_c'$.
4. The *chord modulus* is given by the slope of the line drawn between two points on the stress-strain curve. The line is drawn from a point on the curve corresponding to a strain of 50 μin/in to a point on the stress-strain curve corresponding to 40% of the ultimate load.

The value of modulus of elasticity ($E_c$) to be used for design purposes corresponds to the tangent modulus based on the studies by Pauw [1960], which is specified in the ACI Code [ACI 318-02, Sect. 8.5.1]:

$$E_c = w_c^{1.5} 33\sqrt{f_c'} \text{ psi} \tag{2.2}$$

where $w_c$ is the unit weight of concrete between 90 and 155 lb/ft³. For normal weight concrete, the ACI Code permits the value of $E_c$ to be taken as

$$E_c = 57,000\sqrt{f_c'} \text{ psi} \tag{2.3}$$

### 2.5.3  FLEXURAL STRENGTH OF CONCRETE

The *flexural strength* of concrete, the maximum permitted tensile strength of concrete in bending, is expressed in terms of the *modulus of rupture* $(f_r)$ as specified in ACI 318-02, Sect. 9.5.2.3 (Equation 9.10):

$$f_r = 7.5\sqrt{f_c'} \text{ psi} \tag{2.4}$$

Variation of modulus of rupture with the compressive strength of concrete is shown in Figure 2.5 [Mirza et al. 1979]. Equation 2.4 is used to find $M_{cr}$ (cracking moment), which in turn is used for calculating the effective moment of inertia $(I_e)$, which in turn is used to calculate the deflection of concrete members (Sect. 9.5.2.3, ACI 318-02).

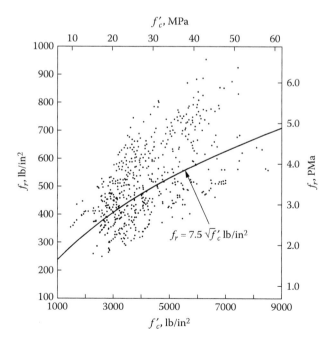

**FIGURE 2.5** Variation of the modulus of rupture with the compression strength of concrete.

The *flexural modulus of elasticity* can be determined from the deflection test of a 6 in × 6 in, 30-in long plain concrete beam, supported on a span of 24 in and subjected to a third-point loading in accordance with ASTM C78 or C293. This is one of the two standard tests to determine the tensile strength of concrete. The stress corresponding to cracking the beam on the tension face, assuming elastic behavior, is computed from Equation 2.5:

$$f_r = \frac{6M}{bh^2} \tag{2.5}$$

where
$\quad$ $M$ = cracking moment
$\quad$ $b$ = width of the beam
$\quad$ $h$ = depth of the beam

Note that the tensile and flexural strengths of concrete (Equation 2.4 and Equation 2.5, respectively) are typically of the order of 10% and 12.5% of its compressive strength.

### 2.5.4 SPLITTING STRENGTH OF CONCRETE

The second standard test to determine the tensile strength of concrete is the *split cylinder test* (ASTM C 496). The test is performed on a standard 6 in × 12 in cylinder placed longitudinally and compressed along its diameter. The value of the tensile strength as determined from the split cylinder test is given by Equation 2.6:

$$f_{ct} = \frac{2P}{\pi LD} \tag{2.6}$$

where
$\quad$ $P$ = compressive splitting load applied on the cylinder
$\quad$ $L$ = length of the cylinder
$\quad$ $D$ = diameter of the cylinder

### 2.5.5 POISSON'S RATIO

For materials subjected to simple axial load in the elastic range, *Poisson's ratio* is defined as the ratio of the lateral strain to the axial strain. For design of common concrete structures such as buildings and bridges, consideration of Poisson's ratio is not critical. However, it is required for analyzing structures such as tunnels, arch dams, flat slabs, and other statically indeterminate structures. The value of Poisson's ratio varies between 0.15 and 0.20 for both the normal weight concrete and the lightweight concrete.

## 2.5.6 SHEAR MODULUS OF CONCRETE

Based on the theory of elasticity, the *shear modulus of concrete* can be calculated from Equation 2.7:

$$G_c = \frac{E_c}{2(1+\mu)} \tag{2.7}$$

where

$G_c$ = shear modulus of concrete
$E_c$ = modulus of elasticity of concrete
$\mu$ = Poisson's ratio

## 2.6 REINFORCING STEEL

### 2.6.1 HOT-ROLLED ROUND BARS

To enhance the strength of concrete members, they are reinforced with steel bars. In beams, the bars are placed near their tension faces because concrete is weak in tension. However, steel bars can also be placed near the compression face to enhance the structural response of concrete in compression. In the form of stirrups, they are used in beams to resist shear; in the form of ties, they are used in columns to confine concrete in compression. The reinforcing bars used in concrete construction are *deformed* bars, which are characterized by ribbed projections (or lugs) rolled onto their surfaces during manufacturing process (Figure 2.6). The reason for having a ribbed bar surface is to enable the bar to develop a better bond with the concrete. Plain bars with smooth surfaces are not used in current construction practices due to poor bond development between steel bars and the surrounding concrete.

**FIGURE 2.6** Examples of deformed bars with varying lug (rib) configurations.

**TABLE 2.1**
**Cross-Sectional Areas of Deformed Bars #9 through #18**

| Bar No. | 9 | 10 | 11 | 14 | 18 |
|---|---|---|---|---|---|
| Square (in) | $1 \times 1$ | $1\ 1/8 \times 1\ 1/8$ | $1\ 1/4 \times 1\ 1/4$ | $1\ 1/2 \times 1\ 1/2$ | $2 \times 2$ |
| Area (in$^2$) | 1.0 | 1.27 | 1.56 | 2.25 | 4.0 |

Bars used as reinforcement are circular in cross-section and are specified by their sizes. Deformed bars are specified by their average diameters expressed in whole numbers such as #3 (read as "number three"), #4, and so on. For bars from #3 to #8, the numbers denote nominal diameters in eighths of an inch. For example, #3 means 3/8-in nominal diameter, #5 means 5/8-in nominal diameter, and so on. Bars #9, #10, #11, #14, and #18 have diameters that provide cross-sectional areas of square bars as shown in Table 2.1. Cross-sectional areas and unit weight (490 lb/ft$^3$) of steel reinforcing bars of various diameters are given in the ACI Code [ACI 318-02]. To specify bars #3 to #11 for design is common practice because the bars in this size group are readily available from suppliers. Bars #14 and #18 can also be used, but these sizes may not be readily available.

Reinforcing bars are classified based on the yield strength of steel from which they are made. They are designated by the word *Grade* (often abbreviated as Gr), followed by a two-digit number expressing the minimum *yield strength* of steel in kilo-pounds (kips) per square inch (kips/in$^2$) and abbreviated as *ksi*. In all, four grades of reinforcing bars are produced by the industry: Grade 40, Grade 50, Grade 60, and Grade 75, the end-two digit numbers indicating the yield strength of the bars (e.g., number 40 for $f_y$ = 40 kips/in$^2$). The industry follows a standard practice of providing size, grade, and producing-mill identification marks on the bars as shown in Figure 2.7 [CRSI 1994].

Although both Grade 40 and Grade 60 bars are commercially available and used in reinforced concrete design, the use of the latter is more economical and common. In the past, Grade 40 billet steel was more economical and readily available; however, today Grade 60 is economical and Grade 40 billet steel is available only in the smaller sized bars #3 through #6. The preferred practice is to use Grade 60 bars.

Note that certain codes and specifications preclude the use of higher strength reinforcement. For example, both the Uniform Building Code [ICBO 1997] and the ACI Code [ACI 318-02] prohibit the use of reinforcement having yield strengths in excess of 60 kips/in$^2$ for shear reinforcement. Similarly, the AASHTO Specifications (Article 8.33) for highway bridges specifies: "design shall not use a yield strength, $f_y$, in excess of 60,000 psi" [AASHTO 1996]. Grade 75 is used in large columns. Both Grade 40 and Grade 60 exhibit the well-defined yield point and elastic-plastic stress-strain behavior, a very important material characteristic that imparts ductility to reinforced concrete members. Grade 40 steel is the most ductile, followed by Grade 60, Grade 75, and Grade 50, in that order [MacGregor 1997].

An important property of reinforcing steel is its modulus of elasticity, which is many times higher than that of concrete. ACI Sec. 8.5.2 specifies the value of the modulus of elasticity of steel, $E_s$, as $29 \times 10^6$ lb/in$^2$ [ACI 318-02]. The ratio of

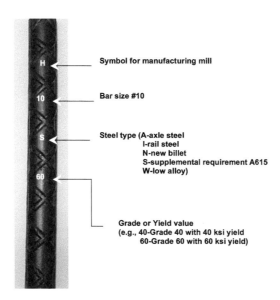

**FIGURE 2.7** Identification marks on reinforcing bars.

modulus of elasticity of steel, $E_s$, to that of concrete, $E_c$, is called *modular ratio* and denoted by $n$:

$$n = \frac{E_s}{E_c} \qquad (2.8)$$

Figure 2.8 shows a typical stress-strain diagram for a tested mild steel bar.

## 2.7 CONSTITUENTS OF FIBER REINFORCED POLYMER (FRP) COMPOSITES

Composites are *manufactured* materials consisting of two or more usually dissimilar materials. However, in some cases these two materials may be the same. For example, carbon-carbon composites are produced with carbon fiber reinforcements and a carbon matrix. The properties of a composite are completely different from those of their constituents. One of the major advantages of composites is the complementary nature of the constituents. For example, glass fibers with high tensile strength are highly susceptible to bending and environmental damage; however, these fibers are complemented by polymers, which are weak in tension but malleable and protective of the fiber surface.

Fiber reinforced polymer (FRP) composites are made of three essential constituents: fibers, polymers, and additives. The additives include plasticizers, impact modifiers, heat stabilizers, antioxidants, light stabilizers, flame retardants, blowing agents, antistatic agents, coupling agents, and others. A discussion of additives used

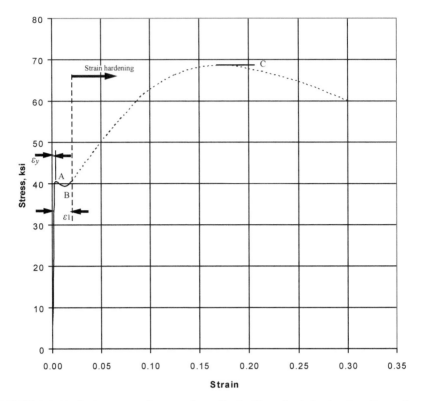

**FIGURE 2.8** Typical stress–strain curve for a Grade 40 steel reinforcing bar. (Legend: $\varepsilon_1$ = elastic strain, $\varepsilon_y$ = yield strain, A = upper yield point, B = lower yield point, C = ultimate strength). (Modified from MacGregor, J. and Wight, J.K., *Reinforced Concrete: Mechanics and Design,* Prentice-Hall, Upper Saddle River, NJ, 2005.)

for FRP composites has been provided by Tullo [2000]. Other constituents present in composites in small quantities are coatings, pigments, and fillers.

FRP composites are manufactured to meet the needs of a specific industry, especially construction, automotive, aerospace, sporting goods, and leisure industries. Structural composites are produced in a variety of forms, such as structural shapes, reinforcing bars, and fabric wraps for structural and nonstructural applications. These include strengthening of beams, columns, and chimneys; production of automotive parts, components for aerospace applications, and sporting goods; and so on, which are manufactured with continuous fiber bundles and polymers. The structural composites are made of different cross-sectional shapes to resist loads in an efficient manner. High strength-to-weight and stiffness-to-weight ratios, high energy absorption, and resistance to de-icing chemicals in relation to metals are some of the important and desirable properties of composites that help enhance the service life of the above structural applications. Furthermore, excellent corrosion and fatigue resistance of FRP composites leads to economy in the life-cycle costs of structures — notably bridges, which must carry moving (i.e., vehicular) loads and are always exposed to environmental hazards.

Different types of fibers, fabrics, and polymers that make up reinforced composites with continuous fibers are discussed in the following sections. The thermomechanical properties of constituents are discussed and information on strength and stiffness properties, specific weight, thermal expansion, and conductivity is provided for a wide range of polymers and glass fibers. Fibrous composites made of metal or ceramic matrixes are not discussed here.

The layers of FRP composite materials (called *laminates*) are manufactured such that the fiber directions of each layer are oriented in different angles to obtain the different strengths and stiffness desired in specific directions [Jones 1999]. This can be accomplished by altering the fiber density in different directions of a fabric or by using three-dimensional fabrics with different fiber orientations and densities obtained through stitching, weaving, or braiding. In addition, resin type and filler material play a significant role in improving certain mechanical properties such as strength and stiffness.

Methods of producing FRP composites differ from those used for conventional materials such as concrete or steel. Composite product manufacturing depends on building up various layers of fibers with polymers through wetting and curing them together, or bonding the layers after laminates are made. Proper combination of the constituents results in a product that is superior to its constituents. In this chapter, a detailed discussion is provided on constituents of composites — polymers, fibers, fillers, sizings, and other additives.

## 2.8 POLYMERS

### 2.8.1 GENERAL CHARACTERISTICS

"Polymers" is an oft-used term in the discussion of composite materials. Essentially, polymers are organic compounds formed by carbon and hydrogen. These compounds can be obtained either from nature or through synthesis of organic molecules in laboratories. A *polymer* is defined as a long-chain molecule having one or more repeating (*poly*) units (*mers*) of molecules joined together by strong covalent bonds These repeating units (subunit) are called *monomers*. Two subunits bonded together form a *dimer*, bonding of three subunits leads to a *trimer*, and bonding of many subunits leads to a polymer. The simplest organic compound containing carbon and hydrogen is methane ($CH_4$). A reaction of such compounds (general term: alkenes) leads to their conversion into high molecular weight compounds commonly known as polymers.

A plastic or polymeric material is a collection of a large number of polymer molecules of similar chemical structure but not necessarily of equal length [Gerig 1974]. The term *polymerization* refers to a chemical reaction or curing, which leads to a formation of a composite in the presence of fibers. The transition period from a liquid state (monomer) to a solid state (matrix), which is a function of the temperature during which curing occurs, is referred to as *cure time*, also commonly referred to as "pot life." After curing, the resulting product attains a solid state. Open or wet lay-up time, known as liquid state, is the amount of time a given mass of mixed resin and hardener will remain in the liquid state at a specific temperature.

For high-quality production of FRP composites, establishing the wet lay-up time of resin systems as accurately as possible is advisable.

Polymers can be in a solid or liquid state, and cured polymer is referred to as *matrix.* "On a submicroscopic scale, various segments in a molecule may be in a state of random excitation. The frequency, intensity and number of segmental motions increase with increasing temperature" [Mallick 1993]. This phenomenon, responsible for temperature-dependent properties of the matrix, plays a key role in the polymer selection process, especially for processing purposes.

Matrixes themselves do not contribute any significant strength (except participating in interlaminar or in-plane force transfer) to a composite as most of the load is taken by the fibers. When a load is applied to a composite, the matrix participates in transferring the loads between the fibers. In addition, the matrix partially protects the fibers against environmental attack and their surface from mechanical abrasion.

Polymeric materials, referred to as "matrixes" after cure, can be classified into two categories, based on their reaction to heating and cooling: thermosets and thermoplastics. Thermoplastics are available in granular form, whereas thermosets are found in liquid form (both are discussed later in this chapter). Choosing the type of polymer to form a matrix material is important because of its important role on the in-plane shear properties and interlaminar shear properties (i.e., between laminae of a laminate). Interlaminar shear strength (i.e., in-plane shear transfer from one laminae to another) is important in structures under bending, whereas the in-plane shear strength is important for structures under torsion, requiring resistance to in-plane shear. The matrix also provides lateral support for fibers against buckling under compression or a combination of forces. The processability and quality of composites depend largely on the physical and thermal properties of matrix materials. The choice of matrix is also important when designing damage-tolerant structures because of the fiber-matrix interaction [Shimp 1998].

The most important difference between thermoset and thermoplastic polymers is their behavior under heat and pressure. Thermoset polymer matrixes are hygrothermally sensitive; they can degrade at moderately high temperatures through moisture absorption (approximately 160° to 180°F in the presence of water). An increase in temperature causes a gradual softening of the polymer matrix material up to a certain point, indicating a transition from a glassy behavior to a rubbery behavior. The temperature at which this occurs is called the *glass transition temperature*, $T_g$, which decreases in the presence of moisture (Figure 2.9). $T_g$ is typically measured using a differential scanning calorimetric (DSC) technique [Gupta 2000; Kajorncheappunngam 1999]. A continued increase in temperature beyond this transition point causes the polymer to undergo a rapid transition from glassy behavior to rubbery behavior. As a result, the matrix-dominated properties such as shear strength and shear stiffness are reduced, and the material becomes too soft for use as a structural material [Gibson 1994].* Hygrothermal changes (expansions and contractions) alter stress-strain distribution in the composite. This change is manifested by the swelling of the composite due to increased temperature or moisture content and contraction

---

* Of major importance is the fact that the polymer undergoes a transition from ductile to brittle behavior on cooling. The term "glass transition temperature" derives from this ductile-brittle transition [Smith 1986].

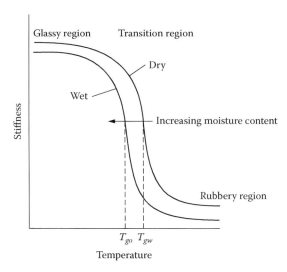

**FIGURE 2.9** Variation of stiffness with temperature for a typical polymer showing the glass transition temperature, $T_g$, and the effect of absorbed moisture in $T_g$. Note: $T_{go}$ = "dry" $T_g$ and $T_{gw}$ = "wet" $T_g$. (From Gibson, R.F., *Principles of Composite Material Mechanics*, New York: McGraw-Hill, 1994. With permission.)

due to a decrease in temperature or moisture content. Softening characteristics of various polymers are often compared by their heat deflection temperature, which is measured from tests according to ASTM D648-72. Note that heat-deflection temperature is not a measure of $T_g$.

Because of their unique properties (which vary with temperature), thermoset and thermoplastic polymers are used for specific structural and nonstructural applications. For example, thermosets are conventionally used for producing structural shapes, whereas thermoplastics are typically used in many nonstructural applications such as automobile panels, dashboards, door frames, and so on. However, recent advances in blends developed from thermosets and thermoplastics are blurring the conventional distinction in terms of their usage. Each resin type offers benefits for particular applications.

Although both types of resins are of long molecular chains, they behave differently during curing because of the arrangement of molecules within these chains. Figure 2.10 schematically shows molecular chains of thermoplastic polymers and thermoset polymers [Mallick 1993]. The notable difference between the two molecular chains is the absence or presence of *crosslinking* (commonly referred to as curing) between them. In thermoplastic polymers, individual molecules are linear (Figure 2.10a) without any chemical linking (crosslinking) between them. In contrast, the molecules of a thermoset resin are chemically crosslinked (Figure 2.10b), which provides a rigid, three-dimensional network structure. As a result, once these crosslinks are formed during the curing process (polymerization), the thermoset polymer cannot be melted and reshaped by the application of heat and pressure. However, softening of these polymers is possible at higher temperatures if the number of crosslinks is low.

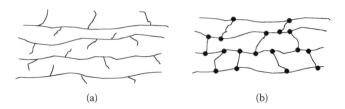

(a)                                                      (b)

**FIGURE 2.10** Schematic representations of (a) thermoplastic polymers and (b) thermosets polymers.

## 2.8.2 Thermoplastic Polymers

### 2.8.2.1 General

Most plastics and virtually all unreinforced plastics are "thermoplastics" by nature [CISPI 1992]. As the name indicates, a thermoplastic polymer becomes soft upon heating and is then capable of being shaped or reshaped while in its heated semi-fluid state. This property can be likened to that of paraffin wax, which upon heating melts and can be poured into any mold for producing the desired shape.

*Thermoplastic* polymers are organic compounds that occur in granular form. They consist of linear molecules that are not interconnected by chemical linking. Instead, they are connected by secondary weak bonds (intermolecular forces), such as van der Waals bonds and hydrogen bonds [Mallick 1993]. These polymers melt upon heating and take the form of resin. With application of heat and pressure, their molecules can be moved into a new position and will freeze in their new position when cooled. This chemical property enables the resin to be reshaped when heated, which is a great advantage. A thermoplastic forms an amorphous (noncrystalline) or a partly crystalline solid. This process can be repeated a few times, but the material slowly degrades and becomes increasingly brittle with the application of each additional thermal cycle [CISPI 1992].

Upon solidification due to temperature reduction, a gradual decrease in volume per unit mass (specific volume) of noncrystalline thermoplastics is observed. An experimental plot of specific volume versus temperature for noncrystalline thermoplastic polymers indicates a slope change at the glass transition temperature, $T_g$, ranging from –18°C for polypropylene to 80°C for PVC, and as high as 300°C in the case of polyimide. The $T_g$ of a thermoplastic depends on variables such as the degree of crystallinity, molecular weight, and rate of cooling. A sudden decrease in specific volume occurs when crystalline thermoplastics solidify and cool. Efficient packing of polymer chains in the crystalline region leads to the decrease in specific volume. An example of a thermoplastic that solidifies to form a partly crystalline structure is polyethylene, whose $T_g$ is about –110°C. The degree of crystallinity of a thermoplastic material affects the tensile strength; the tensile strength increases as the degree of crystallinity increases [Beck et al. 1963]. The mechanical properties of thermoplastics degrade as a result of repeated heating and cooling cycles. However, they provide better impact resistance and toughness and absorb higher levels of moisture than thermosets.

The processing time for thermoplastic resins is shorter than that for thermosets, which lends their use to extended fabrication options, such as injection molding, due to their lower viscosity. However, a high processing temperature of thermoplastics is unsuitable for hand lay-up or spray-up applications. Also, continuous fabrication processes, such as pultrusion with thermoplastic resins, require modification of existing equipment designed for thermoset fabrication. Long fiber-reinforced thermoplastics are used in structural applications because they are considerably less expensive than thermoset compound-based manufacturing. Long-fiber reinforced thermoplastic processing emits no volatile organic compounds, and the compounds often incorporate low-cost recycled materials [Vijay et al. 2000]. These products are easier to recycle than thermoset composites.

### 2.8.2.2 Common Thermoplastic Polymers

A number of thermoplastic polymers are commercially available. Some of the commonly used thermoplastic polymers are acrylonitrile butadiene styrene (ABS), acetal, acrylics, fluoropolymers, polyvinyl chloride (PVC), polycarbonate, polyethylene, polypropylene, polysulfone, and polyether ether ketone (PEEK).

Of those listed above, polyether ether ketone (PEEK), a semicrystalline thermoplastic polymer, is one of the more common thermoplastic polymers. Its outstanding property is manifested in its high fracture toughness, which is rated to be 50 to 100 times higher than epoxies. Another important property is its low water absorption, less than 0.5% at 73°F as compared to 4 to 5% for conventional epoxies.

Chemically, PEEK is a linear aromatic thermoplastic based upon a repeating unit in its molecules (Figure 2.11). Although the crystallinity of PEEK at normal cooling rates is between 30 and 35%, it can be as high as 48% when it is cooled slowly from its melt. However, rapid cooling of molten PEEK results in an amorphous polymer. With the presence of fibers in PEEK, the composite has a tendency to increase the crystallinity to a higher level because the fibers act as nucleation sites for crystal formation, and it has better chemical resistance than epoxy-based composites [Berglund 1998]. Note that even though an increase in the crystallinity leads to an increase in the modulus and yield strength of PEEK, its strain at failure is reduced.

PEEK has a glass transition temperature ($T_g$) of 290°F and a crystalline melting point of 635°F. The temperature range for the melting process of PEEK ranges from 698° to 752°F, and the maximum continuous use temperature is 482°F [Mallick 1993].

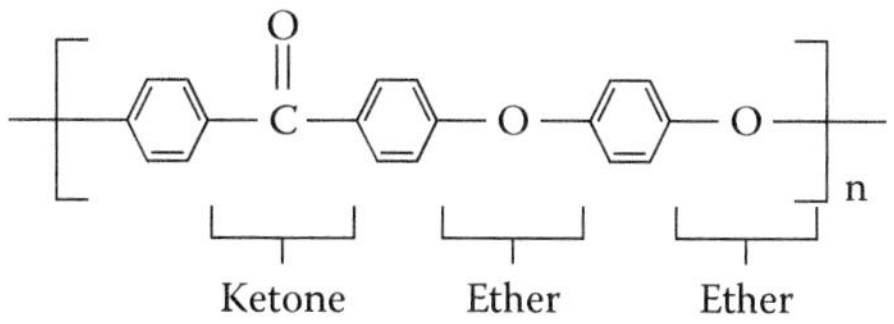

**FIGURE 2.11** Chemical structure of PEEK.

## 2.8.3 Thermoset Polymers

### 2.8.3.1 General

*Thermoset* polymers, sometimes referred to as epoxy resins (or simply resins), are a class of polymers that has been used traditionally as a matrix material for fiber-reinforced composites. Initially, these resins are normally in liquid form and, in a few special cases, also in solid form with low melting points [CISPI 1992]. Most reinforced plastics are thermosetting.

Starting materials used in polymerization of these polymers are formed of low molecular weight liquid chemicals with very low viscosities. They have molecules that are joined together by crosslinks between the linear molecules forming a three-dimensional network structure. Once crosslinks are formed during curing, the resin cannot be melted and reshaped through heat and pressure; this is an irreversible chemical change [Mallick 1993]. This is the main characteristic that sets thermosets apart from thermoplastics. This change can be likened to that of an egg that, once heated or cooked, cannot be converted back into its original liquid form or put back in the shell after hardening [CISPI 1992].

There are three main reasons for using thermoset resins in producing composites [CISPI 1992]:

1. Better bonding between fibers and matrix with compatible sizing
2. Ability to cure at room temperature in the presence of a catalyst
3. Good creep resistance

Curing of thermoset resins at room or elevated temperatures leads to crosslinking. Crosslinking releases heat as the resin cures (exothermic). The cure rate can be controlled in terms of shelf life, pot life, and gel time through proper proportioning with a catalyst.

Special resin formulations can be developed to improve impact and abrasion resistance. A coat of unreinforced resin can be applied to a mold surface, known as gel coat, before laying up fibers. The gel coat helps release the composite from a mold.

### 2.8.3.2 Commercially Available Thermoset Matrixes

Many varieties of thermoset resins are available commercially, mainly polyesters, vinyl esters, and epoxies. Several more, such as acrylics, phenolics, polyurethanes, melamines, silicones, and polymides, are used in small quantities. The principal characteristics of these resins are discussed as follows.

### 2.8.4 Polyester Resins

The starting materials (referred to as precursors) for thermoset polyester resins are unsaturated* polyester resins (also referred to as UP resins or as UPRs) that contain

---

* If one or both of the main constituents are unsaturated (i.e., it contains a reactive C=C double bond), the resulting polyester is called unsaturated. However, saturated acids, such as isophthalic or orthophthalic acid (which do not contain a C=C double bond) are also added to modify the chemical structure between the crosslinking sites [Mallick 1993].

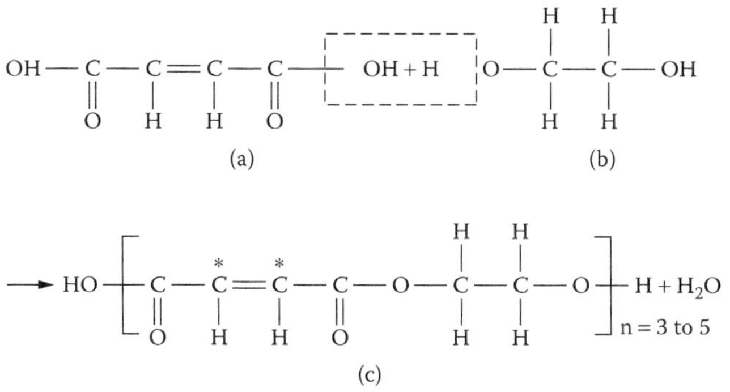

**FIGURE 2.12** Molecules of (a) unsaturated polyester, (b) styrene, and (c) *t*-butyl perbenzoate (tBPB). The asterisk (*) indicates unsaturated/reactive points in the polyester.

a number of C=C double bonds. The fact that a large number of unsaturated polyester resins (i.e., organic acids and glycols) are commercially available and that the constituents can be mixed, and that they are low-cost, makes unsaturated polyesters the most widely used resins, representing almost 75% of the total thermoset resins used by the composites industry.

Polyester resins are often referred to by their ingredients. For example, isopolyester resin builds on isophthalic acids* as its essential ingredient, whereas orthophthalic resin builds on orthophthalic acid as its essential ingredient. Terephthalic resins use terephthalic acids and are used for improved toughness over traditional isopolyesters.

Unsaturated polyesters contain an unsaturated material such as maleic anhydride or fumaric acid as part of the dicarboxylic acid component. The chemical structure for unsaturated polyester is shown in Figure 2.12. A reactive monomer, such as styrene, is used for the finished polymer to obtain a low viscosity. Polyester resins are cured with conventional organic peroxides and the cure is exothermic (i.e., the crosslinking process releases heat as it bonds). Generally, polyester resins are economical and are supplied with a medium to low viscosity (similar to maple syrup or heavy motor oil) with a fast cure time [CISPI 1992]. The cure profile is controlled in terms of shelf life, pot life, gel time, and cure temperature through careful formulation of a catalyst including inhibitor, promoter, or accelerator. Cure formulations are available with reference to fast-cure, ultraviolet (UV) light-cure, and thickenable grades.

Although the strength and modulus of polyester resins are lower than those of epoxy resins, they have a variety of properties that range from hard and brittle to soft and flexible. The major disadvantage of polyester resins is their high volumetric shrinkage (5 to 12%), which can leave sink marks (uneven depressions) in the finished product. These defects can be reduced by partly combining a low-shrinkage polyester resin that contains a thermoplastic polymer [Cossis and Talbot 1998].

Common commercial types of unsaturated polyester resins are made with various acids — such as pthalic anhydride, isophthalic acid, terephthalic acid, and adipic

---

* Phthalic acid can be obtained in *iso*, *ortho* and *tere* forms according to the position of the acid groups around the basic benzene ring of the chemical structure. These acids are solid at room temperature and are distinguished mainly by their melting point.

acid — and glycols — such as ethylene, propylene, and neopentyle. A few of these resins are discussed as follows.

1. *Orthophthalic polyester (OP)*: This was the original form of unsaturated polyester resin. It is made from phthalic anhydride, maleic anhydride, or fumaric acid. This type of resin does not have mechanical properties as good as those of isophthalic resin.
2. *Isophthalic resin*: Also called isopolyester, this resin is made of isophthalic acid and maleic anhydride or fumaric acid. Its mechanical properties, moisture, and chemical resistance are superior to those of orthopthalic resin. The cost of isopolyester resin is greater than that of orthopthalic resin.
3. *Bisphenol A furmerates (BPA)*: These resins offer improved mechanical properties compared to those of orthopthalic or isopthalic resins.
4. *Chlorendics*: These resins are formed from a blend of chlorendic acid and fumaric acid. With the presence of chlorine, these resins provide a degree of fire retardancy and also provide excellent chemical resistance.

### 2.8.5 VINYL ESTER RESINS

*Vinyl ester resins* are unsaturated resins. They are produced by reacting epoxy resin with acrylic or methacrylic acid (unsaturated carboxylic group) and an epoxy, which produces an unsaturated stage and renders them very reactive. This chemical structure is shown in Figure 2.13. The resulting material is dissolved in styrene to give a product that is very similar to a polyester resin. Vinyl ester resins are cured with the same conventional organic peroxides that are used with polyesters. They offer

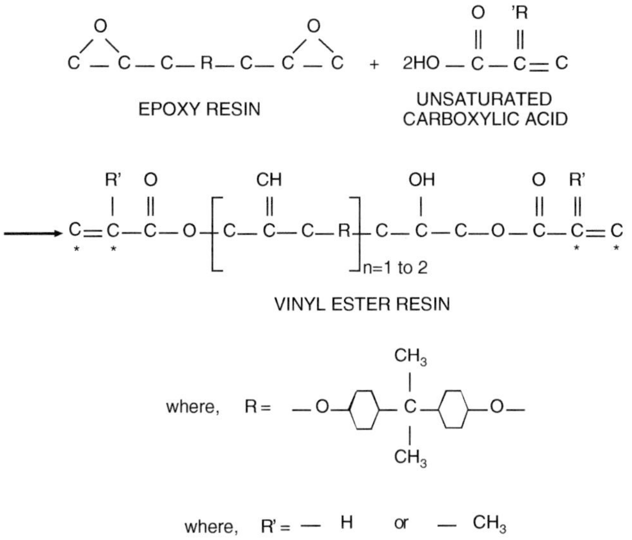

**FIGURE 2.13** Chemistry of vinyl ester resin. The asterisk (*) denotes unsaturation points (reactive sites). (Adapted from Mallick, P.K. *Fiber Reinforced Composite Materials: Manufacturing and Design,* New York: Marcel Dekker, 1993.)

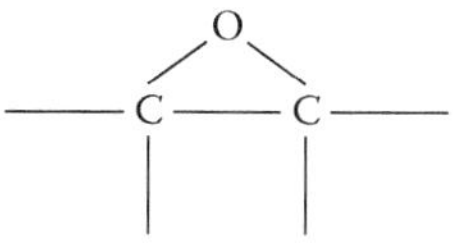

**FIGURE 2.14** Starting materials for an epoxy matrix.

excellent corrosion resistance and higher fracture toughness than epoxy resins without the difficulties experienced with epoxy resins during processing, handling, or special shape fabricating practices [CISPI 1992]. The excellent mechanical properties combined with the toughness and resilience of vinyl esters is due to their high molecular weight and the epoxy resin backbone. Vinyl esters have fewer ester linkages per molecular weight, which combined with acid resistant epoxy backbone gives excellent resistance to acids and caustics. Vinylester resins have a low viscosity and short curing time, characteristics similar to those of polyester resins. The disadvantage of vinyl ester resins is their high volume shrinkage of 5 to 10%. Their adhesive strength is moderate when compared to that of epoxy resins. Vinyl ester resins combined with heat-resistant epoxy resins can improve the heat deflection temperature (a measure of softening characteristics under heat) and thermal stability.

## 2.8.6 EPOXIES

*Epoxies* are characterized by the presence of one or more three-membered rings. These rings are variously known as the epoxy, epoxide, oxirane, or ethoxyline group, and an aliphatic, cycloaliphatic or aromatic backbone [Dewprashad and Eisenbraun 1994]. The starting materials (precursors) for an epoxy matrix are low-molecular-weight organic liquid resins containing a number of epoxide groups, which contain one oxygen atom and two carbon atoms (Figure 2.14) [Penn and Wang 1998]. Other organic molecules are added to the epoxide group to formulate a thermoset resin, which undergoes curing to form a matrix.

Because epoxy resins are mutually soluble, blends of solids and liquids or blends with other epoxy resins can be used to achieve specific performance features or specific properties. The most widely used epoxy resin is diglycidyl ether of bisphenol A and higher-molecular-weight species (Figure 2.15).

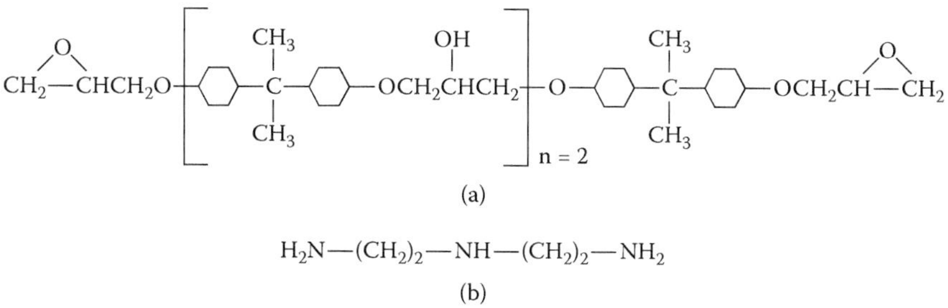

(a)

$$H_2N-(CH_2)_2-NH-(CH_2)_2-NH_2$$

(b)

**FIGURE 2.15** Epoxy matrix preparation molecule of (a) diglycidyl ether of bisphenol A (DGEBA) and (b) diethylene triamine (DETA) for curing.

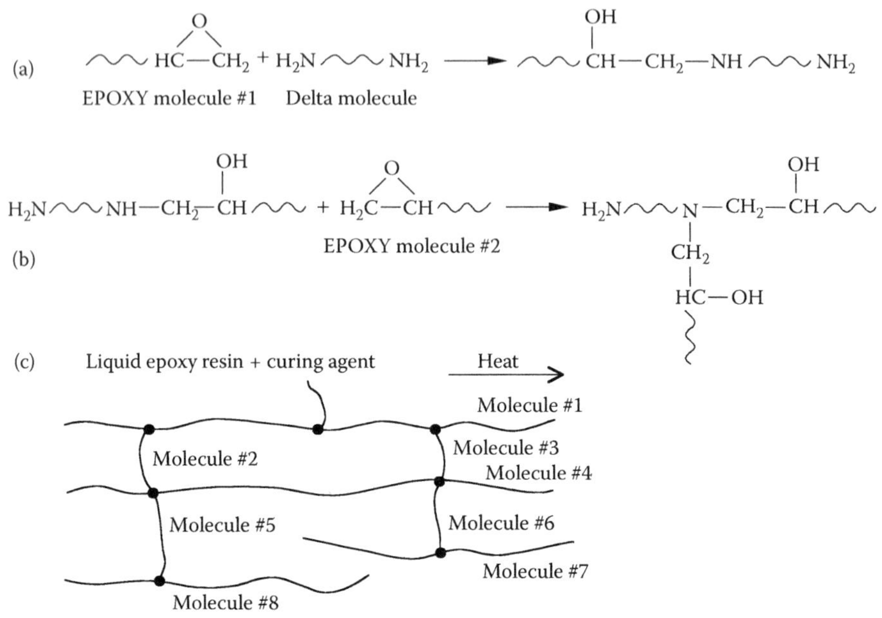

**FIGURE 2.16** Chemical structure of epoxy: (a) reaction of epoxide group with DETA molecule; (b) crosslinking; (c) three-dimensional network structure of cured epoxy. (Adapted from Mallick, P.K. *Fiber Reinforced Composite Materials: Manufacturing and Design,* New York: Marcel Dekker, 1993.)

Epoxies are cured by adding an anhydride or an amine hardener (Figure 2.16). Each hardener (curing agent) produces a different cure profile and imparts different properties to finished products [CISPI 1992]. The cure rates can be controlled through proper selection of hardeners or catalysts to meet process requirements. The network formation through chemical reaction takes place with the aid of curing agents and catalysts, which is an exothermic reaction. This results in the joining of many small molecules and producing a long chain-like (network) structure (polymerization). The curing agents can typically be aliphatic amines or aromatic amines. However, catalysts are added in very small amounts to let the epoxy molecules react directly with each other, a process called *homopolymerization*. A detailed discussion of chemical reactions in epoxy matrix network formation, both when different agents are used and when homopolymerization occurs, has been provided by Penn and Wang [1998].

Epoxies are used in high-performance composites to achieve superior mechanical properties and resistance to a corrosive environment. However, they cost more than other resins. Unless specially formulated otherwise, epoxies also have a much higher viscosity than most polyester resins, which makes them more difficult to use [CISPI 1992]. The many advantages of epoxies over other types of resins are as follows:

1. Wide range of properties allows a greater choice of selection
2. Absence of volatile matters during cure
3. Low shrinkage during curing

4. Excellent resistance to chemicals and solvents
5. Excellent adhesion to a wide variety of fillers, fibers, and other substrates

The major disadvantages of epoxies are their relatively high cost and long cure time.

## 2.8.7 PHENOLICS

A *phenolic* (also called phenolic resin) is a synthetic resin produced by condensation of an aromatic alcohol with an aldehyde, particularly of phenol with formaldehyde (Figure 2.17) [Penn and Wang 1982]. Any reactive phenol or aldehyde can be used to produce phenolic resins. Two-stage (novolac) phenolic resins are typically produced for molding convenience. In the first stage, a brittle thermoplastic resin is produced, which will not crosslink to form a matrix. This resin is produced by reacting less than a mole of formaldehyde with a mole of phenol using an acid catalyst. In the second stage, hexamethylamine tetramine is added

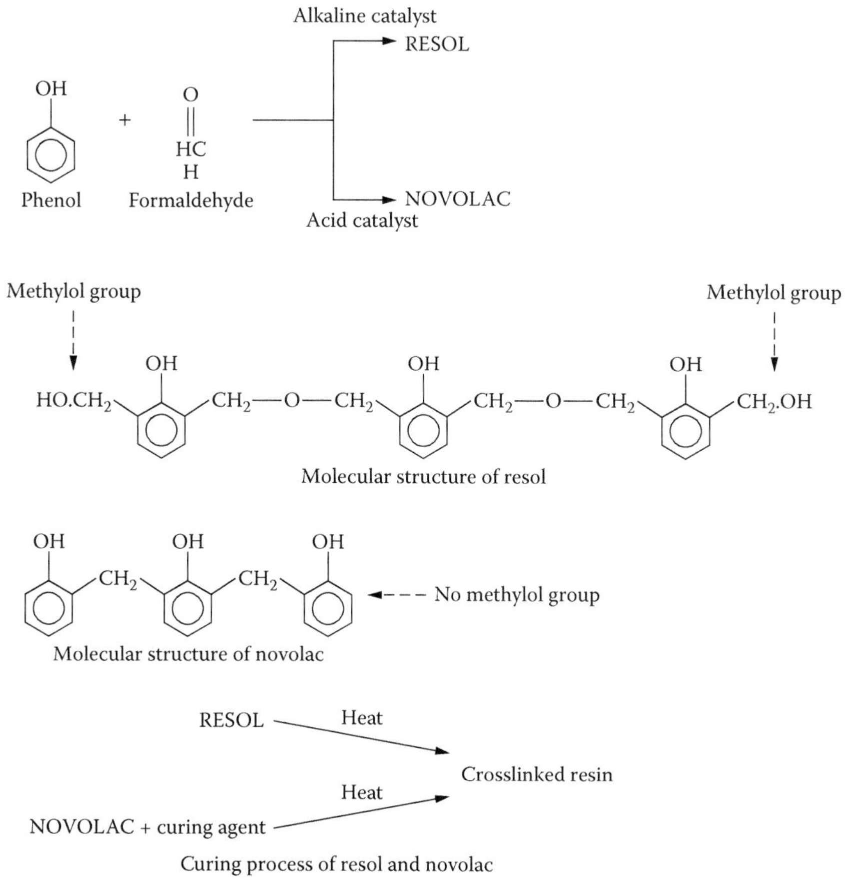

FIGURE 2.17  Preparation of phenolic resins with an acid or alkaline catalyst.

as a basic catalyst to create methylene crosslinkage so as to form a matrix. When heat and pressure are applied, the hexa group decomposes, producing ammonia and leading to methylene crosslinkage [Smith 1990]. During cure, these resins produce water, which should be removed because glass fibers will not absorb water during the short molding period. The water formation can be eliminated by compression molding by releasing the steam generated during the molding cycle by "bumping" the press. The temperature for curing ranges from 250° to 350°F. Molding compounds are made by combining the resin with various fillers. These fillers can weigh up to 50 to 80% of the molding compounds. The high crosslinkage of the aromatic structure produces high hardness, rigidity, and strength combined with good heat and chemical resistance properties. The special properties of phenolic resins are their resistance to fire and their low toxicity and smoke production under fire conditions.

### 2.8.8 POLYURETHANE RESINS

Polyurethanes can be used in thermoset or thermoplastic form. Thermoset polyurethanes are typically used either to bond structural members or to increase Young's modulus (stiffness) of structural components such as automotive bumper facias made from a reaction injection molding (RIM) process. Also, polyimide resins with performance temperatures of the order of 700°F are available with thermoset resin formulation. Similarly, polybutadiene resins have been used in thin-walled glass reinforced radomes in lieu of E-glass reinforced epoxy composites.

The addition of an alcohol to an isocyanate leads to a polyurethane. For example, a reaction between ethylene glycol (alcohol) and isocyanate yields polyurethane. To obtain additional crosslinking (i.e., to increase stiffness and crosslinking), an excess amount of di-isocyanate is used to make sure that some of the polymer chains end in unreacted isocyanate functions. The polymeric di-isocyanate reacts with the urethane linkage in other polymer chains to provide extra crosslinking [Ege 1989]. If a large number of crosslinks are formed with crosslink chains being short and rigid, then the urethane can be hard with higher stiffness, which serves better as a structurally stiff urethane. Similarly, flexible urethanes (commonly found in the form of general-purpose foams) can be obtained by using an extra amount of water for reaction and reducing the isocyanate polymer. However, flexible foams are not of structural significance because of their poor load versus stiffness response, and hence are not discussed further in this book.

### 2.8.9 COMPARISON OF THERMOPLASTIC AND THERMOSET RESINS

Although all thermosets are amorphous, thermoplastics can be amorphous or semicrystalline. The primary advantage of thermoplastic resins over thermoset resins is their high impact strength and fracture resistance, which is exhibited by their excellent damage tolerance property. Thermoplastic resins also provide higher strains-to-failure, which are manifested by better resistance to microcracking in the matrix of a composite. Some of the other advantages of thermoplastic resins are:

1. Unlimited storage (shelf) life at room temperature
2. Shorter fabrication time
3. Postformability (e.g., by thermoforming)
4. Ease of repair by (plastic) welding, solvent bonding, etc.
5. Ease of handling (no tackiness)
6. Recyclability
7. Higher fracture toughness and better delamination resistance under fatigue than thermosets

The disadvantages of thermoplastics are low creep resistance and poor thermal stability. The amorphous polymers show an increased motion in molecules at $T_g$, whereas the molecular motion drops some for semicrystalline thermoplastics. In some crystalline polymers, the processing temperature must exceed the melt temperature. Addition of fiber reinforcement to thermoplastics is more difficult than that of thermosets because of their high solution viscosities [Mallick 1993].

The advantages of thermoset resins over thermoplastic resins are:

1. Better creep resistance
2. Improved stress relaxation
3. Thermal stability
4. Chemical resistance
5. Low-$T_g$ polymers such as polypropylene (PP) have lower-weight molecules and strength

Thermosets achieve good wet-out between fibers and resins with compatible sizing, which results in better mechanical properties than those of thermoplastics. However, they also have some limitations:

1. Low-impact strengths (low strain-to-failure)
2. Long fabrication time in the mold
3. Limited storage life at room temperature (before the final shape is molded)

## 2.8.10 PROPERTIES OF RESINS

The mechanical properties of the resins are much lower when compared to those of the fibers. Properties of the most widely used resins are given in Table 2.2.

## 2.9 REINFORCEMENT (FIBERS)

*Fibers* are the load-carrying constituents of composites and occupy the largest volume in a composite laminate. They are produced in many different forms to suit various industrial and commercial applications. Fibers can be made of different materials such as glass, carbon, aramid, boron, and even natural products such as jute. The most commonly used fibers are the unidirectional fibers, which are produced in the form of single layers of yarn. They are also the types of fibers strongest in tension. Fiber strength is the highest along the longitudinal direction

**TABLE 2.2**
**Structural and Physical Properties and Processing Methods for Representative Engineering Plastics**

| Property Material Type | ASTM Test | Polyvinyl Chloride (PVC) Rigid | Acrylonitile-Butadine-Styrene (ABS) High Impact | Polyethylene (PE) High Density (HDPE) |
|---|---|---|---|---|
| 1. Specific Gravity | D792 | 1.30–1.50 | 1.01–1.04 | 0.94–0.97 |
| 2. Tensile Strength (psi) | D638 | 6000–7500 | 4800–6300 | 3100–5500 |
| 3. Elongation (%) | D638 | 40–80 | 5–70 | 20–13,000 |
| 4. Tensile Elastic Modulus ($10^6$ psi) | D638 | 0.35–0.60 | 0.23–0.33 | 0.06–0.18 |
| 5. Compressive Strength (psi) | D695 | 8000–13,000 | 4500–8000 | 2700–3600 |
| 6. Flexural Strength (psi) | D790 | 10,000–16,000 | 8000–11,000 | — |
| 7. Impact Strength (ft-lb/in), Izod | D256 | 0.4–20.0 | 6.5–7.5 | 0.5–20.0 |
| 8. Hardness (Rockwell) | D785 | D65–D85 (shore) | R85–R105 | D60–D70 (shore) |
| 9. Compressive Elastic Modulus ($10^6$ psi) | D695 | — | 0.14–0.30 | — |
| 10. Flexural Elastic Modulus ($10^6$ psi) | D790 | 0.30–0.50 | 0.25–0.35 | 0.10–0.26 |
| 11. Thermal Conductivity (Btu-in/hr-ft$^2$-°F) | C177 | 1.02–1.45 | — | 3.19–3.60 |
| 12. Thermal Expansion ($10^6$ in/in-°F) | D696 | 27.8–55.6 | 52.8–61.1 | 61.1–72.2 |
| 13. Water Absorption, 24 hr, 1/8-in thick (%) | D570 | 0.04–0.40 | 0.20–0.45 | 0.01 |

| Property Material Type | Polypropylene (PP) Unmodified | Polycarbonate (PC) Unfiled | Polyester Cast Rigid | Epoxy (EP) Cast |
|---|---|---|---|---|
| 1. Specific Gravity | 0.90–0.91 | 1.20 | 1.10–1.46 | 1.11–1.40 |
| 2. Tensile Strength (psi) | 4300–5500 | 8000–9500 | 6000–13,000 | 4000–13,000 |
| 3. Elongation (%) | 200–700 | 100–130 | 5 | 3–6 |
| 4. Tensile Elastic Modulus ($10^6$ psi) | 0.16–0.23 | 0.30–0.35 | 0.30–0.64 | 0.35 |
| 5. Compressive Strength (psi) | 5500–8000 | 12,500 | 13,000–30,000 | 15,000–25,000 |
| 6. Flexural Strength (psi) | 6000–8000 | 13,500 | 8500–23,000 | 13,300–21,000 |
| 7. Impact Strength (ft-lb/in), Izod | 0.5–2.2 | 12.0–18.0 | 0.20–0.40 | 0.2–1.0 |
| 8. Hardness (Rockwell) | R80–R110 | M70–M78 R115–R125 | M70–M115 | M80–M110 |

*Continued.*

**TABLE 2.2** *(Continued)*
**Structural and Physical Properties and Processing Methods for Representative Engineering Plastics**

| Property<br>Material Type | Polypropylene<br>(PP)<br>Unmodified | Polycarbonate<br>(PC)<br>Unfiled | Polyester<br>Cast Rigid | Epoxy (EP)<br>Cast |
|---|---|---|---|---|
| 9. Compressive Elastic Modulus ($10^6$ psi) | 0.15–0.30 | 0.35 | — | — |
| 10. Flexural Elastic Modulus ($10^6$ psi) | 0.17–0.25 | 0.32–0.35 | — | — |
| 11. Thermal Conductivity (Btu-in/hr-ft$^2$-°F) | 0.81 | 1.33 | 1.16 | 1.16–1.45 |
| 12. Thermal Expansion ($10^6$ in/in-°F | 32.2–56.7 | 36.7 | 30.6–55.5 | 25.0–36.1 |
| 13. Water Absorption, 24 hr, 1/8-in thick (%) | 0.01–0.03 | 0.15–0.18 | 0.15–0.60 | 0.08–0.15 |

| Property<br>Material Type | Phenol-Formaldeyhde (PF)<br>Wood Flour and Cotton<br>Flock Filled | Silicone (SI)<br>Glass Fiber Filled<br>Molding Compound |
|---|---|---|
| 1. Specific Gravity | 1.34–1.45 | 1.80–1.90 |
| 2. Tensile Strength (psi) | 5000–9000 | 4000–6500 |
| 3. Elongation (%) | 0.4–0.8 | — |
| 4. Tensile Elastic Modulus ($10^6$ psi) | 0.80–1.70 | — |
| 5. Compressive Strength (psi) | 22,000–36,000 | 10,000–15,000 |
| 6. Flexural Strength (psi) | 7000–14,000 | 10,000–14,000 |
| 7. Impact Strength (ft-lb/in), Izod | 0.24–0.60 | 0.3–0.8 |
| 8. Hardness (Rockwell) | M100–M115 | M80–M90 |
| 9. Compressive Elastic Modulus ($10^6$ psi) | — | — |
| 10. Flexural Elastic Modulus ($10^6$ psi) | 1.00–1.20 | 1.0–2.5 |
| 11. Thermal Conductivity (Btu-in/hr-ft$^2$-°F) | 1.16–2.38 | 2.03–2.61 |
| 12. Thermal Expansion ($10^6$ in/in-°F) | 16.7–25.0 | 11.1–27.8 |
| 13. Water Absorption, 24 hr, 1/8 in thick (%) | 0.30–1.20 | 0.2 |

*Source*: Adapted from the 1975–1976 issue of *Modern Plastics Encyclopedia*. Copyright 1975 by McGraw-Hill, With permission.

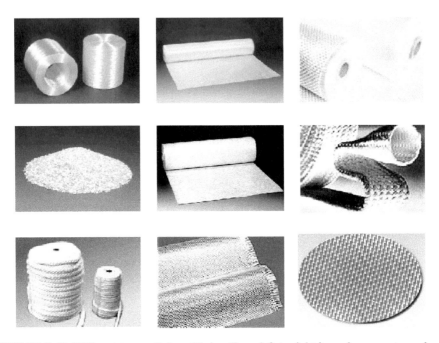

**FIGURE 2.18** Different types of glass fabrics. From left to right in each row — top: glass roving, oven cloth roving, bidirectional cloth; middle: chopped strand, chopped strand mat, braided glass fiber sleeve; bottom: knitted glass fiber rope, glass cloth, woven glass cloth.

and lowest in the transverse or radial direction. Fibers can be continuous or discontinuous; both can be used in producing composites. The strength and modulus of composites produced from continuous fibers are greater than those produced from discontinuous fibers.

A single continuous fiber is called a *filament* and is characterized by its extremely small diameter, which makes it difficult to handle for practical purposes. To obviate this difficulty, a large number of filaments are gathered together into a bundle to produce a commercial form called a *strand*, which can be used as reinforcing material. Note that their high tensile strength is generally attributed to their filamentary form, in which a statistically smaller number of surface flaws are found than in the bulk form. Indeed, measurements have shown the average tensile strength and modulus of fiber strands are smaller than those of single filaments. This phenomenon was discovered first by Griffith [1920], who showed that for very small diameters the fiber strengths approached the theoretical strength between adjacent layers of atoms, whereas for large diameters the fiber strength dropped to near the strength of the bulk glass.

Strands can be bundled together to produce reinforcing elements in a number of forms, such as uniaxial reinforcements (e.g., reinforcing bars and prestressing strands) and fabrics. Typical fabric forms (Figure 2.18) are:

• Chopped strand fiber mats
• Woven, stitched, or braided fabrics

- Bidirectional fabrics
- Short fibers (whiskers)

Unlike the elements with unidirectional fibers, fabrics are produced to meet the strength requirements in different directions. Also, fabrics keep fibers aligned prior to resin impregnation, especially in cases of complex product processing through pultrusion or resin transfer molding (RTM) techniques. Using proper combination of fabrics is important when designing a composite that must resist a complex state of stress.

## 2.9.1 GLASS FIBERS

Glass fibers are the most commonly used fibers for producing FRP composites. Glass fibers are made from molten glass spun from electrically heated platinum-rhodium alloy bushings (or a furnace) at a speed of 200 mph. These filaments cool from a temperature of about 2300°F to room temperature within $10^{-5}$ seconds. Glass fibers have a diameter ranging from 0.000090 to 0.00035 in. (i.e., 90 to 350 μ in). A protective coat (called *sizing*) is applied on individual filaments before they are gathered together into a strand and wound on a drum at speeds of up to 2 miles (3.2 km) per minute. A strand is the basic form of commercially used continuous glass fibers and consists of 204 or more parallel filaments. A group of untwisted parallel strands wound in a cylindrical forming package is called a *roving*.

*Sizing* (the protective coating, discussed in Section 2.10), is a constituent (a mixture of lubricant, antistatic agent, and a binder) used in the production of fibers and performs the following functions:

1. Reduces the abrasive effect of filaments rubbing against one another
2. Reduces static friction between the filaments
3. Packs filaments together into a strand
4. Reduces the damage to fibers during mechanical handling
5. Facilitates the molding process

Several types of commercially available glass fibers are identified below:

1. *E-glass*, which has low alkali content and is the most common type of glass fiber in high-volume commercial use. It is used widely in combination with polyester and epoxy resins to form a composite. Its advantages are low susceptibility to moisture and high mechanical properties.
2. *Z-glass*, which is used for cement mortars and concretes due to its high resistance against alkali attack.
3. *A-glass*, which has a high alkali content.
4. *C-glass*, which is used for applications that require greater corrosion resistance to acids, such as chemical applications.
5. *S-* or *R-glass*, which is produced for extra-high strength and high-modulus applications.
6. Low *K-glass* is an experimental fiber produced to improve dielectric loss properties in electrical applications and is similar to *D-glass* (dielectric glass).

**TABLE 2.3**
**Chemical Composition of E- and S-Glass Fibers**

| | Ingredients (wt%) | | | | | |
|---|---|---|---|---|---|---|
| Type of Fibers | $SiO_2$ | $Al_2O_3$ | CaO | MgO | $B_2O_3$ | $Na_2O_3$ |
| E-glass | 54.5 | 14.5 | 17 | 4.5 | 8.5 | 0.5 |
| S-glass | 64 | 26 | — | 10 | — | — |

*Source:* From Mallick, P.K. *Fiber Reinforced Composite Materials: Manufacturing and Design,* New York: Marcel Dekker, 1993. With permission.

The chemical structure of E- and S-glass fibers indicates silica ($SiO_2$) as the principal ingredient. Oxides such as boric oxide ($B_2O_3$) and aluminum oxide ($Al_2O_3$) are added to modify the network structure of $SiO_2$ and to improve workability. The $Na_2O$ and $K_2O$ content are low to give E- and S-glass fibers a better corrosion resistance against water as well as a higher surface resistivity. Chemical compositions of E- and S-glass fibers are listed in Table 2.3.

Glass fibers offer many advantages, such as:

1. Low cost
2. High tensile strength
3. High chemical resistance
4. Excellent insulating properties

The drawbacks of glass fibers are:

1. Low tensile modulus
2. Relatively high specific gravity
3. Sensitivity to abrasion from handling
4. High hardness
5. Relatively low fatigue resistance

### 2.9.2 CARBON FIBERS (GRAPHITE FIBERS)

#### 2.9.2.1 General Description

Carbon fiber is defined as a fiber containing at least 90% carbon by weight obtained by the controlled pyrolysis of appropriate fibers. The term "graphite fiber" is used to describe fibers that have carbon above 95% by weight. A large variety of precursors are used to manufacture different types of carbon. The most commonly used precursors are polyacrylonitrile (PAN), petroleum or coal tar pitch, cellulosic fibers (viscose rayon, cotton), and certain phenolic fibers. Carbon fibers are distinct from other fibers by virtue of their properties and are influenced by the processing conditions, such as tension and temperature during process. Successful commercial production of carbon composites was started in the 1960s, because of the needs of the aerospace industry, especially military aircraft, for better and lightweight materials. Carbon fiber composites are ideally suited for applications where strength, stiffness, lower weight, and outstanding fatigue characteristics constitute critical requirements. Unlike glass and aramid fibers, carbon fibers do not exhibit stress corrosion or stress rupture failures at room temperature. In addition, they can be used in applications requiring high temperature resistance, chemical inertness, and damping characteristics.

### 2.9.2.2  Classification and Types

On the basis of precursor materials, fiber properties, and final heat treatment temperature, carbon fibers can be classified into the following categories:

1. Based on precursor materials
   PAN-based carbon fibers
   Pitch-based carbon fibers
       Mesophase pitch-based carbon fibers
       Isotropic pitch-based carbon fibers
   Rayon-based carbon fibers
   Gas-phase-grown carbon
2. Based on fiber properties fibers
   Ultra-high-modulus (UHM)-type UHM (> 450 GPa)
   High-modulus (HM)-type HM (325 to 450 GPa)
   Intermediate-modulus (IM)-type IM (200 to 325 GPa)
   Low modulus and high-tensile (HT)-type HT (modulus < 100 GPa and
       strength > 3.0 GPa)
   Super high-tensile (SHT)-type SHT (tensile strength > 4.5 GPa)
3. Based on final heat treatment temperature
   Type I (high-heat-treatment carbon fibers): associated with high-modulus
       type fiber (> 2000°C)
   Type II (intermediate-heat-treatment): associated with high-strength
       strength type fiber (> 1500°C and < 2000°C)
   Type III (low-heat-treatment carbon fibers): associated with low modulus
       and low strength fibers (< 1000°C)

### 2.9.3  MANUFACTURING OF CARBON FIBERS

Carbon fibers are manufactured from synthetic fibers through heating and stretching treatments. Polyacrylonitrile (PAN) and pitch are the two most commonly used precursors (raw material) used for manufacturing carbon fibers. PAN is a synthetic fiber that is premanufactured and wound onto spools, whereas pitch is a by-product of petroleum distillation process or coal coking that is melted, spun, and stretched into fibers. The fibers are subjected to different treatment schemes — thermosetting, carbonizing, and graphitization.

In the thermoset treatment, the fibers are stretched and heated to about 400°C or less. This procedure cross-links carbon chains such that the fibers will not melt in subsequent treatments. During the carbonization treatment, the fibers are heated

TABLE 2.4
PAN Fiber Compositions at Different Treatment Stages

| Treatment Step | Oxygen (wt%) | Hydrogen (wt%) | Nitrogen (wt%) | Carbon (wt%) |
|---|---|---|---|---|
| Untreated | — | 6 | 26 | 68 |
| Thermoset | 8 | 5 | 22 | 65 |
| Carbonize | < 1 | < 0.3 | < 7 | > 92 |
| Graphitize | — | — | — | 100 |

*Source:* Adapted from Bunsell, A.R., *Fibre Reinforcements for Composite Materials*, Amsterdam: Elsevier Science Publishers, 1988.

TABLE 2.5
Effect of Temperature on Modulus of Carbon Fibers

| Carbon Fiber Grade | Low Modulus | Standard Modulus | Intermediate Modulus | High Modulus |
|---|---|---|---|---|
| Carbonization Temperature (°C) | Up to 1000 | 1000–1500 | 1500–2000 | 2000+ (Graphitization) |
| Modulus of Elasticity (GPa) | Up to 200 | 200–250 | 250–325 | 325+ |

*Source:* Adapted from Bunsell, A.R., *Fibre Reinforcements for Composite Materials*, Amsterdam: Elsevier Science Publishers, 1988.

to about 800°C in an oxygen free-environment, which removes noncarbon impurities. During the graphitization treatment, fibers are stretched between 50 to 100% elongation, and heated to temperatures between 1100°C and 3000°C. The stretching results in crystalline orientation and high Young's modulus (300 to 600 GPa). Finally, the last two treatment steps — surface treatment and epoxy sizing — are performed to improve the carbon fiber/epoxy bonding strength. Effects of temperature on carbon fibers are shown in Table 2.4 and Table 2.5.

Carbon fibers offer the following advantages:

1. High tensile strength-to-weight ratio
2. High tensile modulus-to-weight ratio
3. Very low coefficient of linear thermal expansion
4. High fatigue strength

Some of the disadvantages of carbon fibers include high cost; high brittleness; and electrical conductivity, which might limit their application potential.

PAN carbon fibers are produced using higher-cost polymers, whereas isotropic pitch carbon fibers are produced from lower-cost feedstock from petroleum or coal tar. Because of its higher tensile strength, PAN carbon fiber is primarily used as a structural reinforcement. With its lower tensile strength, pitch carbon fiber is customized to meet specific application needs.

## 2.10 ARAMID FIBERS (KEVLAR FIBERS)

### 2.10.1 GENERAL DESCRIPTION

Aramid fiber is the generic name given to aromatic ployamide fibers. The U.S. Federal Trade Commission defines an aramid fiber as "a manufactured fiber in which the filler forming substance is a long chain synthetic polyamide in which at least 85% of the amide linkages are attached directly to two aromatic rings" [Lubin 1982; Schwartz 1992].

Aramis fibers were first introduced by DuPont in 1971 under the trade name Kevlar™. It is an aromatic compound of carbon, hydrogen, oxygen, and nitrogen. The chemical composition of Kevlar is poly para-phenylene-diamine-terephthalamide (PPD-T). It is produced from a condensation reaction of para-phenylene diamine and terephthaloyl chloride. The resultant aromatic polyamide contains aromatic and amide groups. Polymers with high breaking strength often have one or both of these groups. The aromatic ring structure contributes high thermal stability, and the para configuration leads to stiff, rigid molecules that contribute high strength and high modulus [Mallick 1993].

Para-aramid fibers belong to a class of materials called *liquid crystalline polymers*. When PPD-T solutions are extruded through a spinneret and drawn through an air gap during fiber manufacture, the liquid crystalline domains can orient and align in the flow direction. Kevlar can acquire a high degree of alignment of long, straight ploymer chains parallel to the fiber axis. The structure exhibits anisotropic properties, with higher strength and modulus in the fiber longitudinal direction than in the radial direction. The extruded material also possesses a fibrillar structure. This structure results in poor shear and compression properties for aramid composites. Hydrogen bonds form between the polar amide groups on adjacent chains and they hold the individual Kevlar polymer chains together [CISPI 1992; *Modern Plastics Encyclopedia* 1975].

### 2.10.2 CHARACTERISTICS

Aramid fibers have:

1. No melting point
2. Low flammability
3. Good fabric integrity at elevated temperatures
4. Para-aramid fibers, which have a slightly different molecular structure, provide outstanding strength-to-weight properties, high tenacity and high modulus

The many advantages of aramid fibers include very low thermal conductivity; a very high damping coefficient; and a high degree of yielding under compression, which gives superior tolerance to damage against impact and other dynamic loading.

However, aramid fibers also have certain adverse properties. These fibers are hygroscopic, i.e., they can absorb moisture up to about 10% of fiber weight. At high moisture content, they tend to crack internally at pre-existing microvoids and produce longitudinal splitting. They have a low compressive strength and exhibit a loss of strength and modulus at elevated temperatures. Aramid fibers present difficulty in cutting and machining. They also are sensitive to UV lights, a drawback that leads to mechanical property deterioration over time.

### 2.10.3 COMMERCIAL APPLICATIONS OF ARAMID FIBERS

Aramid fibers were originally developed to replace steel in radial tires, which led to its increasing use in the belts of radial car tires and carcasses of radial truck tires, where it saves weight while increasing strength and durability [Schwartz 1992].

The three commercial variants of Kevlar aramid fibers are Kevlar 29, Kevlar 49, and a second-generation fiber called Kevlar 149. Aramid fibers are used for flame-resistant clothing, protective vests and helmets, asbestos replacement, hot air filtration fabrics, tire and mechanical rubber goods reinforcement, ropes and cables, sail cloth, sporting goods, and composites.

## 2.11  BORON FIBERS

Boron fibers are said to be the first advanced fibers that were commercially available [CISPI 1992]. They have a very high tensile modulus, in the range of $50 \times 10^6$ to $60 \times 10^6$ psi. Boron fibers are produced by a chemical vapor deposition (CVD) from the reduction of boron trichloride ($BCl_3$) with hydrogen on fine tungsten wire or carbon monfilament substrate [Agarwal and Broutman 1990].

The tensile strength can be increased by etching part of the outer portion of the filament. This increase in tensile strength is due to some decrease in residual tensile stresses at the inner core, which results from the removal of the outer region of the filament, which contains compressive residual stresses [Agarwal and Broutman 1990]. Advantages of boron fibers include high tensile modulus and good resistance under compressive loads to buckling owing to the large fiber diameter.

## 2.12  ADDITIVES AND OTHER INGREDIENTS

*Additives* and *modifiers* are added during the manufacturing process to modify material properties or the laminate performance to meet design requirements, such as improved processability or product durability. Catalysts, promoters, and inhibitors are used to accelerate or slow the rate of polymerization. Also, release agents are used to facilitate the removal of composites from the mold.

Additives modify the properties of thermoset polymers in the following manner [CISPI 1992]:

1. Low shrink/low profile — used when a smooth surface is desired
2. Fire resistance — to retard or extinguish fire upon contact

3. Viscosity control — to control the flow of the resin where it is needed, depending upon the process of composites
4. Air release agents — which reduce air entrapment and air voids that can cause improper fiber wetting
5. Electrical conductivity agents (shields conductivity from certain fibers)
6. Toughness agents — provides extra toughness to fibers
7. Antioxidants — to keep the polymer from experiencing oxidation
8. Antistatic agents — to reduce a static or electrical charge
9. Foaming agents — to provide lower density, decrease material cost, improve electrical and thermal insulation, increase strength-to-weight ratio, and reduce shrinkage and part warping
10. Plasticizers — to improve processing characteristics and offer a wide range of physical and mechanical properties
11. Slip and blocking agents — to provide lubrication and reduce friction

Additives for thermoplastic resins improve properties, appearance, and processability. Colorants, flame retardants, and mold release agents play a role as discussed for thermosets. Two other additives for thermoplastics are: heat stabilizers, used for protecting polymers from degradation from heat; and UV stabilizers, used to protect from UV degradation [CISPI 1992].

## 2.13  FILLERS

*Fillers* are added to both thermoset and thermoplastic resins to reduce cost, control shrinkage, improve mechanical and physical properties, and to provide UV protection, water resistance, weathering protection, surface smoothness, temperature resistance, impact strength, and fire resistance. Fillers can also provide extra stiffness and dimensional stability. Fillers used for structural applications improve the load transfer and reduce cracking in the unreinforced area.

Fillers are classified as functional and nonfunctional types. Functional types of fillers are alumina trihydrate for flame retardancy; mica, feldspar, and glass milled fibers for reinforcement; and glass microspheres for lower weight. Nonfunctional fillers (such as calcium carbonate) are used to reduce cost; however, special nano-clays are being used to reduce diffusion of harmful materials and corrosion in composites. Fillers may account for about 40 to 65% by weight of the material.

Fillers can also be classified by their particle size: fine and coarse. With their high surface area, fine fillers (with an average particle size less than 5 μm) can produce high viscosities in the resin formulations. Fine fillers provide better cohesiveness and tend to lubricate better during production process. In addition, due to their better distribution in polymers, they help to reduce local shrinkage and improve surface finish.

Fillers must be pure — i.e., free from contamination and uniform — and should contain less than 5% free water content. Contaminants can cause localized reactions leading either to void formations or to the uniformity of coloring of the finished product. Wetting agents are used to add (create) volume to filler material without increasing the viscosity of the resin system. Air release agents are added in the same way as wetting agents, for the purposes of reducing entrapped air in the liquid resin to reduce void ratio in a finished composite. A detailed review of fillers is provided by Monte [1978].

## 2.14 FIBER SURFACE TREATMENT (SIZING)

The surface of fibers is treated primarily to improve the fiber surface wettability with the matrix; create a stronger bond at the fiber-matrix interface; and protect the fiber from certain environmental attacks. Sizing is necessary for the effective transfer of stress from the matrix to the fiber and vice versa. To accomplish this objective, glass, carbon, and aramid fibers are surface-treated with compounds called "coupling agents." The surface treatment of filaments is carried out immediately after their formation to ease processing and protect fibers from breakage during processing. Sizing comprises only 0.25 to 6% of total fiber weight.

Sizing chemistry determines the grade of fiber for pultrusion, filament winding, and other processes. Sizing chemistry varies from one application to another. For example, sizing for polyester resin is different from one used with phenolic resin. Sizing formulations evolve continually for improving fiber processing characteristics (to reduce fuzz, to improve wet-art, and to chop glass fibers cleanly); as such their constituents are closely guarded trade secrets. A description of these surface treatments is summarized from Mallick [1993] as follows.

### 2.14.1 GLASS FIBERS

The chemical coupling agents commonly used with glass fibers are organofunctional silicon compounds called "silanes," which are used in aqueous solution. Before treating the glass fibers with coupling agents, cleaning their surface from sizing applied at the time of manufacture is necessary. This is accomplished by burning the sizing through heating the fiber in an air-circulating oven at 644°F (340°C).

Treating glass fibers with chemical coupling agents helps to improve the fiber/matrix interfacial strength through physical and chemical bonds and to protect the fiber surface from moisture and reactive fluids, an extremely important consideration for prevention of fiber strength degradation.

### 2.14.2 CARBON FIBERS

The primary need for the surface treatment of carbon fibers arises from the fact that their surfaces are chemically inactive and must be treated to form surface functional groups that promote good chemical bonding with the matrix. The surface treatment also increases the surface area by creating surface pits on the potentially porous carbon fiber surface. The porous surface provides a large number of contact points for fiber-matrix interfacial bonding.

Two types of surface treatments for carbon fibers are in commercial use: oxidative and nonoxidative. The oxidative surface treatment produces acidic functional groups, such as caboxylic, phenolic, and hydroxilic, on the surface of carbon fibers.

The nonoxidative treatments are of several types. One of these treatments involves coating the carbon fiber surface with an organic polymer that has functional groups capable of reacting with the resin matrix. A discussion on the surface treatments of fibers has been provided by Donnet and Bansal [1970].

### 2.14.3 ARAMID FIBERS

The commercially available aramid fibers, Kevlar 49, also suffer from weak interfacial adhesion, which makes the surface treatment a necessity. Two methods of surface treatments are in use: filament surface oxidation, also called plasma etching, and formation of reactive groups on the fiber surface. A discussion of these surface treatments has been provided by Morgan and Allread [1989].

## 2.15  PROPERTIES OF FIBERS

The stress-strain plots of fibers subjected to tension indicate a linear relationship. In contrast with steel, fibers do not exhibit a plastic region in the stress-strain relationship, and the fiber failure is brittle. Fibers are produced in different diameters of 4 to 10 µm from glass, carbon, aramid, and boron. Fibers can be produced with a wide range of physical and mechanical properties. For example, carbon fibers are manufactured with different:

1. Moduli (low, intermediate, high, and ultrahigh)
2. Strength (low and high)
3. Cross-sectional areas
4. Shapes
5. Twists
6. Number of fiber ends

Recognizing that properties of carbon fibers produced by different manufacturers can be different is very important. This is due to the subtle differences in precursor types and the carbon fiber processes. Different manufacturers may not have the same number of fibers or twists, which could affect the composite properties. Also, some manufacturers may have better processing techniques over other manufacturers [Mallick 1993]. General fiber properties of glass, carbon, aramid, and boron fibers are listed in Table 2.6.

**TABLE 2.6a**
**General Properties of Fibers**

|  | Specific Gravity | Tensile Strength (ksi) | Tensile Modulus ($10^6$ psi) | Coefficient of Thermal Expansion ($10^{-6}/°C$) | Strain to Failure (%) |
|---|---|---|---|---|---|
| Glass | 2.48–2.62 | 217–700 | 10.2–13.0 | 2.9–5.0 | 4.8–5.0 |
| Carbon PAN | 1.76–1.96 | 220–820 | 33–70 | −0.60 to −0.75 | 0.38–1.81 |
| Carbon Pitch | 2.0–2.15 | 275–350 | 55–110 | −1.30 to −1.45 | 0.32–0.50 |
| Aramid | 1.39–1.47 | 435–525 | 10.1–19.0 | −2.0 to −6.0 | 1.9–4.4 |
| Boron | 2.7 | 450 | 57 | 5 | 0.2 |

*Source:* Data from Mallick 1993, ACI 440.1R-03.

**TABLE 2.6b**
**Specific Strength and Stiffness Properties of Fiber Reinforcements**

| Type of Fiber Reinforcement | (1) Specific Gravity | (2) Density (lb/in³) | (3) Tensile Strength ($10^3$ psi) | (4)* Specific Strength ($10^6$ in) | (5) Tensile Elastic Modulus ($10^6$ psi) | (6)† Specific Elastic Modulus ($10^8$ in) |
|---|---|---|---|---|---|---|
| **Glass** | | | | | | |
| E monofilament | 2.54 | 0.092 | 500 | 5.43 | 10.5 | 1.14 |
| 12-end roving | 2.54 | 0.092 | 372 | 4.04 | 10.5 | 1.14 |
| S monofilament | 2.48 | 0.090 | 665 | 7.39 | 12.4 | 1.38 |
| 12-end roving | 2.48 | 0.090 | 550 | 6.17 | 12.4 | 1.38 |
| **Graphite** | | | | | | |
| High strength | 1.80 | 0.065 | 400 | 6.15 | 38 | 5.85 |
| High modulus | 1.94 | 0.070 | 300 | 4.29 | 55 | 7.86 |
| Intermediate | 1.74 | 0.063 | 360 | 5.71 | 27 | 4.29 |

* Strength/density.
† Elastic modulus/density.

*Source:* Adapted from *Structural Plastics Design Manual,* ASCE Manuals and Reports on Engineering Practice, No. 63, American Society of Civil Engineers, pp. 66, 1984.

## REFERENCES

ACI 440.1R-03: Guide for the Design and Construction of Concrete Reinforced with FRP Bars, Detroit, Michigan: American Concrete Institute, 2003.

Agarwal, B. and Broutman, L., *Analysis and Performance of Fiber Composites*, New York: John Wiley & Sons, 1990.

American Association of State Highways and Transportation Officials (AASHTO), *Standard Specifications for Highway Bridges*, 16th ed., Washington, DC, 1996.

American Concrete Institute (ACI), State-of-the-report on fiber-reinforced concrete, *An International Symposium: Fiber-Reinforced Concrete*, ACI Committee 544, Special Publication, SP-44, Farmington Hills, MI, 1974.

American Concrete Institute (ACI), State-of-the-report on fiber-reinforced concrete, ACI Committee 544, *Report ACI 544-1R-82, Concrete International: Design and Construction* (re-approved 1986), Farmington Hills, MI, 1982.

American Concrete Institute (ACI), *Building Code Requirements for Structural Concrete (ACI 318-99) and Commentary (ACI 318R-99)*, ACI 318-99, Farmington Hills, MI, 1999.

American Concrete Institute (ACI), *Cement and Concrete Technology*, ACI 116R-78, ACI Committee 116, Farmington Hills, MI, 1978.

American Concrete Institute (ACI), *Concrete International*, ACI Committee 212, Chemical Admixtures, Farmington Hills, MI, 3, 5, 1981.

American Concrete Institute (ACI), *Proc. ACI Journal*, ACI Committee 223, Farmington Hills, MI, 73(6), 1976, 319–339.

American Society of Testing Materials (ASTM) C125-06 Standard Terminology Relating to Concrete and Concrete Aggregates, Philadelphia, 2006.

American Society of Testing Materials (ASTM) C150-05 Standard Specification for Portland Cement, Philadelphia, 2005.

American Society of Testing Materials (ASTM) C260-01 Standard Specification for Air-Entraining Admixtures for Concrete, Philadelphia, 2001.

American Society of Testing Materials (ASTM) C330-05 Standard Specification for Light-weight Aggregates for Structural Concrete, Philadelphia, 2005.

American Society for Testing and Materials (ASTM) C 469-94, Test Method for Static Method of Elasticity and Poisson's Ratio of Concrete in Compression, Philadelphia, 1994.

American Society of Testing Materials (ASTM) C494/C494M-05a Standard Specification for Chemical Admixtures for Concrete, Philadelphia, 2005.

American Society of Testing Materials (ASTM) C567-05 Standard Test Method for Determining Density of Structural Lightweight Concrete, Philadelphia, 2005.

American Society of Testing Materials (ASTM) C595-05 Standard Specification for Blended Hydraulic Cements, Philadelphia, 2005.

Aziz, M.A., Paramsivam, P., and Lee, S.L., Concrete reinforced with natural fibers, in *New Reinforced Concretes*, R.N. Swamy (Ed.), Glasgow: Surrey University Press, 1984.

Balguru, P.N., and Shah, S.P. Alternative reinforcing materials for developing technology, *Intl. J. Dev. Tech.*, 2, 1985, 87–105.

Balguru, P.N. and Shah, S.P., *Fiber-Reinforced Cement Composites*, New York: McGraw-Hill, 1992.

Beck, D.L., Hiltz A.A., and Knox, J.R., Experimental data of specific volume vs. temperature to determine $T_g$ of polypropylene, *Soc. Plast. Eng. Trans.*, March, 1963, 279.

Berglund, L.A., Thermoplastic resins, in *Handbook of Composites*, S.T. Peters (Ed.), New York: Chapman & Hall, 1998.

Brunauer, S. and Copeland, L.E., The chemistry of concrete, *Scientific American*, April 1964.

Bunsell, A.R., *Fibre Reinforcements for Composite Materials*, Amsterdam: Elsevier Science Publishers, 1988.

Canadian Society for Civil Engineers (CSCE), *Advanced Composite Materials with Application to Bridges*, Montreal, 1992.

Composites Institute of the Society of the Plastics Industry (CISPI), *Introduction to Composites*, Washington, DC, 1992.

Concrete Reinforcing Steel Institute (CRSI), *Manual of Standard Practice*, Chicago, 1994.

Cornish, E.H., *Materials and the Designer*, New York: Cambridge University Press, 1987.

Cossis, F.A. and Talbot, R.C., Polyester and vinyl ester resins, in *Handbook of Composites*, S.T. Peters (Ed.), New York: Chapman & Hall, 1998.

Davis, H.E., Troxell, G.E., and Wiskosil, C.T., *The Testing and Inspection of Engineering Materials*, New York: McGraw-Hill, 1964.

Dewprashad, B., and Eisenbraun, E.J. Fundamentals of epoxy formulation, *J. Chem. Educ.*, 71, 4, April, 1994, 290–294.

Donnet, J.B. and Bansal, R.C., *Carbon Fibers*, New York: Marcel Dekker, 1984.

Ege, S., *Organic Chemistry*, 2nd ed., Lexington, MA: D.C. Heath & Co., 1989.

Gerig, J.T., *Introductory Organic Chemistry*, New York: Academic Press, 1974.

Gibson, R.F., *Principles of Composite Material Mechanics*, New York: McGraw-Hill, 1994.

Griffith, A.A., The phenomenon of rupture and flow in solids, *Phil. Trans. Royal Soc.*, 221A, 1920, 163–198.

Gupta, R.K., *Polymer and Composite Rheology*, 2nd ed., New York: Marcel Dekker, 2000.

Gustafsson, C.G., Initiation of growth of fatigue damage in graphite/epoxy and graphite/PEEK laminates, Ph.D. dissertation, Royal Institute of Technology, Stockholm, 1988.

Hassoun, M.N., *Design of Reinforced Concrete Structures*, Boston: PWS Engineering, 1997.

International Conference of Building Officials (ICBO), *Uniform Building Code-1997, Volume 2*, Whittier, CA: 1997.

Jones, R.M., *Mechanics of Composite Materials*, New York: Hemisphere Publishing, 1975.

Kajorncheappunngam, S., The effect of environmental aging on durability of glass/epoxy composite, Dissertation submitted to the College of Engineering and Mineral Resources, EVU, 1999.

Kosmatka, S.H. and Panarese, W.C., *Design and Control of Concrete Mixtures*, 13th ed, Chicago: Portland Cement Association, 1992.

Lubin, G. (Ed.). *Handbook of Composites,* New York: Van Nostrand Reinhold Co., 1982.

MacGregor, J.G., *Reinforced Concrete: Mechanics and Materials*, Upper Saddle River, NJ: Prentice Hall, 1997.

Mallick, P.K., *Fiber Reinforced Composite Materials: Manufacturing and Design,* New York: Marcel Dekker, 1993.

McConnel, V.P., Bridge column retrofit, *High-Performance Composites*, September/October, 1993, 62–64.

McConnel, V.P., Infrastructure update, *High-Performance Composites*, May/June, 1995, 21–25.

Mehta, P.K., *Concrete: Structure, Properties, and Materials*, Upper Saddle River, NJ: Prentice-Hall, 1986.

Mirza, S.A., Hatzinkolas, M., and MacGregor, J.G., Statistical description of strength of concrete, *Proc. ASCE J. Struct. Div.*, 105, ST6, June, 1979.

*Modern Plastics Encyclopedia*, New York: McGraw-Hill, 1975.

Monte, *Handbook of Fillers and Reinforcements*, 1978.

Morgan, R.J. and Allred, R.E., *Aramid Fiber Reinforcements, Reference Book for Composites Technology,* Vol. 1, Lee, S.M., ed., Lancaster, PA: Technomic Pub. Co., 1989.

Munley, E., FHWA structures research program: fiber reinforced polymer (FRP) composite materials, *Summary of Remarks for 10th ASM/ESD Advanced Composite Conference and Exposition*, Dearborn, MI, November 8, 1994.

Nielsen, E.L. and Landel, R.F., *Mechanical Properties of Polymers and Composites*, 2nd ed., New York: Marcel Dekker, 1994.

Pauw, A., Static modulus of elasticity of concrete as affected by density, *Proc. ACI Journal*, 57, 6, December, 1960, 679–687.

Penn, L.S. and Wang, H., Epoxy resins, in *Handbook of Composites*, S.T. Peters (Ed.), New York: Chapman & Hall, 1998.

Schwartz, M., *Composites Materials Handbook*, 2nd ed., New York: McGraw-Hill Co., 1992.

Shimp, D.A., Specialty matrix resins, in *Handbook of Composites*, S.T. Peters (Ed.), New York: Chapman & Hall, 1998.

Smith, W.F., *Principles of Material Science and Engineering*, 2nd ed., New York: McGraw-Hill, 1990.

*Structural Plastics Design Manual*, ASCE Manuals and Reports on Engineering Practice, No. 63, American Society of Civil Engineers, pp. 66, 1984.

Tullo, A.H., Plastics additives' steady evolution, *C & EN*, December 2000, 21–31.

Vijay, P.V., GangaRao, H.V.S., and Bargo, J.M., *Mechanical Characterization of Recycled Thermoplastic Polymers for Infrastructure Applications,* ACMBS-3, Ottawa, Canada, pp. 55–60, Aug., 2000.

# 3 Manufacturing of Composite Components

## 3.1 INTRODUCTION

Composites consisting mainly of fibers/fabrics and resins (polymers) are manufactured using automated and manual processes. Over 20 different methods of composite component manufacturing are in practice with each having its own advantages, limitations, and capability to produce specific end products such as FRP bars, bridge deck shapes, automobile parts, utility poles, airplane components, and others. Depending on the manufacturing process, fibers/fabrics are wetted and cured with resins by different techniques. During manufacturing, resins are mixed with additives and modifiers (e.g., accelerators, pigments, UV ray inhibitors, fire retardants and others) to achieve proper curing characteristics, viscosity, durability, appearance, and finish. Some composite products are manufactured by using readily available inserts such as metal bolts, brackets, and other accessories. Some of the important parameters that affect short- and long-term properties of composites are:

- Fiber/fabric properties and configurations
- Resin properties
- Additive and modifier properties
- Percent cure of resins and fiber volume content
- Process parameters (temperature, pressure, cure time, and surface finish requirements)

Several commonly used manufacturing methods are listed in Section 3.2. Among the manufacturing processes listed, pultrusion is used for manufacturing FRP bars and hand lay-up is used for wrapping or bonding FRP fabrics to concrete member surfaces for external strengthening. Other manufacturing methods described in this chapter provide details on their capabilities, advantages, and limitations. Additional details on composite manufacturing methods are available in handbooks and publications related to a particular manufacturing method [ASM 1989, 2001; CSCE 1992; Mallick 1997; Gutowski 1997; Holloway 1990; Peters 1998].

## 3.2 MANUFACTURING METHODS

Several commonly used manufacturing methods are listed below and briefly described in Section 3.2.1 through Section 3.2.9:

- Hand (wet) lay-up/automated lay-up
- Pultrusion
- Filament winding
- Resin transfer molding (RTM)
- Sheet molding compound (SMC)
- Seemann composite resin infusion molding process (SCRIMP)
- Injection molding
- Compression molding
- Extrusion

### 3.2.1 Hand (Wet) Lay-up/Automated Lay-up

The *hand (wet) lay-up* and *automated lay-up* processes have been used to produce a significantly large number of fiber reinforced polymer composite products. More than half of structural composites in the aerospace industry are made from hand lay-up processes. In the hand lay-up process (Figure 3.1), fibers in the form of chopped or continuous fabrics are impregnated with resin using handheld rollers and brushes. Several fabric layers with required fiber/fabric orientation can be stacked on each other, with each layer being coated or sprayed with resin. To minimize tooling costs, open molds used for hand lay-up process can be easily modified to manufacture products with different shapes and surface textures [Anderson et al. 2000; Branco et al. 1995].

In an open mold process, the mold surface is treated with several layers of release agent (wax) and then spray coated with a pigmented polyester resin called a *gel coat*. On top of the gel coat, fiber/fabric (reinforcement) layers saturated (wet) with resin and catalyst at the desired room temperature are positioned. Each fiber/fabric layer is pressed with hand rollers to ensure proper and uniform wetting of the reinforcement. Some of the structural parts are sandwiched with honeycomb or rigid foam blocks as core material. Hand lay-up processes can utilize resin pre-impregnated reinforcement called "prepreg" to provide consistent control over reinforce-

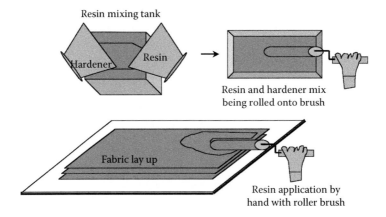

**FIGURE 3.1** Schematic representation of a hand lay-up process.

ment to resin ratio by weight or volume. Several variations in the hand/automated lay-up manufacturing process are possible. Common steps used in a hand lay-up manufacturing process are described below.

- *Mold preparation* — A mold of the part to be manufactured is created and a release film is applied to the mold surface. Mold design depends on the product shape, dimensions, reinforcing materials, resin type, additives, and the magnitude of required pressure application during production. Mold design also depends on required life, tooling, and labor costs.
- *Gel coating* — This step consists of applying a specially formulated resin layer called gel coat on the mold surface, which will become the outer surface of the manufactured product. Gel coats are used only when a good surface appearance is required.
- *Hand lay-up* — Reinforcement in the form of chopped strand mat, fabric, or woven roving is applied to the gel coat surface, which is precoated with resin mix. Resin mix consisting of resin and catalyst (hardener) with a specified ratio is applied to the fiber/fabric surface using rollers and gently pressed against the mold to attain air removal, resin saturation, and consolidation. The composite is completely cured under ambient conditions or with the aid of external heaters (e.g., infrared heaters).
- *Finishing* — Desired machining and assembly work is carried out on the cured composite part [Davima et al. 2004].

In automatic lay-up procedure, multiaxis Computer Numeric Control (CNC) machines are used to lay prepreg tape or prepreg fibers. Prepreg curing is typically carried out with the use of ovens, heated-platen presses, or autoclaves. Room temperature curing can take as long as two hours, whereas heat-assisted curing can take about one half-hour or less. Curing can also be carried out with vacuum bag molding, wherein a nonadhering plastic film — usually polyester — is sealed around the mold plate and lay-up assembly. Vacuum force is applied through the bag covering the whole fiber/fabric lay-up to draw out both excess resin and entrapped air. For vacuum bag and autoclave assisted curing lay-up processes, some of the accessories needed include separator films, bleeder plies, vent cloth, vacuum lines and fittings, edge seals, thermocouples, and autoclave units [Gutowski 1997].

The hand lay-up process is simple but slow, and involves several labor-intensive steps. It requires good ventilation and protective equipment for workers. To attain consistent quality of the end product, special care is necessary. Typical production rates using hand lay-up are about 1 lb/hr [Gutowski 1997].

### 3.2.2 PULTRUSION

The *pultrusion* process derives its name by using the word "pul" from the pulling force applied to fibers passing through a heated die and combining it with the word "trusion" from the extrusion process that consists of extruding (pushing) hot molten material (metal or polymer) through a die. Though the "pultrusion" and "extrusion" processes have some similarities, molten material in the extrusion process is pushed out through the die as opposed to the beneficial pulling force applied on fibers exiting a heated die in a pultrusion process.

In pultrusion processes, reinforcement in the form of fiber rovings or fabrics are continuously pulled from creels and preshaped with a series of guides for producing the required products. Figure 3.2 through Figure 3.5 show pultrusion manufacturing steps involved in pultruding a composite FRP deck that was designed and developed by the Constructed Facilities Center (CFC) of West Virginia University. Bedford Reinforced Plastics Inc. of Pennsylvania performed the manufacturing of those FRP decks that have been field-installed in many bridge structures constructed by several state Department of Transportation agencies in the United States. The fibers/fabrics are often pulled through preheaters to dry them and remove any undesirable condensation that might inhibit wetting. The fibers/fabrics used can be of single material types or hybrids (e.g., a combination of glass, kevlar, or carbon fibers).

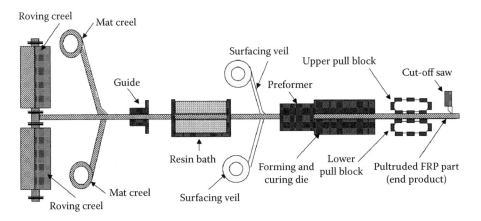

**FIGURE 3.2** Schematic of a pultrusion process.

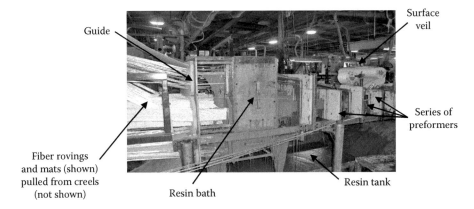

**FIGURE 3.3** Pultrusion process: fibers/fabrics passing through guides and into the resin bath (left and middle part); resin tank at the bottom of the resin bath (middle portion); addition of surface veils (right part).

Heated
die

Pultruded composite
deck being pulled

Pulley
assembly

**FIGURE 3.4** Pultrusion process: composite shape being pulled through the die.

Pulley blocks (top and bottom) pulling composite shape

**FIGURE 3.5** Pultrusion process: unit with caterpillar-type pullers being used for pulling the composite shape.

During pultrusion, the fibers/fabrics are wetted in a resin bath before feeding them through fiber preformers or a heated die [Paciornik et al. 2003; Suratno et al. 1998]. The resin bath consists of required accelerators, filler materials, catalysts, and wetting agents. Following curing, the hardened FRP product is cooled while being gripped and pulled by the pull mechanism (typically made of durable urethane foam) and cut to the required length. A pultrusion process is used for manufacturing FRP rebars used in the construction industry. Some FRP rebars may be manufactured by a two-stage process, where the bars are pultruded to a partially cured state during the first stage, followed by a second stage consisting of compressing an additional layer on top of the first stage product to provide surface lugs similar to steel bars. The average production output of pultrusion process is about 1 to 5 linear ft/min [Gutowski 1997; Miller et al. 1998].

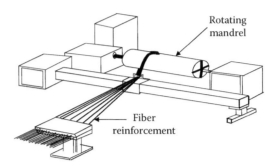

**FIGURE 3.6** Schematic of a filament winding process.

### 3.2.3 FILAMENT WINDING

*Filament winding* is performed by winding a series of continuous fibers from uni-directional rovings or continuous strand mats around a rotating mandrel (Figure 3.6) [Koussios et al. 2004]. The fiber wetting is done by passing fibers through a resin bath or by resin pre-impregnation (tow-preg). After the fiber winding process is complete, the part is hot cured using a set of heating lamps fitted to the machine or by passing the mandrels along a curing oven. Some processes cure the lay-up while it is still attached to the mandrel using external infrared heaters or internal electrical/steam heating sources [Korotkov et al. 1993]. For some applications, banding or tape-winding techniques with FRP fabrics have been employed using conventional winding machines.

Filament winding processes can be setup in several ways and a few of them are [Chen and Chiao 1996; CISPI 1992; Henninger and Friedrich 2002; Mantell and Springer 1994]:

- Continuous winder — linked nonrotating mandrels move through a winding station
- Polar winder — mandrels rotate on two axes to create a closed cylinder or sphere
- Continuous mandrels — reinforcements and resin are applied to a rotating removable mandrel to create continuous tubular products
- Braider winder — linked nonrotating mandrels move through a rotating ring

Some of the different material types used as mandrels are [CISPI 1992]:

- Steel mandrel — used for high volume repetitive products
- Disposable mandrels (plywood, sheet aluminum, and others) — used for low-volume products or limited prototype parts
- Air pressure bottles (flexible bladder) — used for closed products, where the bladder is inflated during the winding process and then deflated for removal or else the bottle can become an integral liner in the final product

Pressure- and shrink-wraps are used during curing to minimize voids in a filament winding process, which could be as high as 3% as compared to about 1% or less in a hand lay-up. The production rate through filament winding is as high as 100 lb/hr [Gutowski 1997].

### 3.2.4 RESIN TRANSFER MOLDING (RTM)

*Resin transfer molding* (RTM) (Figure 3.7 and Figure 3.8) is a closed-mold process in which a composite is shaped between male and female molds under low-pressures (less than 100 psi) [Lawrence et al. 2005; Le Riche et al. 2003]. During manufacturing, preformed fiber reinforcement is placed in the mold and infused with resin and catalyst, which are pumped into the closed mold under low pressure (40 to 50 psi). The mold is then heated and cured to create a composite part. Mold surfaces utilized in the RTM process can be coated with a gel coat or covered with a veil to create a "Class A" smooth surface of the final product.

In a RTM process, complex shapes can be made in one operation with or without inserts [Ferret et al. 1998; Holmberg and Berglund 1997]. RTM processes can be automated with limited void content, i.e., less than 1%. One major advantage of RTM is the release of fewer hazardous emissions. The resins used in RTM are polyesters (most common), vinyl esters, epoxies, urethanes, phenolics, and acrylic/polyester hybrids. Mineral fillers (e.g., nanoclays) can be added to the resin and exfoliated to enhance fire retardancy, mechanical properties, durability, and surface finish. Glass fibers consisting of continuous strand mats and chopped strands are commonly used for product manufacturing [CISPI 1992]. Carbon, aramid, and other synthetic fibers are also used either individually or in combination with each other. The whole RTM production process can be well-controlled to obtain high product strength with minimal fiber damage under low pressure. An RTM process is used extensively by the automobile industry because it offers advantages such as a high production rate, the flexibility to include inserts, and an ability to produce complex shapes [Kim and Lee 2002; Suh and Lee 2001].

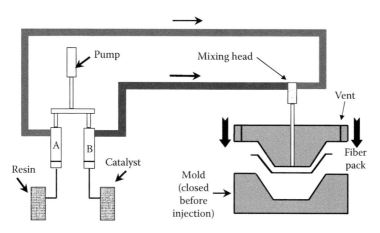

**FIGURE 3.7** Resin transfer molding process.

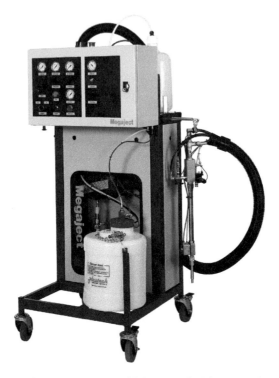

**FIGURE 3.8** RTM equipment used at the CFC, West Virginia University.

### 3.2.5 SHEET MOLDING COMPOUND (SMC)

Manufacturing with *sheet molding compound* (SMC) consists of utilizing SMC sheets in a compression press and squeezing them between heated male and female molds to a desired shape [ASM 1989; Dumont et al. 2003]. A SMC sheet consists of required ingredients mixed and packed as a flexible sheet for further use in a molding process. SMC ingredients consist of fibers (from 1/2 to 2 in length), resin, fillers, chemical thickeners, catalysts, mold release agents, and others. The SMC material (sheet) is cut to the necessary dimensions and stacked or oriented in the mold as per the design configuration. Curing is achieved through a heated die under pressure. The process temperature varies from 250° to 350°F, whereas the molding pressure ranges from 250 to 3000 psi [Hollaway 1990]. The time for each molding cycle can vary from 1 to 4 min and depends upon the part thickness, mold temperature, fiber and resin quantity, and the amount of catalyst. After removing the molded part, it may be subjected to secondary operations such as stud insertion, piercing, bonding, deflashing, and others. SMC consists of any one of the following fiber types:

- Random chopped fibers
- Continuous unidirectional fibers
- Combined random chopped fibers
- Discontinuous unidirectional fibers of approximately 4 in length.

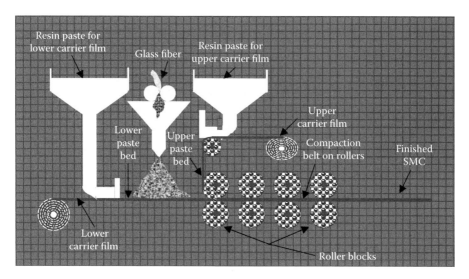

**FIGURE 3.9** Schematic of a sheet molding compound (SMC) process.

The SMC process is schematically represented in Figure 3.9. To prepare a SMC sheet, the ingredients consisting of thermosetting resin, additives, and fillers — except reinforcement — are mixed together in either a batch mixer or a continuous mixer. The resulting paste will have a viscosity similar to pancake batter. The paste material is taken to a SMC machine and metered through doctor blades onto separate upper and lower plastic "carrier films," usually made of polyethylene [CISPI 1992]. The fibers are distributed over the lower paste bed. The upper and lower paste beds are "sandwiched" together to form a sheet, which goes through a compaction belt. The sheet is cut into convenient lengths and packaged into nylon sleeves to reduce the evaporation of resin volatiles and then stored in a temperature-controlled area until molding. The nylon sleeves and carrier films are removed before using SMC sheets in a molding process for manufacturing FRP parts.

## 3.2.6 SEEMANN COMPOSITE RESIN INFUSION MOLDING PROCESS (SCRIMP)

The *Seemann composite resin infusion molding process* (SCRIMP) is used for co-molding composite skins and core in one piece without the need for an autoclave. The SCRIMP process is similar to RTM and offers the benefit of preforming parts using dry fabrics and core; however, it does not require a two-sided mold and resin pressure as in the RTM process. Parts molded using the SCRIMP process require a one-sided vacuum-tight surface. Utilizing vacuum pressure and a SCRIMP resin infusion system, complete wetting of fiber/fabric stack is carried out as shown in Figure 3.10. The entire fiber/fabric stack including any core is saturated in one infusion step, thus eliminating the weaker secondary bonds and relatively longer times that are associated with the RTM process. Very thick fabrics, mats, and cores can be used to greatly speed the lay-up process [Kopf 1995].

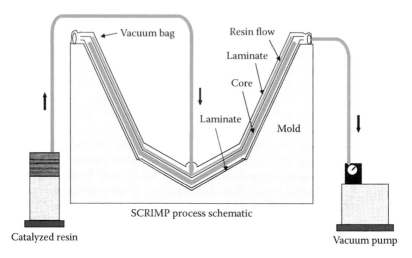

**FIGURE 3.10** Schematic representation of SCRIMP process-resin wetting of laminates.

As compared to many other composite manufacturing processes, the SCRIMP process is inherently repeatable. The resin infusion is automatically stopped once an equilibrium resin content is achieved. Typically, a maximum of 55 to 60% fiber volume fraction is used in the SCRIMP process. Air voids are eliminated prior to resin infusion by using fabric preform as an effective breather layer. Resin infused through the fiber/fabric layers travels in controlled waves and completely saturates them, thereby eliminating possible voids that could be created by the volatile organic compounds (VOC) emitted by the resin during the cure cycle [Boh et al. 2005; Han et al. 2000].

Large parts weighing up to 3000 lbs and measuring up to 2000 ft² can be manufactured using the SCRIMP process. Composite parts such as those with a single skin and cored construction, complex three-dimensional truss parts, and others are manufactured with a SCRIMP process and have mechanical properties comparable to those manufactured by highly controlled, expensive autoclave processes. Parts processed by SCRIMP can be about 20 to 30% less expensive as compared to the ones obtained by hand lay-up and the closed molding manufacturing systems [Kopf 1995].

### 3.2.7 INJECTION MOLDING

The *injection molding* process is mainly used for making parts from thermoplastic polymers (Figure 3.11). Suitable modifications are necessary for making parts from an injection molding process using thermosetting polymers.

In injection molding, a thermoplastic polymer in the form of pellets and flakes is placed in a vertical hopper that feeds into a heated horizontal injection unit. A typical injection unit consists of reciprocating or co-rotating screws that push the polymer through a long heated chamber, where the material is softened into a fluid state [Bickerton and Abdullah 2003; Engineering Materials Handbook 2001]. Melted

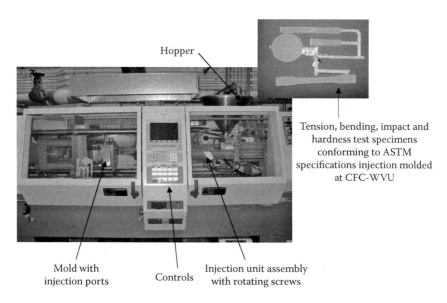

Hopper

Tension, bending, impact and
hardness test specimens
conforming to ASTM
specifications injection molded
at CFC-WVU

Mold with
injection ports

Controls

Injection unit assembly
with rotating screws

FIGURE 3.11 Injection molding equipment. (Courtesy of CFC, West Virginia University.)

polymer from the heated chamber enters a cooler mold through a nozzle under high pressure. High clamping force is applied to keep the mold sections closed during polymer cooling. Following cooling, the mold sections are opened and the finished part is pushed out from the mold by ejector pins. Pressure, temperature, and molding time play an important role in the manufacturing of parts, in addition to the properties of polymers, fibers, and other additives.

### 3.2.8 COMPRESSION MOLDING

*Compression molding* can be used to manufacture parts from both thermoset and thermoplastic polymers. In compression molding, polymer materials with fibers, additives, modifiers, and other additions are placed directly in the mold cavity and compressed under pressure and heat [Hulme and Goodhead 2003; Odenberger et al. 2004]. Under pressure and heat, the charge (polymers, fibers, and other ingredients) flows throughout the mold [Ankermo and Angstrom 2000; Lee and Tucker 1987; Wakeman et al. 2000]. Thermoset polymer cures under applied heat, whereas a thermoplastic polymer will soften and take the mold cavity shape with heat and pressure. Temperature, pressure, and time of residence in a compression mold depend on the properties of the charge (polymer and other constituents placed in the mold) and the type of the finished product (shape, dimensions, surface finish). Figure 3.12 shows compression molding equipment and Figure 3.13 shows a field-installed highway offset block made of a recycled polymer shell with glass fabric reinforcement and discarded rubber tires or wood as a core material. The offset block was manufactured using a compression molding machine at the Constructed Facilities Center (CFC) laboratory of West Virginia University.

**FIGURE 3.12** Compression molding equipment. (Courtesy of CFC, West Virginia University.)

**FIGURE 3.13** Field-installed highway guardrail offset blocks made of recycled polymers manufactured using a compression molding machine.

### 3.2.9 EXTRUSION

An *extrusion* process is used for making continuous parts with thermoplastic polymers. In this process, thermoplastic polymer in the form of pellets and flakes is placed in the vertical hopper of the extrusion machine (Figure 3.14 and Figure 3.15) that feeds into a long horizontal chamber — typically consisting of four or more heating zones from start to exit — with continuously revolving screws. Screw movement forces the softened resin out of the heating chamber through a die. Hot polymer extruded from the die is fed onto a conveyor belt and quenched (cooled)

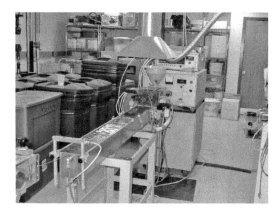

**FIGURE 3.14** Extrusion equipment used by the CFC, West Virginia University.

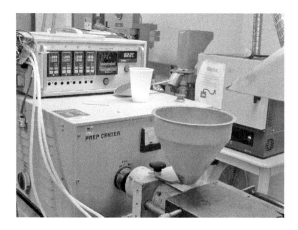

**FIGURE 3.15** Temperature control for four different heat zones in the extrusion equipment used by the CFC, West Virginia University.

through water immersion or air-blowing techniques. The die is formed in the shape of the final product, e.g., tubes, rods, continuous films, and shapes. Biaxial polymer geogrids used for soil strengthening are manufactured from extrusion process with the use of high-density polyethylene or polypropylene. Extrusion equipment used by the Constructed Facilities Center (CFC) of West Virginia University is shown in Figure 3.14 and Figure 3.15.

## 3.3 SIGNIFICANCE OF QC AND QA IN MANUFACTURING

To obtain consistent physical, thermal, and mechanical properties in FRP composite parts manufactured by various processes, good quality control (QC) and quality assurance (QA) are necessary. Manufacturing at a production plant with a controlled environment can provide products with higher consistency in mechanical properties with fewer defects as compared to field-prepared composites. The strength properties of composites for design derived through different standard tests must appropriately reflect manufacturing type and location, which are discussed in Chapter 4.

## REFERENCES

Anderson, F.R., Reichhold, A.S., and Norway, Open molding: hand lay-up and spray-up, *ASM Handbook Composites*, 21, 2000, 450–456.
Ankermo, M. and Angstrom, B.T., Modeling component cost in compression molding of thermoplastic composite and sandwich components, *Composites: Part A*, 31, 2000, 319–333.
ASM International, *ASM Handbook*, Vol. 21, Cleveland, Ohio: ASM International Publications, 2001.
ASM International, *Engineering Materials Handbook-Composites*, Vol. 1, Boca Raton, FL: CRC Press, 1989.
Bickerton, S. and Abdullah, M.Z., Modeling and evaluation of the filling stage of injection/compression moulding, *Composites Sci. Tech.*, 63, 2003, 1359–1375.
Boh, J.W., Louca, L.A., Choo, Y.S., and Mouring, S.E., Damage modeling of SCRIMP woven roving laminated beams subjected to transverse shear, *Composites Part B: Eng.*, 36, 2005, 427–438.
Branco, C.M., Ferreira, J.M., Faelt, P., and Richardson, M.O.W., A comparative study of the fatigue behaviour of GRP hand lay-up and pultruded phenolic composites, *Int. J. Fatigue*, 18, 4, 1995, 255–263.
Canadian Society for Civil Engineers (CSCE), *Advanced Composite Materials with Application to Bridges*, Montreal, 1992.
Caulk, D.A., General results on the flow of chopped fiber compounds in compression molding, *J. Non-Newt. Fluid Mech.*, 28, 1998, 333–347.
Chen, H.C. and Chiao, S.M., Fiber consolidation in the filament winding process: modeling with undulating channels, *Composites Sci. Tech.*, 56, 1996, 1161–1169.
Composites Institute of the Society of the Plastics Industry (CISPI), *Introduction to Composites*, Washington, DC, 1992.
Davima, J.P., Reis, P., and António, C.C., Experimental study of drilling glass fiber reinforced plastics (GFRP) manufactured by hand lay-up, *Composites Sci. Tech.*, 64, 2004, 289–297.
Dumont, P., Orgea, L., Le Corre, S., and Favier, D., Anisotropic viscous behavior of sheet molding compounds (SMC) during compression molding, *Intl. J. Plasticity*, 19, 2003, 625–646.
Ferret, B., Anduze, M., and Nardari, C., Metal inserts in structural composite materials manufactured by RTM, *Composites Part A: Appl. Sci. Manuf.*, 29, 1998, 693–700.
Gutowski, T.G. (Ed.), *Advanced Composites Manufacturing*, New York: John Wiley & Sons, 1997.
Han, K., Jiang, S., Zhang, C., and Wang, B., Flow modeling and simulation of SCRIMP for composites manufacturing, *Composites Part A: Appl. Sci. Manuf.*, 31, 2000, 79–86.
Henninger, F. and Friedrich, K., Thermoplastic filament winding with online-impregnation. Part A: process technology and operating efficiency, *Composites: Part A*, 33, 2002, 1479–1486.
Holloway, L. (Ed.), *Polymers and Polymer Composites in Construction*, London: Thomas Telford, 1990.

Holmberg, J.A. and Berglund, L.A., Manufacturing and performance of RTM U-beams, *Composites Part A: Appl. Sci. Manuf.*, 28, 1997, 513–521.

Hulme, A.J. and Goodhead, T.C., Cost effective reprocessing of polyurethane by hot compression moulding, *J. Mat. Proc. Tech.*, 139, 2003, 322–326.

Kim, P.J. and Lee, D.G., Surface quality and shrinkage of the composite bus housing panel manufactured by RTM, *Composite Struct.*, 57, 2002, 211–220.

Kopf, S., *SCRIMP Systems*, New Castle, DE: Hardcore DuPont, 1995.

Korotkov, V.N., Chekanov, Y.A., and Rozenberg, B.A., The simultaneous process of filament winding and curing for polymer composites, *Composites Sci. Tech.*, 47, 1993, 383–388.

Koussios, S., Bergsma, O. K., and Beukers, A., Filament winding. Part 1: Determination of the wound body related parameters, *Composites: Part A*, 35, 2004, 181–195.

Lawrence, J.M., Fried, P., and Advani, S.G., Automated manufacturing environment to address bulk permeability variations and race tracking in resin transfer molding by redirecting flow with auxiliary gates, *Composites Part A: Appl. Sci. Manuf.*, 36, 2005, 1128–1141.

Le Riche, R., Saouab, A., and Bréard, J., Coupled compression RTM and composite layup optimization, *Composites Sci. Tech.*, 63, 2003, 2277–2287.

Lee, C.C., and Tucker, C.L. III, Flow and heat transfer in compression mold filling, *J. Non-Newt. Fluid Mech.*, 24, 1987, 245–264.

Mallick, P.K. (Ed.), *Composites Engineering Handbook*, New York: Dekker Publishing, 1997.

Mantell, S.C. and Springer, G.S., Filament winding process models, *Composite Struct.*, 27, 1994, 141–147.

Miller, H., Dodds, N., Hale, J.M., and Gibson, A.G., High speed pultrusion of thermoplastic matrix composites, *Composites Part A: Appl. Sci. Manuf.*, 29, 1998, 773–782.

Odenberger, P.T., Andersson, H.M., and Lundström, T.S., Experimental flow-front visualization in compression molding of SMC, *Composites: Part A*, 35, 2004, 1125–1134.

Paciornik, S., Martinho, F.M., de Mauricio, M.H.P., and d'Almeida, J.R.M., Analysis of the mechanical behavior and characterization of pultruded glass fiber-resin matrix composites, *Composites Sci. Tech.*, 63, 2003, 295–304.

Peters, S.T. (Ed.), *Handbook of Composites*, 2nd ed., New York: Chapman & Hall, 1998.

Suh, J.D. and Lee, D.G., Manufacture of composite screw rotors for air compressors by RTM process, *J. Mat. Proc. Tech.*, 113, 2001, 196–201.

Suratno, B.R., Ye, L., and Mai, Y.M., Simulation of temperature and curing profiles in pultruded composite rods, *Composites Sci. Tech.*, 58, 1998, 191–197.

Wakeman, M.D., Rudd, C.D., Cain, T.A., Brooks, R., and Long, A.C., Compression moulding of glass and polypropylene composites for optimized macro- and micro-mechanical properties. 4: Technology demonstrator — a door cassette structure, *Composites Sci. Tech.*, 60, 2000, 1901–1918.

# 4 Durability: Aging of Composites

## 4.1 INTRODUCTION

Fiber reinforced polymer (FRP) composite structures are finding broader acceptance from end users through a wider range of applications in civil infrastructure as an alternate to conventional concrete, steel, and timber structures [Arockiasamy et al. 2003; Demers et al. 2003; Fyfe, Watson, and Watson 1996; GangaRao et al. 2001; Gerritse, 1998; JCI 1998; Lampo, Hoy, and Odello 1996; Porter and Barnes 1998; Priestly, Seible, and Fyfe 1992; Roll 1991; Taerwe 1993; Uomoto 2001]. This acceptance is attributed primarily to the noncorrosive properties of FRP composite structural components and systems. Such acceptance should be viewed in the context that direct and indirect costs of maintenance, rehabilitation, replacement of systems, and loss of productivity due to steel corrosion are of the order of $297 billion (U.S.) per year or 3% of the gross domestic product [NACE, CCT, and FHWA 2002]. Structural FRP composites provide similar or superior mechanical, thermal, and chemical properties when compared to conventional steel, concrete, and timber materials, thus leading to lower life-cycle costs. However, the thermomechanical properties of FRP composites decrease during their service life — especially under harsh environments — similar to conventional structural materials.

The durability of FRP composites consisting of different fibers such as glass, carbon, or aramid is the main focus of this chapter. Effects of chemical, thermal, and mechanical loading on the durability of composites during their service life and their safety have been reflected in the design approach provided by ACI 440.1R-03 and ACI 440.2R-02. The *durability* of a material or structure is defined as "its ability to resist cracking, oxidation, chemical degradation, delamination, wear, and/or the effects of foreign object damage for a specified period of time, under appropriate load conditions and specified environmental conditions" [Karbhari et al. 2003]. Several studies conducted over the last decade indicate that the mechanical properties of FRPs used as internal and external reinforcement for concrete structural members, including their bond strength (with concrete), decrease with time [ACI 440.1R-03 and ACI 440.2R-02; Bakht et al. 2000; GangaRao et al. 1996; GangaRao and Vijay 1997; JCI 1998; Sonobe et al. 1997; Vijay and GangaRao 2001]. This chapter presents details on property change processes (durability) and mechanisms in FRP composites that affect their hygro-thermomechanical properties under varying moisture and temperature conditions, chemical exposure, and sustained stresses.

Environmental factors affecting FRP properties and the reduction factors of different FRP systems under varying exposure conditions, as suggested by ACI 440.2R-02 and ACI 440.1R-03, are provided in Section 4.2 and Section 4.3.

Fiber and resin types play an important role in the durability of FRP bars and fabrics that are used to reinforce or strengthen concrete members. External exposure conditions, also referred to as *environmental loading*, can vary widely from one location to another and also from one season to another. The exposure conditions consist of moisture fluctuations, freeze-thaw effects, temperature cycles (elevated to ambient), pH variations (dependent on chemical nature of the surrounding media or substrate), and ultraviolet (UV) radiation. Strength, stiffness, and bond properties of FRP reinforcements that influence the response of concrete members are affected by environmental exposure [Coomarasamy and Ipe 1998; Litherland et al. 1981; Nanni et al. 1998; Vijay and GangaRao 1999a]. Some of the responses of FRP reinforced or strengthened concrete members that are influenced by environmental exposure include ultimate moment, deflection, crack width, energy absorption, and failure mode.

The degree of reduction in FRP properties under varying environmental conditions — or even their damage — depends on various factors such as type of fibers and resins, fiber-sizing chemistry, cure conditions, quality control during manufacturing, and the severity of the external environmental agents such as temperature and pH. The type and quality of concrete influence the durability of FRP reinforced concrete members.

Durability of FRP composites in the form of FRP bars, external wraps, and bonded strips is discussed in several publications such as ACI 440.1R-03 and ACI 440.2R-02 design guidelines, ISIS-Canada design guidelines, Japanese Society of Civil Engineers (JSCE), Canadian Highway Bridge Design Code (CHBDC), Military Handbook MIL-HDBK-17, and design guidelines proposed by Fyfe Company, Sika Carbodur Company, the Society of the Plastics Industry (SPI), and many others.

## 4.2  ENVIRONMENTAL FACTORS AFFECTING FRP PROPERTIES

The fibers in FRP composites are the main load-carrying elements. The matrix (cured resin) protects fibers and fabrics from damage, preserves fiber alignment, and facilitates load distribution to individual fibers. Fiber selection criteria are application dependent, which may be governed by strength, stiffness, durability, cost, and other requirements. For example, depending on the grade, the cost of carbon fibers is four to eight times that of glass fibers. Therefore, glass fibers are commonly used for strength-dependent applications, whereas carbon fibers are found to be appropriate for stiffness-dependent applications. Resins are selected on the basis of application and type of processing, i.e., wet lay-up versus factory production. The addition of modifiers and additives to the resin enhances durability against UV radiation and resistance to moisture transport [GangaRao et al. 1995; Kato et al. 1998]. The durability of FRP fiber/fabric composite bonded concrete beams depends on properties of the concrete substrate (e.g., alkaline agents in

concrete, presence of cracks and spalls, surface texture that affects bonding characteristics) and primer compatibility with concrete, fiber, and resin [Malvar 1998;
Marshall and Busel 1996, Vijay and GangaRao 1999a]. Fibers typically used in the
construction industry are carbon and glass along with thermoset resins such as
epoxy, vinyl ester, polyester, and urethane. Usage of aramid fibers in the construction
industry is limited because of their potentially high moisture pickup (greater than
10%) over the service life of a structure.

Some of the different environmental conditions that affect durability (e.g., over
75 years) of FRP composites in terms of their strength, stiffness, fiber/matrix interface integrity, micro- and macro-cracking are [GangaRao et al. 1995; Schutte 1994;
Seible and Karbhari 1996]:

- Water/sea water
- Chemical solutions (salt, alkaline, and acid)
- Prolonged freezing
- Thermal cycling (freeze-thaw)
- Elevated temperature exposure
- Thermal gradient (high and low temperature along the depth or across the
  section)
- Oxidation
- UV radiation
- Creep and relaxation
- Aging (chemical and physical under accelerated/natural conditions)
- Fatigue
- Fire
- Others

The durability (aging) of FRP composites and associated effects in structural
members made of FRP composites are further discussed below.

## 4.3  DURABILITY AND SAFETY FACTORS

The ACI 440.1R-03 guide for internal FRP reinforcement and ACI 440.2R-02 guide
for external FRP strengthening have recommended an environmental reduction
factor, $C_E$, to represent the reduction in strength and strain properties of FRP
materials during their service life (Table 4.1). The environmental reduction factor
$C_E$ depends on the location and severity of exposure conditions. For example, the
$C_E$ for carbon FRP varies from 0.85 to 0.95, whereas the $C_E$ for glass FRP varies
from 0.75 to 0.50. Note that overall reductions when using $C_E$ for carbon FRP are
lower than those with glass FRP, which indicates a better durability of carbon FRP
systems. Reduction factors are higher for external exposure conditions such as
bridge decks, beams, and columns. Lower reduction factors are suggested for interior
exposure conditions such as building beams and slabs because of the reduced
severity and exposure to environmental elements (e.g., moisture, temperature fluctuations, and others).

**TABLE 4.1**
**Environmental Reduction Factor ($C_E$) for Different FRP Systems and Exposure Conditions**

| Exposure Condition | Fiber and Resin Type | Environmental Reduction Factor ($C_E$) |
|---|---|---|
| Interior exposure | Carbon/epoxy | 0.95 |
| | Glass/epoxy | 0.75 |
| | Aramid/epoxy | 0.85 |
| Exterior exposure (bridges, piers, and unenclosed parking garages) | Carbon/epoxy | 0.85 |
| | Glass/epoxy | 0.65 |
| | Aramid/epoxy | 0.75 |
| Aggressive environment (chemical plants and waste water treatment plants) | Carbon/epoxy | 0.85 |
| | Glass/epoxy | 0.50 |
| | Aramid/epoxy | 0.70 |

*Source:* ACI 440.1R-03 and ACI 440.2R-02.

The British Standard [BS 4994, 1987] for design and construction of vessels and tanks in reinforced plastics with wet lay-up technique (see Chapter 3) suggests a design factor

$$K = 3k_1 k_2 k_3 k_4 k_5.$$

The actual values of factors $k_1$ through $k_5$ vary depending upon the application type. Factor $k_1$ represents the manufacturing method, $k_2$ represents the chemical environment, $k_3$ represents the heat distortion temperature, $k_4$ represents loading cycles, and $k_5$ is related to the curing procedure. The product of several independent factors might lead to overly conservative design and unrealistic representation of interactions between different factors [Karbhari et al. 2003; Vijay and GangaRao 1999a]. The Concrete Society of the United Kingdom (U.K.) defines the design strength property $f_{fd}$ to be equal to

$$f_{fd} = \frac{f_{fk}}{\gamma_{mf}\gamma_{mn}\gamma_{mE}},$$

where the characteristic material strength property $f_{fk}$ is divided by the product of partial safety factors $\gamma$ representing ultimate strength (e.g., $\gamma_{mf} = 1.4$ for CFRP), manufacturing method (e.g., $\gamma_{mn} = 1.1$ to 1.2), and design modulus at ultimate (e.g., $\gamma_{mE} = 1.1$ for CFRP). Characteristic material strength represents 95% of test results meeting or exceeding that value. Characteristic material strength is determined as

$$f_{fk} = f_{fm} - 2\sigma,$$

where $f_{fm}$ is the material strength at ultimate and $\sigma$ is the standard deviation. Similar to $f_{fk}$, ACI 440.1R-03 uses an analogous term called *design strength*. To account for variations in mechanical properties of FRP composite bars manufactured through different techniques, a $3\sigma$ reduction from $f_{fm}$ has been suggested [Vijay and GangaRao 2001]. Manually bonded FRP composite systems used for highway bridge applications are likely to have more variations and defects than the factory manufactured composites.

## 4.4  PHYSICAL, CHEMICAL, AND THERMAL AGING

Continuous changes and potential degradation (refer to Section 4.5) in physical, mechanical, and chemical properties of composites are expected under service environments leading to "aging." Aging can occur in polymer composites without the application of external (mechanical or environmental) loads. In addition to several other parameters, mechanical properties and durability of polymer composites are influenced by:

- Chemical bonds between polymer chains through permanent crosslink and cure density (percent cure of polymer chains)
- van der Waals and valence forces
- Physical state of the material including morphology, sample size, fiber orientation, etc.

The aging-related degradation rate depends on the fiber-resin interface (degree of bonding, sizing, and curing), moisture ingress and temperature variations, reaction of composites with chemicals, and others [Porter and Barnes 1998; Pritchard and Speake 1988]. Additional details about sizings or coupling agents are given in Section 4.6. Some of the interface characteristics of composites that affect durability are [GangaRao et al. 1995; Kelen 1983; Pritchard and French 1993]:

- Interfacial imperfections
- Heat and pressure curing
- Surface pretreatment of fibers
- Thermal properties of adhesive versus adherend

Aging and degradation of structural composites typically occur through molecular interactions between water and the polymer network. Hygrothermal action on FRP composites may lead to material cracking, fiber-matrix debonding, loss of fiber structural network (e.g., silica network in glass), and an increase in induced (e.g., residual) FRP stresses [Parkyn 1985; Rao et al. 1981]. Aging is a complex phenomenon because of interaction among different factors that influence mechanical properties of polymer composites. Aging can be significant during the service of constructed facilities over a life span of about 75 years.

Aging can be chemical or physical. *Chemical aging* involves a change in the chemical or molecular structure of the polymer such as chain scission, oxidation, and crosslinking [GangaRao et al. 1995; Haskins 1989; Janas and McCullough 1987; Vijay and GangaRao 1999a]. *Physical aging* involves an attempted regrouping of macromolecules to a new state below the glass transition temperature ($T_g$).

Mechanical properties are related to the amount of free volume contained in the bulk polymer, which corresponds to the unoccupied regions accessible to segmental (chain) motions. A change in the bulk polymer temperature produces a thermodynamic driving force for the polymer chains to rearrange themselves to a new equilibrium of free volume state. At temperatures above the glass transition $T_g$, polymer molecules have sufficient instantaneous mobility to regain equilibrium during temperature changes. When a polymer is quenched (i.e., cooled from a high to a low temperature, such as from above to below $T_g$), the lack of instantaneous mobility results in free volume in the system. This change in free volume during the movement of polymer molecules toward an equilibrium state results in altering the mechanical properties of the bulk polymer, including the build up of residual stresses.

As quenching is commonly employed for polymer processing, changes in mechanical properties depend on the processing history, i.e., the aging temperature and the difference between $T_g$ and the quench temperature, $T_q$. When an aged polymer is raised above $T_g$, it re-establishes its free volume equilibrium with no trace of past thermal history. Quenching it back to $T_q$ will again result in physical aging. If a polymer is subjected to cycles of thermal history (raising above $T_g$ and quenching to $T_q$), the mechanical properties due to physical aging follow identical time-dependent paths during each cycle. Tensile creep and dynamic mechanical analysis (DMA) tests can be employed to characterize physical aging factors [Haskins 1989; Janas and McCullough 1987].

## 4.5  MECHANISMS OF POLYMER DEGRADATION

Hygrothermal and mechanical properties of FRP composites are dependent upon the primary and secondary chemical bonds in the polymer chain. Polymer degradation can take place through several mechanisms, such as random chain scission, depolymerization, crosslinking, side group elimination, and reaction of side groups among themselves [Kelen 1983]. Polymer degradation depends on:

- Chemical and physical structure of polymers (dislocation energy of primary and secondary bonds, and other components of the chemical structure)
- Additives (lubricants, plasticizers, nanoclays, reinforcing fillers, UV inhibitors) and modifiers
- Moisture
- Stress/pressure
- Temperature
- Physical and chemical aging
- Others (e.g., contaminants, biological ingredients, etc.)

## 4.6  COUPLING AGENT AND INTERFACE

The *coupling agent* (sizing or coating on fibers) influences the long-term durability of composites. The structure and selection of the coupling agent (coating) mainly depend on the fiber-resin chemistry and the substrate to which they are bonded. The interface between fiber-sizing-resin (e.g., siloxane-based sizings for glass) has two boundaries where sizing (couplant or lubricant) assists in the bundling of fibers and preventing abrasive contact. Typically, failure occurs at the interface between fiber and coupling agent, though it may be initiated by either alkaline reaction or polymer plasticization due to moisture, or even localized residual stress build-up. Water penetrates faster at the fiber-matrix interface than through matrix (hardened resin) and significantly influences interface degradation [Hogg and Hull 1983; Rao et al. 1981]. Coupling agents improve the bond between fibers and resin, leading to a reduction in moisture attack at the interface.

The chemical resistance of FRP composites can be severely reduced because of "wicking," i.e., the ingress of liquids due to capillary action along the fiber strands containing hundreds of filaments [Parkyn 1985; Springer 1981; Vijay et al. 1998; Zheng and Morgan 1993]. Even with a perfect interfacial bond, capillaries along individual filaments pave the way for water ingress. Resins of poorer chemical resistance with chopped strands exhibit better chemical resistance than the resins with continuous filament roving or woven roving because of limited wicking in the former case. A thin glass monofilament or a scrim (woven or nonwoven acrylic or polyester fibers) is used during manufacturing to eliminate the protrusion of glass fibers from the outer surface and avoid paths for moisture ingress [GangaRao et al. 1996]. Under highly alkaline environments (e.g., hydration of concrete), glass fibers can react to form expansive silica gels that may lead to cracking of concrete. Long-term resistance of FRP against aggressive solutions varies with respect to fibers and type of products [Rostasy 1996]. Carbon FRP reinforcements are more durable in nearly all the environments relevant to concrete construction as compared to aramid or glass FRP [ACI 440.1R-03; GangaRao et al. 1998; Hahn and Kim 1988; Odagiri et al. 1997].

The chemical composition of resin influences the hydrolysis of FRPs. As an example, moisture-related damage in glass FRP has been found to decrease in this order: isophthalic, epoxy, and vinyl ester resins. Similarly, the magnitude of damage in glass FRP was higher at 90°C as compared to 25°C [Kajorncheappungam et al. 2002]. A reduction in tensile strength of glass FRP bars conditioned under alkaline and freeze-thaw temperature for 203 days decreased in this order: isocyanurate vinylester (49.1% loss), isophthalic unsaturated polyester (19.7% loss), and low viscosity urethane modified vinyl ester (15.6% loss) [Vijay 1999]. Therefore, resin optimization for environmental durability would require a hygrophobic response of resins to minimize their equilibrium water content, leading to a reduction in susceptibility to hydrolysis.

## 4.7 FACTORS AFFECTING FRP PROPERTIES

### 4.7.1 EFFECT OF MOISTURE

Moisture ingress in composites affects the performance of FRP composites [Adams 1984; Springer 1981]. Water penetrates FRPs through two processes: diffusion through the resin, and flow through cracks or other material flaws. During diffusion, absorbed water is not in the liquid form but consists of molecules or groups of molecules that are linked together by hydrogen bonds to the polymer. Water molecules are dissolved in the surface layer of the polymer and migrate into the bulk of the material under a concentration gradient. Water penetration into cracks or other flaws occurs by capillary flow [Parkyn 1985; Rao et al. 1981].

Matrix softening due to hydrolysis leads to a reduction in matrix-dominant properties of a composite, such as shear strength, glass transition temperature, and composite strength and stiffness. Hydrolysis-induced mechanical property reduction is accentuated in the presence of stress and temperature [Chateauminois et al. 1993]. Relevant mathematical models (e.g., Fickian and non-Fickian laws of moisture absorption) and design factors related to moisture and temperature effects can be found in the literature [GangaRao et al. 2001; Springer 1981; Vijay et al. 1998].

### 4.7.2 Effect of Alkaline/Acid/Salt Solutions

Alkaline, acid, and salt reaction of FRP composites used for reinforcing concrete is a major durability issue to be considered in design. Concrete environment can be highly alkaline (~ pH = 12.8 or higher) and may lead to a combination of chemical interactions with fibers, particularly glass, leading to reductions in FRP composite strength, stiffness, and toughness [GangaRao and Vijay 1997]. Chemical reactions may also cause fiber embrittlement.

The changes in mechanical, thermal, and chemical properties of glass are influenced by its composition, homogeneity, temperature, stress, and corrosion media [Adams 1984]. Silica forms a major part (54.5%) of glass composition and it is the silica network that gets attacked during environmental or chemical agent exposure. Exposure of glass composites to various environmental and chemical agents (e.g., rain water, deicing salts, and other solutions that are alkaline, acidic, and pH-neutral) results in degradation of different chemical bonds in glass. These degradations are referred to as *glass corrosion* [Adam 1984; Vijay 1999] and are briefly described below.

#### 4.7.2.1 Alkaline Effects

Reduction in mechanical properties of glass FRP composites under alkaline environments is widely reported in the literature [Benmokrane et al. 1998; Benmokrane and Rahaman 1998; GangaRao et al. 1996; Katsuki, and Uomoto 1995; Porter and Barnes 1998]). Alkaline attack on glass is attributed to: (1) etching and (2) hydroxylation and dissolution, which have been summarized by Adams [1984], Al Cheikh and Murat [1988], and Vijay [1999]. During etching, silica network is attacked and the constituents of the glass are released. If there is no further accumulation of reaction products on the remaining glass surface and no change in the activity of the surrounding solution, reaction proceeds at a constant rate. However, any accumulation of reaction products in the solution suppresses the reaction rate such that saturated silica will reduce the reaction rate to zero.

$$2 \text{ x NaOH} + (SiO_2)_X \rightarrow \text{x Na}_2SiO_3 + \text{x H}_2O$$

Hydroxylation and dissolution is caused by chemical hydroxylation of silica in the glass. Deposition of the hydroxylation product on the glass surface slows down the reaction. Hydroxylation is associated with dissolution and is characterized by leaching of calcium from the glass. The leached calcium when combined with water, deposits calcium hydroxide on the surface of the glass and reduces the rate of reaction. Following hydroxylation and dissolution, notching is caused by the formation of calcium hydroxide crystals on the glass surface [Al Cheikh and Murat 1988].

### 4.7.2.2  Acid Effects

Acid attack leads to leaching process where hydrogen or hydronium ions exchange for alkali and other positive mobile ions in the glass. The remaining glass network, mainly silica, retains its integrity. It may become hydrated if the network is relatively unstable; or it may become more dense and stable than original glass. Unless the leached layer is removed or altered, reaction rate reduces even to zero. Acid reacts slowly with glass in comparison to alkali [Adams 1984].

$$Na^+ + HCl \rightarrow H^+ + NaCl$$

### 4.7.2.3  Salt Effects

Water, salt and other solutions of neutral pH produce attack on glass similar to those of acids. Ajjarapu, GangaRao, and Faza [1994] suggested the rate of degradation of glass FRP composite under salt environment using a simple relationship $\sigma_t = \sigma_o e^{-\lambda t}$, where $t < 450$ days, $\lambda = 0.0015$, $\sigma_o$ = tensile strength of FRP at time $t = 0$, and $\sigma_t$ = tensile strength of FRP at time $t$. From the texts, it was concluded that maximum reduction in tensile strength of glass FRP was 50% in 450 days. However, beyond 450 days, the strength did not change considerably.

### 4.7.3  Effect of Temperature

Temperature affects the rate of moisture absorption and the mechanical properties of FRP composites [Allred 1984; Devalapura et al. 1998; Katz et al. 1998; Pritchard and Speake 1988]. Mechanical properties of FRP composites decrease when the material is exposed to elevated temperatures (37° to 190°C). An increase in temperature accelerates creep and stress relaxation, which become pronounced when the temperature reaches a value close to glass transition (particularly beyond $(T_g - 30°F)$ [GangaRao et al. 2001]. A decrease in mechanical properties of FRP composites is not as severe in cold region structures [Dutta, Hui, and Prasad 1994]. Strength and stiffness variations (both increase and decrease, depending on the temperature range) are noted in polymers at low temperatures, resulting in premature brittle failure. The flexibility and toughness of polymers at low temperatures are related to their glassy state of molecular movements [Allred 1984; GangaRao et al. 1995; Kelen 1983]. A decrease in temperature can lead to possible increases in:

- Modulus
- Tensile and flexural strength
- Fatigue strength and creep resistance

Also, a decrease in temperature can lead to possible reductions in:

- Elongation and deflection
- Fracture toughness and impact strength
- Compressive strength
- Coefficient of linear expansion

Temperature variation produces residual stresses in FRP due to the lower longitudinal coefficient of thermal expansion of fibers in relation to resin [Gentry and Hudak 1996]. In cold regions, the difference in curing and operating temperatures of the composite material can be as high as 200°F, and the resulting residual stresses can be high enough to cause microcracking within the matrix and at matrix-fiber interfaces [Dutta et al. 1994].

Vijay and GangaRao [1999a] studied the effects of freeze-thaw temperature cycles (–20° to 150°F), elevated temperature (150°F), and sustained stress (up to 40% of ultimate stress) in terms of failure strength on two types of pultruded glass FRP bars. The two types of bars were sand coated bars and bars with lugs similar to steel bars called C-bars, which were made of urethane modified vinylester resins. Maximum strength reductions due to alkaline environment were twice those of a salt environment under freeze-thaw conditioning. An increase in temperature and sustained stress resulted in larger strength reductions.

### 4.7.4 EFFECT OF STRESS

The strength of FRP composites decreases when exposed to varying environmental conditions under sustained stress. Moisture participates as an active agent for fiber-resin bond breakage, and the rate of degradation is influenced by the sustained stress and temperature [Lagrange et al. 1991]. In glass fibers, such a phenomenon is partly due to the oxidation of metallic particles present in fibers. Generally, hydrolysis reactions are reversible in nature and the regaining of lost properties upon drying is possible. However, if the hydrolysis reaction produces fragments that are attached to different networks such as those in glass, then the fragments can have a tendency to be separated by interfacial stress beyond a certain threshold stress level. Thus, the hydrolytic cleavage may become irreversible. The rate of hydrolysis is a function of many parameters and can be very significant during the service life of glass FRP composites [Pritchard and French 1993; Pritchard and Speake 1988].

Vijay and GangaRao [1999a] have studied the effect of moisture with different pH and stress levels on glass FRP bars called *C-bars* made of urethane modified vinylester resins. C-bars were manufactured using a two-stage manufacturing process with a pultruded core and a compression-molded shell. Maximum strength reductions in C-bars over a 30-month duration in salt and alkaline conditioning at room temperature were 24.5% and 30%, respectively. With sustained stress, maximum strength reductions in salt and alkaline conditioning at room temperature were 25.2% (10 months of 32% applied stress) and 14.2% (8 months of 25% applied stress), respectively. The fiber/matrix interface region plays a critical role in the durability of composites based on observations from C-bar response. The interface controls the rate of moisture ingress and therefore the rate of stress corrosion of the fibers.

Damage due to sustained stress in FRP composites consists of random fiber fracture, leading to relaxation of the matrix around those fractures and causing a reduction in stiffness [Chateauminois et al. 1993]. The second phase involves matrix cracking, interfacial debonding, and more fiber fractures, causing a rapid decrease in stiffness. The final phase due to stress rupture consists of total failure. This damage development process is very similar to that of cyclic fatigue [Dolan et al. 1997; GangaRao and Kumar 1995; GangaRao et al. 2001; Talreja 1987]. The time to fracture is strongly dependent on the initial strain and a minimum strain is required for the onset of fracture.

Clarke and Sheard [1998] studied the durability of FRP-reinforced concrete in alkali, wet/dry conditions at different temperatures and stress levels. They suggested a 100-year life threshold stress limit of about 25% for E-glass, 50% for aramid, and 75% for carbon fiber [Sheard et al. 1997].

## 4.7.5 CREEP/RELAXATION

The creep behavior of FRP composites is influenced by the type of fibers and resins, fiber orientation, fiber volume fraction, and loading conditions [Arockiasamy et al. 1996; Brown and Bartholomew 1996; Vijay and GangaRao 1998]. Poor matrix properties, including cure percent, can significantly increase creep strains. Both creep and stress relaxations are manifestations of resin (polymer) viscoelasticity, which is attributed to the presence of long-chain molecules in polymers. Some molecules in a chain and sometimes part of a chain tend to rearrange and slide past others under applied stress. This is especially significant above the polymer glass transition temperature $T_g$ but it is also possible in the glassy state below $T_g$. Crosslinks in thermosets restrict polymer chain mobility. Large chain deformations result in rupture. The presence of fillers and reinforcements in polymers further restrict creep. The creep coefficients are lower for loading along the fiber axis, whereas the creep behavior is significantly dependent on creep properties of the matrix for off-axis loading.

The ACI-440 guide documents the use of a conservative factor $\lambda$ to account for creep. Carbon shows less creep as compared to glass and aramid [Karbhari et al. 2003; Machida 1993; Sen et al. 1998; Vijay and GangaRao 1998]. The stress rupture level for glass, aramid, and carbon fibers under ambient conditions with 10% failure probability is stated to be 50%, 60%, and 75%, respectively [Karbhari et al. 2003]. ACI 440.R1-03 recommends a design stress rupture threshold of 20%, 30%, and 55% of the ultimate for glass, aramid, and carbon fibers, respectively, for infrastructural applications as discussed in Chapter 5 and Chapter 6.

## 4.7.6 Creep Rupture

*Creep rupture*, also called *static fatigue*, refers to the tensile fracture of a material subjected to sustained stress during the service life of a structural element when the material reaches its strain rupture limit. The time required for rupture under creep loads (endurance time) decreases with the increasing ratio of the sustained tensile stress to the short-term strength of FRP composites. Carbon fibers exhibit better creep characteristics as compared to glass or aramid fibers [Dolan et al. 1997].

## 4.7.7 Fatigue

The long-term behavior and damage mechanisms of composite materials subjected to fatigue loading have been an active area of research during the past two decades, indicating its significance [Lopez-Anido, Howdyshell, and Stephenson 1998]. Many of the fatigue load-related test results available on FRP reinforced concrete correspond to a frequency range of 4 to 5 Hz at low amplitudes [GangaRao et al. 2001; Kumar and GangaRao 1998; Vijay and GangaRao 1999]. Unlike homogeneous materials, FRP composites accumulate damage with increasing number of cycles even at a low fatigue stress range (about 15 to 20% of ultimate stress) rather than developing localized damage in the form of a single macroscopic crack that propagates and results in fracture [Mandell and Meier 1983; Natarajan, GangaRao, and Shekar 2005]. The damage accumulation in FRP composites is microstructural, which includes fiber/matrix debonding, matrix microcracking, delamination, and fiber fracture [Talreja 1987]. The fatigue behavior of composite materials depends on the fabric lay-up sequence, temperature, moisture content, frequency, and maximum to minimum stress or strain ratio. S-N (stress vs. number of cycles-to-failure) data from fatigue testing are typically fit with an exponential law such as

$$N / N_b = (S / S_b)^m,$$

where $N$ is the number of cycles and $S$ is the strength. The subscript $b$ refers to the baseline value used in the equation.

The fatigue damage in composites is measured by the variable $D$, which is a function of the number of cycles applied on a composite and of other parameters. Cyclic loading causes the damage to increase from $D_i$ to $D_f$ after $N$ cycles, at which point catastrophic failure of a composite laminate occurs [GangaRao et al. 2001]. The lifetime ($N_f$), which is the number of cycles to increase damage $D$ from $D_i$ (corresponding to the initial state) to $D_f$ (corresponding to the failure state) is given by

$$N_f = \int_{D_i}^{D_f} dD / f(\Delta\sigma, R, D).$$

Note that the fatigue life of composites is dependent on the strain in the matrix and interfacial characteristics rather than fiber strength [Natarajan et al. 2005]. Matrix and interfacial properties become more critical as the thickness of composites increases. Residual strength is used as a damage metric in many of the mathematical models related to fatigue. One such model is

$$d\sigma/dN = -(1/\gamma)f\sigma_a^\gamma\sigma^{1-\gamma},$$

where, $\sigma$, $\sigma_a$, $f$, and $\gamma$ are, respectively, instantaneous strength, maximum applied cyclic stress, and dimensionless functions that do not depend on $\sigma$. Other methods of expressing damage due to fatigue include stiffness as the damage metric. Fatigue studies conducted on FRP bonded wood and FRP composite deck panels have been modeled using damage energy concepts by comparing the total energy to the energy loss per fatigue cycle [Natarajan et al. 2005]. Several studies indicate excellent fatigue performance of FRP composites under fatigue [GangaRao et al. 2001; Odagiri et al. 1997]. FRP bar-reinforced concrete decks have shown excellent fatigue performance similar to those of steel bar-reinforced decks [Kumar and GangaRao 1998].

### 4.7.8 Ultraviolet (UV) Radiation

FRP composites exposed to UV radiation undergo photochemical damage near the exposed surface, leading to discoloration and reductions in molecular weight that results in the degradation of composites [Kato et al. 1998]. Long UV exposure durations can lead to resin erosion that may lead to fiber exposure, moisture penetration, and matrix cracking, causing a reduction in the thermomechanical properties of composites [GangaRao et al. 1995]. Carbon fibers are less susceptible to UV damage in comparison to glass or aramid. Strength and stiffness reduction due to UV exposure is greater in thin composites than in thick composites. UV inhibitors are mixed with resins during FRP manufacturing to resist damage caused by UV radiation. External FRP reinforcement bonded to concrete beams is protected from UV radiation with aesthetically pleasing special coatings that contain UV inhibitors.

## 4.8  ACCELERATED AGING

Information on the durability of field-installed FRP applications is limited and not available for a variety of resin-fiber-process combinations. Typically, long-term strength and stiffness values of FRP reinforcement for concrete applications are extrapolated based on short-term accelerated aging test results. *Accelerated aging* tests consist of subjecting FRP composites and FRP reinforced or bonded concrete beams to elevated temperatures or freeze-thaw cycling under water, salt, alkaline, or acidic solution immersion. Based on accelerated aging test results, charts are prepared using time-temperature-stress superposition principles (Section 4.8.1 and Section 4.8.3). Using those charts, accelerated aging test data are correlated to natural aging results of FRP composites. Natural aging consists of exposing FRP specimens to natural environmental weathering in open areas consisting of some or all elements such as sunlight, rain, snow, freeze-thaw cycling, humidity changes, and temperature variations. The Arrhenius temperature dependence concept described below (Section 4.8.2) is used for correlating tension test data obtained from accelerated aging tests with those from natural aging (weathering). Additional details on accelerated aging methodology are available in the literature and a brief summary of accelerated aging methodology — along with its limitations — is provided in Section 4.8.3 [Litherland et al. 1981; Proctor et al. 1982; Porter and Barnes 1998].

### 4.8.1 TIME–TEMPERATURE–STRESS SUPERPOSITION PRINCIPLE

A polymer composite material property such as time-dependent stress at one temperature can be used to find those properties at another temperature (with certain limitations), which is referred to as the *time-temperature-stress superposition principle*. This principle is employed to calibrate naturally aged results of FRP at ambient temperature with accelerated aging results. A procedure employing the above principle to predict the service life of an FRP composite is described in Section 4.8.3 along with a brief description of the Arrhenius principle.

### 4.8.2 THE ARRHENIUS PRINCIPLE

The *Arrhenius principle* states that rate at which chemical degradation occurs is dependent on temperature. This principle is employed to exploit the temperature dependence of polymers subjected to environmental aging consisting of several temperature levels.

$$k = Ae^{-E_a/RT}$$

where $k$ is the reaction rate constant with respect to a temperature $T$, $A$ is a "preexponential factor," $E_a$ is the activation energy for the reaction, $R$ is a constant, and $T$ is the temperature in Kelvin.

### 4.8.3 ACCELERATED AGING METHODOLOGY

The following procedure is used to correlate natural aging to accelerated aging [Litherland et al. 1981; Vijay and GangaRao 1999; Vijay 1999].

*Step 1*: Consists of subjecting the composite specimens immersed in cement representative pH solution conditioning schemes to 6 or 7 evenly spread different temperature aging from –20°F (low temperature may slow down aging but causes brittle failures) to 180°F (below glass transition temperature).

*Step 2*: Consists of plotting strength loss curves (which are typically nonlinear curves conforming to some power law, e.g., $C = C_o + mt^n$) with respect to an aging

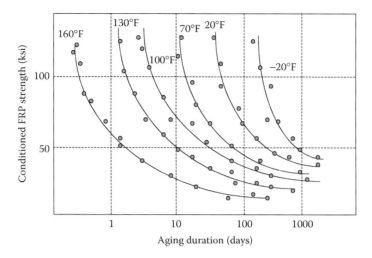

**FIGURE 4.1** Strength retention of aged FRP at different temperatures.

period (number of days). Strength loss is plotted along the vertical axis and the aging period is plotted along the horizontal axis (Figure 4.1).

*Step 3*: Consists of plotting the curves in Step 2 for an Arrhenius-type relationship, i.e., $A = A_o \exp(-\Delta E/RT)$. The log (time to reach a particular strength value, i.e., 90, 80 ksi) is plotted along the vertical axis and the inverse of temperature (°K) is plotted along the horizontal axis (Figure 4.2).

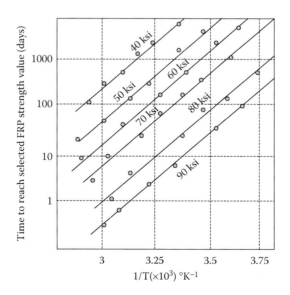

**FIGURE 4.2** An Arrhenius plot for temperature- and time-dependent strength retention.

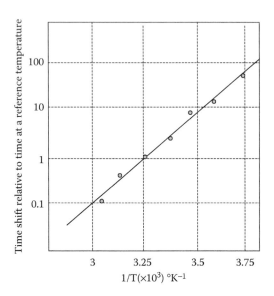

**FIGURE 4.3** Normalized time displacement curve (Arrhenius plot) relative to a reference temperature.

*Step 4*: Involves normalizing the curves in Step 3 into a single curve by plotting the logarithm of the time (for a given strength loss) at different aging temperatures ($T = 273 + t_0$, selected in Step 1) along the vertical axis (relative to the time at some REFERENCE temperature), against the inverse of temperature along the horizontal axis (Figure 4.3).

The normalization procedure is as follows:

- Select a REFERENCE temperature, e.g., 70°F.
- Plot the logarithm of the ratio of the time taken for the composite strength to reduce to a given value at $T = 273 + t_0$ (pick all the temperatures individually as selected in Step 1) relative to the time to reduce to that value at 70°F (reference temperature) versus the inverse of the absolute temperature corresponding to $t_0$ (where the time is read from the fitted curves plotted as per Step 2).

*Step 5*: A normalized Arrhenius plot gives one overall picture of the relative acceleration of strength or stiffness loss at different temperatures. From the known time-scale shift (i.e., plot of Step 4), changes expected over a long period under lower service temperatures is predicted by considering following calibration.

- Strength loss data from naturally weathered samples (Figure 4.4)
- Using the mean annual temperature and other factors (i.e., moisture, freeze-thaw, and pH level) as a basis for calibration

Litherland, Oakley, and Proctor [1981] have correlated their accelerated aging data of glass fibers with natural weathering samples of about 10 years. In their tests,

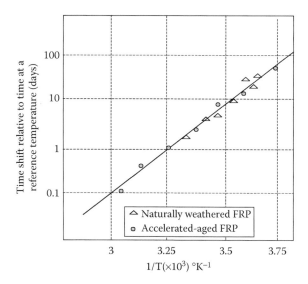

**FIGURE 4.4** Normalized time displacement curve (Arrhenius plot) with natural weathering data.

the media surrounding the glass was cement representative, so as to correlate natural and accelerated aging. Some of the factors to be considered before using Litherland, Oakley, and Proctor's method described above are:

- The mean annual temperature is taken as the sole criteria for determining the accelerating factors. The identical mean annual temperature at different locations does not necessarily account for the geographical variations in magnitude and distribution of temperature, humidity, and precipitation throughout the year.
- The correlation of natural and accelerated weathering is carried out on samples without stress.
- Present-day manufacturing methods and durable resins offer a better degree of protection against water, salt, or alkaline attack, thus taking more time to reduce the strength to a selected value under identical aging conditions considered by Litherland et al. [1981]. In effect, the shift of the time-scale factor is necessary while interpreting Litherland's data.

A study by Vijay and GangaRao [1999b] correlated accelerated and natural weathering on GFRP bars. Calibration charts developed for the nonstressed GFRP bars show that one day of chamber conditioning (accelerated) in their study was equivalent to 34 days of natural weathering at Morgantown, West Virginia, or 36 days of typical U.K. weather. These calibrations were developed similar to the accelerated aging results of Litherland, Oakley, and Proctor [1981] on glass-reinforced composites. Chamber weathering (freeze-thaw between 12.2° to 120.2°F or −11° to 49°C) of 30 months in alkaline conditioning (pH = 13) carried out in this study corresponds to natural weathering of 1020 months (85 years). However, under a sustained stress of 20%, the natural weathering of GFRP bars was equivalent to 704 months (58.67 years) instead of 1020 months. Concrete cover was found to provide a beneficial effect of slowing down the aging duration (time to reach a particular strength loss value) of FRP bars embedded in cracked concrete beams.

## 4.9  MANUFACTURING AND DURABILITY

The manufacturing techniques discussed in Chapter 3 play an important role in the durability of FRP composites. The possibility of high void content exists in a composite during manufacturing. For example, variation in pull speeds can provide some differences in void fractions, fiber wet-out, and degree of cure, which affect strength, stiffness, and durability of a composite. The presence of voids in FRP composites increases the moisture absorption and diffusion coefficient, which eventually lead to degradation in strength and stiffness. Hence, the void contents in FRP composites should be kept to a minimum during the manufacturing process, i.e., less than 0.5% if possible and no more than 1%. Good quality control (QC) and quality assurance (QA) during manufacturing is essential to obtain FRP composites with consistent and durable properties.

## 4.10  CURRENT GAPS IN DURABILITY ANALYSIS

Data related to durability of composites is still evolving. As composites are beginning to find increased usage in the construction industry, note that several of the available durability studies are validated over short-terms (typically from 1980 and beyond). Most of the durability data have not been validated over a long-term duration, such as 50 to 100 years under field conditions. Available data on composite properties is influenced by durability test methods employed (type of exposure, concentration of salts and alkali, exposure methods), test conditions (temperature, humidity, specimen type and dimensions, rate of testing), variation in constituents (resins, fibers, additives, cure conditions), and manufacturing issues (QA and QC). Data on durability is sometimes controversial due to the complexity of the synergistic action of a set of environmental loads and corresponding composite responses that may indicate conflicting trends in results. The Civil Engineering Research Foundation (CERF) has identified current gaps in durability analysis of composites used for construction based on the effects from exposure to moisture, alkali solution, temperature, creep/relaxation, fatigue, ultraviolet radiation, and fire [Karbhari et al. 2003]. The CERF document mainly presents information as a roadmap to generate and coordinate future efforts to establish an integrated knowledge system; a methodology for test protocols for generation of test data, collection of test data and validation; and the implementation of plans for correlating field data with laboratory test results.

## REFERENCES

Adams, P.B., Glass corrosion, a record of the past? A predictor of the future? *J. Non-Cryst. Solids*, 67, 193–205, 1984.

Ajjarapu, S., Faza, S., and GangaRao, H.V.S., *Strength and Stiffness Degradation of Glass Reinforced Polyester and Vinyl Ester Structural Plates*, Morgantown, WV: Constructed Facilities Center, West Virginia University, 1994.

Al Cheikh, A. and Murat, M., Kinetics of non-congruent dissolution of E-glass fiber in saturated calcium hydroxide solution, *Cement Concrete Res.*, 18, 943–950, 1988.

Allred, R.E., The effects of temperature and moisture content on the flexural response of Kevlar/epoxy laminates: Part II [45/0/90] Filament orientation, *Environmental Effects on Composites Materials*, v. 2, Springer, G. (Ed.), Lancaster, PA: Technomic Publishing Company, 1984.

American Concrete Institute (ACI), ACI 440.1R-03: Guide for the design and construction of concrete reinforced with FRP bars, Detroit, Michigan.

American Concrete Institute (ACI), ACI 440.2R-02: Design and construction of externally bonded FRP systems for strengthening concrete structures, Detroit, Michigan.

Arockiasamy, M., Amer, A., Shahawy, M., and Chidambaram, S., Long-term behaviour of concrete beams reinforced with CFRP bars under sustained loads, *ACMBS-II*, El-Badry, M. (Ed.), Montreal, August 1996.

Arockiasamy, M., GangaRao, H.V.S., Vijay, P.V., Benmokrane, B., and Shahawy M., Fiber reinforced polymer composites for concrete bridge reinforcement, Final Report, Project 10-55, FY'99, Submitted to National Cooperative Research Highway Program (NCHRP), Transportation Research Board, National Research Council, May 2003.

Bakht, G.A., Banthia, N., Cheung, M., Erki, M., Faoro, M., Machida, A., Mufti, A., Neale, K.W., and Tadros, G., Canadian bridge design code provisions for fiber-reinforced structures, *J. Composites Construction*, ASCE, 4(1), 3–15, 2000.

Benmokrane, B. and Rahman, H. (Eds.), Durability of fiber-reinforced polymer (FRP) composites for constructions, *Proc. CDCC '98*, Benmokrane, B. and Rahman, H. (Eds.), Sherbrooke, Québec, Canada, August 5–7, 1998.

Benmokrane, B., Rahman, H., Ton-That, T.M., and Robert, J.F., Improvement of the durability of FRP rebars for concrete structures, *CDCC '98*, Benmokrane, B. and Rahman, H. (Eds.), Sherbrooke, Québec, Canada, August 5–7, 1998.

Brown, V.L. and Bartholomew, C.V., Long-term deflections of GFRP-reinforced concrete beams, *Proc. ICCI '96*, Saadatmanesh, H. and Ehsani, M.R. (Eds.), Tucson, AZ, January 1996.

BS 4994:1987, Specification for Design and Construction of Vessels and Tanks in Reinforced Plastics, 1987, U.K.

Chateauminois, A., Chabert, B., Soulier, J.P., and Vincent, L., Effects of hygrothermal aging on the durability of glass/epoxy composites. Physico-chemical analysis and damage mapping in static fatigue, *Proc. ICCM '93*, 1993.

Civil Engineering Research Foundation (CERF), Gap analysis for durability of fiber reinforced polymer composites in civil infrastructure, CERF-01, Research Affiliate of American Society of Civil Engineers, 2001.

Clarke, J.L. and Sheard, P., Designing durable FRP reinforced concrete structures, durability of fibre reinforced polymer (FRP) composites for construction, *Proc. CDCC '98*, Benmokrane, B. and Rahman, H. (Eds.), Sherbrooke, Québec, Canada, August 5–7, 1998.

Coomarasamy, A. and Ipe, A.K.C, Evaluation of fiber reinforced plastic (FRP) materials for long term durability in concrete structures, *Proc. CDCC '98*, Benmokrane, B. and Rahman, H. (Eds.), Sherbrooke, Québec, Canada, August 5–7, 1998.

Demers, M., Popovic, A., Neale, K., Rizkalla, S., and Tadros, G., FRP retrofit of the ring-beam of a nuclear reactor containment structure, *ACI Spec. Publ.*, SP-215, November 2003.

Devalapura, R.K., Greenwood, M.E., Gauchel, J.V., and Humphrey, T.J., Evaluation of GFRP performance using accelerated test methods, *Proc. CDCC '98*, Benmokrane, B. and Rahman, H. (Eds.), Sherbrooke, Québec, Canada, August 5–7, 1998.

Dolan, C.W., Leu, B.L., and Hundley, A., Creep-rupture of fiber-reinforced-plastics in a concrete environment, *Proc. 3rd Intl. Symp. Non-Metallic (FRP) Reinforcement for Concrete Structures*, v.2, Japan Concrete Institute, 1997.

Dutta, P.K., Hui, D., and Prasad, Y., Influence of subfreezing temperatures on the flexural behavior of thick composites, *2nd Biennial European Joint Conf. Eng. Sys. Des. Analy. (ESDA)*, Queen Mary and Westfield College, University of London, July 4–7, 1994.

Fyfe, E.R., Watson, R.J., and Watson, S.C., Long term durability of composites based on field performance and laboratory testing, *Proc. ICCI '96*, Saadatmanesh, H. and Ehsani, M.R. (Eds.), Tucson, AZ, January 1996.

GangaRao, H.V.S. and Kumar, S., Design and fatigue response of concrete bridge decks reinforced with FRP rebars, *Proc. 2nd Intl. RILEM Symp. (FRPPRCS-2), Non-Metallic (FRP) Reinforcement for Concrete Structures*, Gent, Belgium, 1995.

GangaRao, H.V.S. and Vijay, P.V., Aging of structural composites under varying environmental conditions, *Proc. 3rd Intl. Symp. Non-Metallic (FRP) Reinforcement for Concrete Structures*, v.2, Japan Concrete Institute, October 1997.

GangaRao, H.V.S, Vijay, P.V., Altizer, S.D., Douglass, N., Pauer, R., and Faza, S., Thermoset polymer performance under harsh environments to evaluate glass composite rebars for infrastructure applications, Advanced Composite Materials, State-of-the-Art Report, First Middle East Workshop on Structural Composites, Sponsored by the Egyptian Society of Engineers, Sharm El-Shiekh, Egypt, June 1996.

GangaRao, H.V.S., Vijay, P.V., and Burdine, E., Durability of steel stringers reinforced with CFRP plates, *Proc. CDCC '98*, Benmokrane, B. and Rahman, H. (Eds.), Sherbrooke, Québec, Canada, August 5–7, 1998.

GangaRao, H.V.S., Vijay, P.V., and Dutta, P.K., Durability of composites in infrastructure, CORROSION-95, Paper No. 550, NACE International Annual Conference and Corrosion Show, March 1995.

GangaRao, H.V.S., Vijay, P.V., Gupta, R.K., and Barbero, E., Mechanical-hygrotheramal responses and predictive models for FRP composites-state-of-the-art review submitted to US Army Corps of Engineers, Engineer Research and Development Center, Construction Engineering Research Laboratory, CFC-Report No. 02-100, June 2001.

Gentry, T.R. and Hudak, C.E., Thermal compatibility of plastic composite reinforcement and concrete, *ACMBS-II*, El-Badry, M. (Ed.), Montreal, August 1996.

Gerritse, A., Assessment of long term performance of FRP bars in concrete structures, *Proc. CDCC '98*, Benmokrane, B. and Rahman, H. (Eds.), Sherbrooke, Québec, Canada, August 5–7, 1998.

Hahn, H.T. and Kim, K.S., Hygroscopic effects in aramid fiber/epoxy composite, *J. Eng. Matls. Tech.*, 110, 153–157, April 1988.

Haskins, J.F., Thermal aging, *SAMPE J.*, 25, 2, 29–31, March-April 1989.

Hogg, P.J. and Hull, D., Corrosion and environmental deterioration of GRP, *Developments in GRP Technology-1*, New York: Applied Science Publishers, 1983.

Janas, V.F. and McCullough, R.C., The effects of physical aging on the viscoelastic material of a thermoset polyester, *Composite Sci. Tech.*, Elsevier Applied Science, 1987.

Japan Concrete Institute (JCI), Technical report on continuous fibre reinforced concrete, *Technical Committee on Continuous Fibre Reinforced Concrete (JCI TC952)*, Japan, 1998.

Kajorncheappunngam, S., Gupta, R.K., and GangaRao, H.V.S., Effect of aging environment on degradation of glass reinforced epoxy, *J. Composites Construction*, ASCE, Vol. 6, No. 1, Feb. 2002.

Karbhari, V.M., Chin, J.W., Hunston, D., Benmokrane, B., Juska, T., Morgan, R., Lesko, J.J., Sorathia, U., and Reynaud, D., Durability gap analysis for fiber-reinforced polymer composites in civil infrastructure, *J. Composites Construction*, 73, 238–247, August 2003.

Kato, Y., Nishimura, T., Uomoto, T., and Yamaguchi, T., The effect of ultraviolet rays to FRP rods, *Proc. CDCC '98*, Benmokrane, B. and Rahman, H. (Eds.), Sherbrooke, Québec, Canada, August 5–7, 1998.

Katsuki, F. and Uomoto, T., Prediction of deterioration of FRP rods due to alkali attack. Non-metallic (FRP) reinforcement for concrete structures, *FRPCS-2*, Gent, Belgium, 1995.

Katz, A., Berman, N., and Bank, L.C., Effect of cyclic loading and elevated temperature on the bond properties of FRP rebars, *Proc. CDCC '98*, Benmokrane, B. and Rahman, H. (Eds.), Sherbrooke, Québec, Canada, August 5–7, 1998.

Kelen, T., *Polymer Degradation*, New York: Van Nostrand Reinhold Co., 1983.

Kumar, V.S. and GangaRao, H.V.S., Fatigue response of concrete decks reinforced with FRP rebars, *J. Struct. Eng.*, 124, 1, 11–16, January 1998.

Lagrange, A., Melennec, C., and Jacquemet, R., Influence of various stress conditions of the moisture diffusion of composite in distilled water and in natural sea-water, *Durability of Polymer Based Composite Systems for Structural Applications*, St. Louis, MO: Elsevier Applied Science, 1991.

Lampo, R.G., Hoy D.E., and Odello, R.J., Field performance of FRP composite prestressing cables, Materials for the New Millennium, *Proc. 4th Matls. Eng. Conf.*, ASCE, November 1996.

Litherland, K.L., Oakley, D.R., and Proctor, B.A., The use of accelerated aging procedures to predict the long term strength of GRC composites, *Cement Concrete Res.*, 11, 455–466, 1981.

Lopez-Anido, R., Howdyshell, P., and Stephenson, L.D., Durability of modular FRP composite bridge decks under cyclic loading, *Proc. CDCC '98*, Benmokrane, B. and Rahman, H. (Eds.), Sherbrooke, Québec, Canada, August 5–7, 1998.

Machida, A. (Ed), State-of-the-art report on continuous fiber reinforcing materials, *Concrete Engineering Series 3, Research Committee on Continuous Fiber Reinforcing Materials*, Tokyo: Japan Society of Civil Engineering, 1993.

Malvar, L.J., Durability of composites in reinforced concrete, *Proc. CDCC '98*, Benmokrane, B. and Rahman, H. (Eds.), Sherbrooke, Québec, Canada, August 5–7, 1998.

Mandell, J.F. and Meier, U., Effects of stress ratio, frequency and loading time on the tensile fatigue of glass reinforced epoxy. Long-term behaviour of composites, *ASTM STP 813*, O'Brien, T. K. (Ed.), Philadelphia, PA: American Society for Testing and Materials, 1983.

Marshall, O.S. Jr. and Busel, J.P., Composite repair/upgrade of concrete structures, Materials for the New Millenium, Chong, K.P. (Ed.), *Proc. 4th Matls. Eng. Conf.*, Washington, DC, November 1996.

NACE, CCT, and FHWA, Cost of Corrosion Study (CCS), July 2002, http://nace.org/nace/content/nacenews/news_win/html/costcorr.htm.

Nanni, A., Bakis, C.E., and Mathew, J.A., Acceleration of FRP bond degradation, Durability of Fiber Reinforced Polymer Composites for Construction (FRP), *Proc. CDCC '98*, Benmokrane, B. and Rahman, H. (Eds.), Sherbrooke, Québec, Canada, August 5–7, 1998.

Natarajan, V., GangaRao, H.V.S., and Shekar, V., Fatigue response of fabric reinforced poly-meric composites, *J. Composite Materials*, 39, 17, 1541–1559, 2005.

Odagiri, T., Matsumoto, K., and Nakai, H., Fatigue and relaxation characteristics of continuous aramid fiber reinforced plastic rods, *Proc. 3rd Intl. Symp. Non-Metallic (FRP) Rein-forcement for Concrete Struct.*, v.2, Japan Concrete Institute, 1997.

Parkyn, B., Glass reinforced plastics, In: *Encyclopedia of Polymer Science and Engineering*, 2nd Ed., London: CRC Press, 1970; New York: John Wiley & Sons, 1985.

Porter, M.L. and Barnes, B.A., Accelerated durability of FRP reinforcement for concrete structures, *Proc. CDCC '98*, Benmokrane, B. and Rahman, H. (Eds.), Sherbrooke, Québec, Canada, August 5–7, 1998.

Priestley, M.J.N., Seible, F., and Fyfe, E., Column seismic retrofit using fiberglass/epoxy jacket, advanced composite materials in bridges and structures, Neale, K.W. and Labossiere, P. (Eds.), Canadian Society for Civil Engineers, Canada, 1992.

Pritchard, G. and French, M.A., The fracture surface of hybrid fiber composites, *Composite Sci. Tech.*, 47, 217–223, 1993.

Pritchard, G. and Speake, S.D., Effects of temperature on stress-rupture times in glass/poly-ester laminates, *Composites*, 19, 1, 29–35, 1988.

Proctor, B.A., Oakley, D.R., and Litherland, K.L., Development in the assessment and per-formance of GRC over 10 years, *Composites*, 173–179, April 1982.

Rao, R.M.G.K., Balasubramanian, N., and Chanda, M., Factors affecting moisture absorption in polymer composites, *Part-II, Influence of External Factors, Environment Effects on Composite Materials*, v. III, Springer, G. (Ed.), Lancaster, PA: Technomic Pub-lishing Company, 1981.

Roll, R.D., Use of GFRP rebar in concrete structures, Advanced Composite Materials in Civil Engineering Structures, Iyer, S.L. (Ed.), *Proc. Soc. Specialty Conf.*, New York: ASCE, January 1991.

Rostasy, F.S., Frp: The European Perspective, *Proc. ICCI '96*, Saadatmanesh, H. and Ehsani, M.R. (Eds.), Tucson, AZ, January 1996.

Schutte, C.L., Environmental durability of glass fibre composites, *Matl. Sci. Eng.*, 13. 7, 265–323, November 1994.

Seible, F. and Karbhari, V., Advanced composites for civil engineering applications in the U.S, *Proc. ICCI '96*, Saadatmanesh, H. and Ehsani, M.R. (Eds.), Tucson, AZ, January 1996.

Sen, R., Shahawy, M., Rosas, J., and Sukumar, S., Durability of AFRP pretensioned elements in a marine environment, *ACI Struct. J.*, 95, 5, 578–587, 1998.

Sheard, P.A., Clarke, J.L., Dill, M.J., Hammersley, G.P., and Richardson, D.M., Eurocrete — Taking account of durability for design of FRP reinforced concrete structures, *Proc. 3rd Intl. Symp. Non-Metallic (FRP) Reinforcement for Concrete Struct.*, v.2, Japan Concrete Institute, October 1997.

Sonobe, Y., Fukuyama, H., Okamoto, T., Kani, N., Kimura, K., Kobayashi, K., Masuda, Y., Matsuzaki, Y., Nochizuki, S., Nagasaka, T., Shimizu, A., Tanano, H., Tanigaki, M., and Teshigawara, M., Design guidelines of FRP reinforced concrete building struc-tures, *J. Composites Construction*, 90–115, August 1997.

Springer, G.S. (Ed.), *Environmental Effects on Composite Materials*, v.I–III, Lancaster, PA: Technomic Publishing Company, 1981.

Taerwe, L., FRP developments and applications in Europe, *Fiber-Reinforced-Plastic (FRP) Reinforcement for Concrete Structures: Properties and Applications*, New York: Elsevier Science Publishers, 1993.

Talreja, R., *Fatigue of Composite Materials*, Lancaster, PA: Technomic Publishing Company, 1987.

Uomoto, T., Durability considerations for FRP reinforcements, *Proc. FRPRCS-5*, Burgoyne, C.J. (Ed.), Cambridge, 2001.

Vijay, P.V., Aging Behavior of Concrete Beams Reinforced with GFRP Bars, Ph.D. dissertation, Department of Civil and Environmental Engineering, West Virginia University, Morgantown, May 1999.

Vijay, P.V. and GangaRao, H.V.S., Creep behavior of concrete beams reinforced with GFRP bars, *Proc. CDCC '98*, Benmokrane, B. and Rahman, H. (Eds.), Sherbrooke, Québec, Canada, August 5–7, 1998.

Vijay, P.V. and GangaRao, H.V.S., Development of fiber reinforced plastics for highway application: aging behavior of concrete beams reinforced with GFRP bars, *CFC-WVU Report No. 99-265 (WVDOH RP #T-699-FRP1)*, 1999a.

Vijay, P.V. and GangaRao, H.V.S., Accelerated and natural weathering of glass fiber reinforced plastic bars, *4th Intl. Symp., Fiber Reinforced Polymer Reinforcement for Reinforced Concrete Structures*, ACI Fall Conference, SP-188, October-November 1999b.

Vijay, P.V. and GangaRao, H.V.S., Bending behavior and deformability of glass fiber-reinforced polymer reinforced concrete members, *ACI Struct. J.*, 98(6), 834–842, November-December 2001.

Vijay, P.V., GangaRao, H.V.S., and Kalluri, R., Hygrothermal response of GFRP bars under different conditioning schemes, *Proc. CDCC '98*, Benmokrane, B. and Rahman, H. (Eds.), Sherbrooke, Québec, Canada, August 5–7, 1998.

Zheng, Q. and Morgan, R.J., Synergistic thermal-moisture damage mechanisms of epoxies and their carbon fiber composites, *J. Composite Matls.*, 27, 15, 1465–1478, 1993.

# 5 Strengthening of Structural Members

## 5.1 INTRODUCTION

Conventional reinforced concrete members consist of Portland cement concrete and steel reinforcement. In such beams, concrete resists compressive forces and steel reinforcing bars embedded in concrete — typically on the tension side — resist tensile and dowel (shear) forces. Such an arrangement is structurally efficient because of concrete's inherent resistance to compression, whereas that of steel's in tension and partially in shear. In some reinforced concrete members, steel reinforcement is also used to partially resist compression and enhance the flexural strength and stiffness of members with limited depth and even to limit crack widths. Recent advances in fiber reinforced polymer (FRP) composite technologies have resulted in alternative reinforcing materials that can be used efficiently as supplemental, externally bonded reinforcement [Crasto et al. 1996; Dolan 1993; Machida 1993; Neale and Labossière 1997]. A class of such materials comprises fiber reinforced polymeric (FRP) materials, commercially available in the form of fabrics or sheets, that can be bonded to the outer surface of concrete members (hence the term, "external reinforcement") to accomplish a number of desired objectives.

Two types of FRP fabrics/sheets are in common use:

1. Glass fiber reinforced polymer (GFRP) laminates or wrap systems
2. Carbon fiber reinforced polymer (CFRP) laminates or wrap systems

Although physically both GFRP and CFRP are in laminate form or wraps, in this book they will be referred to as *FRP external reinforcement systems*, or simply FRP-ER. The word "system" is used in conjunction with FRP external reinforcement to indicate that these laminates (i.e., FRP wraps) can be used as external reinforcement for not only new steel-reinforced flexural members but also for various other applications such as:

1. Repairing damaged/deteriorated concrete beams and slabs to restore their strength and stiffness, assuming that debonding the FRP wrap would not cause member failure
2. Limiting crack width under increased (design/service) loads or sustained loads

(a)                                            (b)

**FIGURE 5.1** Strengthening columns of Pond Creek bridge, West Virginia: (a) wet lay-up system, *in situ* hand wrap; (b) pre-cured system, GFRP shells prior to installation.

3. Retrofitting concrete members to enhance the flexural strength and strain-to-failure of concrete elements necessitated by increased loading conditions such as earthquakes or traffic loads
4. Designing new concrete members having depth limitations or needing high demands on ductility
5. Rectifying design and construction errors
6. Enhancing the service life of concrete members
7. Increasing the shear strength of in-service concrete members
8. Providing confinement for concrete members, such as concrete columns, as an alternative to steel jacketing
9. Restoring or retrofitting structures built with masonry and wood
10. Repairing old and historic structures

Since the late 1980s, glass and carbon fiber composite wraps have seen a dramatic increase in their applications [Alkhrdaji and Thomas 2005; Dolan 1993; JSCE 1992; Machida 1993; Marshall et al. 1999; Mufti et al. 1991; Neale and Labossière 1997; Nikolaos et al. 1995]. Used initially as demonstration projects, FRP wraps are now being used as routine construction material for retrofitting columns (Figure 5.1) and the strengthening of beams and slabs (Figure 5.2). This chapter presents the use of FRP-ER for design and construction of externally bonded FRP systems, particularly for the strengthening of concrete structures. Theory and design principles are presented to illustrate the use of FRP-ER for enhancing the strength and stiffness of concrete members. Several examples conforming to ACI 440 guidelines are provided on the flexural and shear design of concrete beams bonded with FRP-ER [ACI 440.2R-02].

## 5.2 BONDING CONCRETE BEAMS WITH FRP-ER

The technique of bonding FRP-ER to a prepared concrete (substrate) surface consists of several sequential steps. The first step is to prepare the concrete surface for achieving a proper FRP-ER bondable surface, which should be even and uniform.

**FIGURE 5.2** Library building floor beams rehabilitated with FRP, San Antonio, Texas, 1993. (Courtesy of Constructed Facilities Center, West Virginia University.)

For concrete substrate having excessive cracking or uneven surfaces, a high-viscosity polymer putty (e.g., heavy-duty methacrylate or epoxy) can be used as a filler for cracks, bug holes, surface pores, and irregularities up to 1/4-in. width or depth. The second step is to apply a primer coat to obtain a uniform bondable substrate. The primer is a low-viscosity polymer used as a first coat, essentially for filling concrete pores. Primer bonds to both concrete substrate and to the resin applied on the FRP fabric. The third step involves the application of a resin coat to the concrete/primer substrate. The resin is a polymer that wets the fabric and chemically bonds to the primer applied on the concrete substrate and the fabric. When necessary, a protective coating is applied over the composite wrap to resist environmental exposure effects due to moisture ingress, leaching of salt water and alkali solution, ultraviolet (UV) radiation, chemical emissions, and other external ingress. Protective coatings may be used to enhance fire rating of FRP-ER.

## 5.3  TYPES OF FRP SYSTEMS

The FRP-ER systems are commercially available in several forms to suit different application needs, such as load conditions, member dimensions, shape, and field environment. Wet lay-up FRP-ER systems are commonly used in the field. Precured FRP-ER systems (called "shells") have also been used in the field (Figure 5.1). Machine-applied FRP-ER systems are not common due to complexities in field applications.

In the wet lay-up system, the fibers of FRP-ER are saturated with resin at the site and cured *in situ*, typically at ambient temperatures, resulting in a composite wrap or laminate. The FRP-ER can be in the form of dry fiber tows, unidirectional fiber sheets, or fabrics consisting of multidirectional fibers. A variation of a wet lay-up system, commonly referred to as "prepreg," consists of unidirectional fiber sheets or fabrics (having multidirectional fibers) that are pre-impregnated and saturated with a resin system. Prepreg sheets bond to a concrete surface with or without further resin application. Prepreg systems are cured *in situ*; extra heat may be used to achieve a full cure, if necessary. For example, extra heat can be generated *in situ* through a heating blanket which can be wrapped over the external reinforcement.

The pre-cured FRP composite systems are commercially available in a variety of shop-manufactured shapes. They can be in the form of unidirectional thin ribbon strips, laminated sheets, or pre-cured shells. As in wet lay-up systems, concrete substrate must be properly finished to achieve smoothness and primed before installing the pre-cured composite system.

Presently, FRP composite wrap technology has matured to a point where the potential for field applications depends primarily on the availability of validated design information and guidelines. In the United States, guidelines for the strengthening of structural elements using FRP systems have been developed by ACI Committee 440 [ACI 440.2R-02]. These guidelines are based on the knowledge acquired from worldwide research, analytical work, and data from many field applications [GangaRao and Vijay 1999; Karbhari and Seible 1999; NCHRP Report 514 2004; Sonobe et al. 1997; Vijay and GangaRao 1999]. A few field applications are summarized in Section 5.5.

The design strength of concrete members using externally bonded FRP-ER depends on the design strength of commercially available proprietary FRP systems (for field installation details, see Section 5.5). As these systems have been developed by the industry through material characterization and testing, their recommended guidelines should be strictly followed [Sika Design Manual 1997]. Consulting experienced engineers while specifying primer or adhesive for field applications would be prudent. This suggestion is made to ensure the compatibility of primers, resins, and sizing (i.e., coatings on fabrics) necessary to obtain the optimal service life of a wrapped concrete member.

## 5.4 ADVANTAGES AND LIMITATIONS OF FRP COMPOSITE WRAPS FOR REINFORCED CONCRETE MEMBERS

Conventional steel reinforced concrete has been extensively used world-wide as one of the most important structural materials. Structural applications include buildings, bridges, retaining walls, tunnels, tanks, underground pipes, and so on. However, concrete members with steel reinforcement corrode due to environmental exposures such as deicing salts, chemicals, and moisture ingress due to micro-cracking in concrete. Premature cracking in steel-reinforced concrete due to the corrosion of steel leads to reduced strength, stiffness, and service life as well as concrete failure, which in turn can lead to structural failure.

In many cases, the corrosion of steel-reinforcing bars weakens concrete structures as a result of tension caused by the expansion of corroded steel. These concrete members need rehabilitation to restore their strength and stiffness after controlling corrosion rates through cathodic protection or other conventional means. The application of FRP composite wrap technology to concrete members has been found to be an excellent solution to this problem [NCHRP Report 514 2004; Oehlers 1992].

To date, successful applications of FRP composite wrap technology have been associated with rehabilitating and retrofitting concrete structures [Malvar et al. 1995]. The success of these applications is attributed to the many advantages this technology has to offer over steel [ACMBS-III 2000; CDCC 1998; ICCI 1998]. Some of the advantages of using FRP-ER over steel jacketing are:

1. Higher strength-to-weight ratio ($\approx$ 15 and 35, respectively, for glass and carbon, as compared with that of steel)
2. Higher stiffness-to-weight ratio ($\approx$ 1 and 3, respectively, for glass and carbon, as compared with that of steel)
3. Higher corrosion resistance
4. Lighter unit weight, resulting in less-expensive equipment for economical handling, shipping, and transportation as well as lighter erection equipment
5. Higher durability, leading to lower life-cycle costs [ACMBS-III 2000]
6. Greater ductility, providing ample warning before collapse [ACMBS-III 2000]
7. Easier-to-reinforce microcrack zones
8. Easier-to-control tension crack growth by the confining concrete
9. Better customization for specific needs
10. Faster field installation, resulting in more economical procedures for the confinement of concrete in columns than steel jacketing
11. Simpler field corrections in case of installation defects of bonding of FRP with concrete substrate

However, some limitations exist to FRP composite wrap applications:

1. Uncertainties about the durability of FRPs, as their long-term performance data are limited
2. Concerns of fire resistance, adverse effects from smoke and toxicity, and poorer resistance of resins to UV rays
3. Limited knowledge of material properties and application procedures, and possible continuation of corrosion of steel reinforcing bars in wrapped concrete members
4. Lack of adequate laboratory and field data with respect to various structural actions, including the shear-lag phenomenon due to an increase in the number of fiber composite wrap layers

## 5.5  FIBER WRAP TECHNOLOGY APPLICATIONS: CASE STUDIES

Several structures in the United States, Japan, and Europe have been successfully rehabilitated using the fiber wrap technology. A few case studies of fiber wrap applications are presented as follows.

### 5.5.1 KATTENBAUSCH BRIDGE, GERMANY

The Kattenbausch Bridge in Germany is a continuous, prestressed concrete box girder with 45-m spans and two 36.5-m side spans. Several spans of this bridge exhibited cracking around their working joints located near the points of contraflexure where the tendons were coupled. Transverse cracks were predominantly noticed at the bottom slab of the box girder at the joint area, which widened significantly due to inadequate reinforcement in the bottom slab and a combination of other stresses. Instead of employing a traditional steel plate bonding technique to strengthen such joints, 20 glass-fiber reinforced plates were used to restore one of the working joints. Each plate was 3.2 m long, 0.15 m wide, and 0.003 m thick. This rehabilitation technique resulted in a reduction of 50% in crack widths and 36% lower stress amplitude due to fatigue [Meier et al. 1993].

### 5.5.2 IBAACH BRIDGE, LUCERNE, SWITZERLAND

Built in 1969 near Lucerne, Switzerland, the Ibaach Bridge consists of a multispan, pre-stressed concrete box beam with a total length of 228 m. One of the spans (39 m long and 16 m wide) crossing a national highway was damaged accidentally during a core-boring operation for mounting new traffic signals. Several wires of the prestressing tendon in the outer web were completely severed by an oxygen lance. The span was repaired satisfactorily by using 2-150 mm × 5000 mm × 1.75 mm and 1-150 mm × 5000 mm × 2 mm CFRP sheets [Meier 1992].

### 5.5.3 CITY HALL OF GOSSAU, ST. GALL, SWITZERLAND

An elevator needed to be installed as a part of a renovation program at the City Hall of Gossau, St. Gall, Switzerland. To accommodate the elevator, a rectangular cut was made in one of the slabs. The sides of the rectangular hole were strengthened with CFRP sheets. For aesthetic reasons, thin CFRP sheets rather than thick steel plates were used [Meier et al. 1993].

### 5.5.4 COLUMN WRAPPING AT HOTEL NIKKO, BEVERLY HILLS, CALIFORNIA, USA

Thirty-four rectangular columns in the parking structure of the Hotel Nikko, Beverly Hills, California — located within 12 mi of an earthquake fault — were retrofitted with fiber wraps before the 1994 Northridge earthquakes. As a result of this retrofitting, these columns did not suffer any damage during the earthquake [McConnell 1995]. In contrast, a number of taller columns that were not wrapped before the earthquake suffered damage during the event.

### 5.5.5 COLUMN WRAPPING PROJECTS BY CALTRANS FOR SEISMIC RESISTANCE, USA

The California Department of Transportation (Caltrans) has retrofitted columns supporting many highway bridges in California using composite fabrics. For retrofitting columns, typically a mat of woven unifabric made of glass or aramid is applied in a wet lay-up system using epoxy as a resin. The mat cures in place at ambient temperature. An expansive grout is injected beneath the mat to assure good contact with the original concrete. The mats utilized about 15% aramid fibers to resist the high shear strength. Fibers were primarily oriented along the column circumference (95%), whereas the remaining 5% were parallel to the column axis (this fiber configuration is expressed as 0°/90°). Laboratory testing of this system showed that this wrapping procedure helped increase the ductility factor from 1 to 8. The effectiveness of this system was evident from the response of two of the bridge columns that were located 20 mi from the epicenter of the 1994 Northridge earthquake and retrofitted with this scheme, which suffered no damage [McConnell 1995].

### 5.5.6   COLUMN WRAPPING FOR CORROSION PREVENTION, USA

Many Departments of Transportation (DOTs) in the United States (e.g., Nevada, Pennsylvania, and West Virginia DOTs) have adopted fiber wrapping as a reliable and cost-effective method for column retrofitting. For example, the use of wrap saved the Vermont Department of Transportation $8000 (U.S.) as compared to using the conventional steel jacketing method [Tarricone 1995].

### 5.5.7   FLORIDA DEPARTMENT OF TRANSPORTATION, USA

Several bridges involving beam strengthening were undertaken by the Florida Department of Transportation. Techniques included the use of carbon/epoxy, aramid/epoxy, and a hybrid of glass/carbon woven materials to provide beam strengthening. Two layup procedures were used: dry carbon fiber fabric and prepreg. The wet system (prepreg) was found to be easier to apply to beams (tension side) in the field compared to dry carbon fabric [McConnell 1995].

### 5.5.8   SINS WOODEN BRIDGE, SWITZERLAND

The Sins wooden bridge, a historic structure near Sins, Switzerland, was built in 1807. Damaged in a civil war, the bridge was rehabilitated several times through various techniques [Meier et al. 1993]. Bonding 0.04–in. thick CFRP laminates/wraps strengthened two of its most highly loaded crossbeams and helped preserve this national monument. One crossbeam was strengthened by bonding 10-in. and 8-in. wide high modulus fiber wraps/laminates to, respectively, the bottom faces of the beam. A second crossbeam was strengthened with high-strength fiber sheets, 12–in. wide at the top of the beam and 8–in. wide on the bottom. The retrofitted bridge is satisfactorily carrying 45kN-category traffic.

### 5.5.9   SOUTH TEMPLE BRIDGE, INTERSTATE 15 (I-15), UTAH, USA

Built in 1962, South Temple Bridge on Interstate 15, Utah, was retrofitted with externally bonded CFRP composites. It involved retrofitting a reinforced concrete (RC) bent consisting of a cap beam and three columns that were supported on three

**FIGURE 5.3** View of the viaduct with a train running on top (East Street Viaduct, West Virginia).

pile caps [Pantelides et al. 1999]. Retrofitting with CFRP composite laminates/wraps greatly enhanced the displacement ductility of the RC bridge bent. The existing bent had developed substantial flexural cracks in the upper region of the columns, which had a high lateral load capacity and a maximum displacement ductility of 2.8. After retrofitting, the bent achieved the target displacement ductility of 6.3, more than twice the capacity that of as-built bent capacity. The increase in peak lateral load capacity was found to be 16%. An increase of 35% in shear strength was also achieved at the cap beam-column joints using CFRP laminates.

### 5.5.10 EAST STREET VIADUCT, WEST VIRGINIA, USA

The East Street viaduct that supports the CSX goods railway operation (Figure 5.3) is located in Wood County, West Virginia. This 130-ft long and 55-ft wide viaduct carries two lanes of traffic with a 4-ft wide walkway on each side. The middle portion of the viaduct is 12 ft wide and consists of columns on the edges. The viaduct had moisture-induced damage in the wing and abutment walls in addition to corrosion-induced spalling of concrete (Figure 5.4). Rehabilitation of this viaduct was carried out by the West Virginia Department of Transportation (Division of Highways) in cooperation with the Constructed Facilities Center (CFC) of West Virginia University. The rehabilitation scheme consisted of bonding glass FRP wraps to the abutment, wing walls, and beams.

To alleviate the moisture problem, the bottom of the slab was cleaned and shotcreted. Grooves were cut in the slab at a 4- to 5-ft spacing in the longitudinal and transverse directions. The grooves were aligned with the columns to drain water. The grooves were plugged with inverted plastic T-shapes and caulked. In addition, the following field operations were carried out at the bottom portion of the concrete slab:

1. Heating of wet areas and drying moisture patches
2. Application of primer before bonding FRP wraps

**FIGURE 5.4** View of the damaged wing wall (East Street Viaduct, West Virginia).

Similar to the underside of slabs, wing wall surfaces were cleared of loose concrete and shotcreted. Weep drains were installed in the wing and abutment walls. Steel columns were cleaned and painted, whereas concrete pedestal surfaces were cleaned and made even with putty application. After shotcreting, the abutments, wing walls, and beams were strengthened using glass FRP wraps.

The concrete surface was prepared by filling the cracks with epoxy putty to obtain an even surface. Primer (low-viscosity resin with a viscosity of 500 to 1000 cps) was applied to the prepared concrete surface to obtain a bond between the base concrete (substrate) and glass fabric composite. After primer curing, a saturant (resin) was applied to the primed surface, and then fabrics were placed on the substrate and gently pressed to remove the trapped air. After this, a second coat of saturant was applied on the bonded FRP wrap. After curing, a UV- and moisture-protective coating that matched the color of the concrete surface was applied to enhance the appearance. Figure 5.3 through Figure 5.9 illustrate a sequence of operations ranging from the damaged to restored state of the viaduct.

### 5.5.11 MUDDY CREEK BRIDGE, WEST VIRGINIA, USA

Located in Preston County, West Virginia, the Muddy Creek Bridge was rehabilitated in October 2000 using carbon FRP laminates/wraps. The West Virginia Department of Transportation carried out the bridge rehabilitation in cooperation with Specialty Group Inc., Bridgeport, West Virginia. The Constructed Facilities Center (CFC) of West Virginia University provided the necessary rehabilitation design. The bridge consisted of three 42-ft wide spans, each supported on four T-beams. The 8 1/4-in thick flanges of the exterior and interior T-beams were, respectively, 84 and 90 in.; the webs were 21 in. wide and 36 3/4 in. deep.

The bridge experienced several problems such as lateral movement in the interior spans, hairline vertical cracks on piers and abutments, and localized concrete spalling and cracking due to steel corrosion. The surface of the T-beams was prepared after

**FIGURE 5.5** View of the wing wall and concrete girder being restored with GFRP fabric and the abutment wall with moisture-trap channels (East Street Viaduct, West Virginia).

**FIGURE 5.6** Wing wall and abutment being restored with GFRP (East Street Viaduct, West Virginia).

**FIGURE 5.7** View of the restored wing wall (East Street Viaduct, West Virginia).

**FIGURE 5.8** Concrete pedestals pretreated (left) and wrapped with a FRP fabric (right) (East Street Viaduct, West Virginia).

**FIGURE 5.9** View of the restored wing wall, girder, and the abutment walls (East Street Viaduct, West Virginia).

**FIGURE 5.10** Muddy Creek Bridge (West Virginia) before wrapping.

applying putty and primer. Following the surface preparation, exterior T-beams were strengthened with CFRP fabrics by bonding them to the sides and bottom. Rehabilitation of this bridge included a protection scheme to resist galvanic corrosion between carbon and steel [Tavakkolizadeh and Saadatmanesh 2001; Sen et al. 2006; Wheat et al. 2006]. Various stages of strengthening the Muddy Creek Bridge are shown in Figure 5.10 through Figure 5.16.

FIGURE 5.11 Primer application on Muddy Creek Bridge (West Virginia) beams before wrapping.

FIGURE 5.12 Wrapping of the beam side with CFRP fabrics (Muddy Creek Bridge, West Virginia).

**FIGURE 5.13** Preparations for wrapping the beam bottom with CFRP fabrics (Muddy Creek Bridge, West Virginia).

**FIGURE 5.14** Removal of air entrapment from the bonded CFRP fabric (Muddy Creek Bridge, West Virginia).

FIGURE 5.15 Application of finishing coat (Muddy Creek Bridge, West Virginia).

FIGURE 5.16 Finished interior of the external beam with carbon fabric and protective coating (Muddy Creek Bridge, West Virginia).

## 5.6  COMPATIBILITY OF STEEL-REINFORCED CONCRETE AND FRP-ER

In steel-reinforced concrete structures, concrete and steel work together in a synergistic sense that the advantages of one material compensate for the limitations of the other. The structural contribution of external fiber composites (FRP-ER) to concrete members, particularly beams, is somewhat similar as summarized below:

1. Similar to internal steel bars acting as reinforcement on the tension side, FRP-ER also acts as reinforcement on the tension side.
2. FRP-ER provides protection to the concrete surface and the reinforcing steel from chemical and moisture ingress.
3. FRP-ER enhances flexural strength and stiffness, thereby delaying the formation of tension cracks. For the structural contribution of FRP-ER, the bonding material (primer and resin) between fiber composites and concrete is essential as the application is bond-critical.

The bond between the fiber wraps and concrete substrate is achieved by the use of primer, which can penetrate into concrete pores [Hollaway 1990; Mallick 1993]. Chemical bonds are created between the primed substrate and FRP-ER through a resin system. The resin system can be modified with fire retardants to enhance the fire safety of the FRP-ER [Hollaway 1993; Mallick 1993]. The FRP-ER can be engineered for compatibility with reference to strength, stiffness, and thermal coefficients so that its ability to work in unison with steel-reinforced concrete members can be assured (Master Builders: Design Guidelines). Air pockets formed between the wrap and the concrete surface during construction can be eliminated through compatible resin injection (Figure 5.14).

Composite laminates/wraps can be applied easily to concrete structures through *in situ* adhesive application and wrapping by following the contour of the member surface. Extreme care should be taken to provide special detailing while applying the composite wraps (Tonen Corporation: Design Guidelines) to minimize the potential for structural failures resulting from wrap delamination. For example, when a composite wrap (FRP-ER) is used in a beam-slab system, the wrap should be anchored under the slab with a minimum of 6 in. of extension from the beam-slab joints. Similar to steel cover plates used for steel girders, longitudinal curtailment (cut-off) of composite wraps can be achieved in a stepwise manner to minimize excessive stress concentration at a single cross-section. Step-wise FRP-ER curtailment provides a gradual change in the beam stiffness along the length and prevents a peeling-off of FRP-ER at the ends [Bazaa, Missihoun, and Labossiere 1996].

## 5.7  THE ACI GUIDE SPECIFICATIONS

### 5.7.1  INTRODUCTION

American Concrete Institute (ACI) Committee 440 has developed guide specifications for the design and construction of externally bonded FRP systems [ACI 440.2R-

02]. It is a consensus document based on the knowledge gained from world-wide research and field applications and the expertise of many professionals. It is divided into six parts:

1. Introduction and background information of externally bonded FRP systems, including historical developments
2. Identification of constituent materials and their physical, mechanical, and viscoelastic properties and durability considerations
3. Requirements for construction, which include shipping, storage, handling, installation, inspection, evaluation, acceptance, maintenance, and repair
4. Design recommendations for flexural strengthening, shear, axial compression, tension and ductility enhancement, reinforcement details, drawings, and specifications
5. Design examples
6. Appendices (material properties and summary of standard test methods)

Items under part 3 of the guide are discussed in the following sections, whereas design aspects and examples are presented in later sections of this chapter.

### 5.7.2 SHIPPING AND STORAGE

Generally, resins used for bonding fiber wraps are thermoset. ACI 440.2R-02 recommends the shipping of thermoset resins to comply with the Code of Federal Regulations (CFR), No. 49, dealing with transportation. The resins should be shipped under the provisions for "Hazardous Materials Regulations," and stored in compliance with the Occupational Safety and Health Administration's (OSHA) recommendations provided by the manufacturers, e.g., Material Safety Data Sheets (MSDS) provided by the Tonen Corporation, Master Builders, Mitsubishi Corporation, Ashland Chemicals, and other manufacturers. Properties of the constituents of a resin system can change with time, temperature, and humidity. Any material that has exceeded its shelf life should not be used because of the deterioration in its chemical composition. Disposal of constituents should be carried out in compliance with manufacturer's specifications and state and federal environmental regulations.

### 5.7.3 HANDLING

Thermoset resins must be handled carefully. MSDSs of constituent materials that are obtained from the manufacturers should be accessible at the job site [ACI 440.2R-02]. The handling of resin constituents can be a health and safety hazard that can cause:

1. Skin irritation such as burns, rashes, and itching
2. Breathing problems from organic vapors
3. Fire when exposed to sparks, cigarettes, flames, and other sources of ignition, or even exothermic reactions of mixtures and constituent materials

Wearing disposable plastic gloves and safety eyeglasses is recommended when handling resin systems. Dust masks or respirators should be used if dust or organic vapors are present or anticipated.

The workplace should be well ventilated. The floor should be properly covered to protect against spills and consequent chemical reactions. Uncontrolled reactions (including fire) when curing (after the mixing of constituents) may occur in mixing containers. Hence, containers should be monitored to prevent possible workplace hazards such as explosions, fire, or even atmospheric contamination with chemical fumes.

The clean-up of resin systems may require the use of flammable solvents; hence, appropriate precautions are necessary. All waste materials should be contained and disposed of in compliance with applicable environmental regulations and information provided by the MSDS.

### 5.7.4 INSTALLATION

FRP composite wraps should be installed by skilled workers and authorized contractors in accordance with the procedures developed and recommended by manufacturers [ACI 440.2R-02]. Contractor competency can be evidenced by proof of training or by demonstrated experience in the surface preparation and field installation of FRP-ER.

To attain a complete mixing, all resins should be mixed in the recommended volumetric ratio for the correct length of time and within the specified temperature ranges. Temperature, relative humidity, and moisture at installation time can affect the performance of the cured resin (matrix) in the composite wrap. Prior to and during installation, the surface temperature of the concrete should not be less than 50°F or as specified by the manufacturer. In addition, the relative humidity should not be above 60% or as specified by the manufacturer. If necessary, an auxiliary heat source such as heat blankets or space heaters can be used to raise the ambient and surface temperature during installation. Some resin systems are formulated to attain full cure only when in contact with water. However, most of the resin systems do not cure well when applied to a damp surface; therefore, the bonding of fabrics to damp areas should be avoided.

Equipment to apply resin systems includes resin impregnators, sprayers, mixing blades, pumping devices, and hand-stirring units. The equipment should be operated by trained personnel and must be kept clean at all times. In the case of equipment failure, back-up parts for equipment should be available at the job site for uninterrupted work.

### 5.7.5 SUBSTRATE PREPARATION

The ACI 546R report discusses problems associated with the conditions of concrete substrate and the methods of repairs and surface preparation. In addition to following the ACI 546R guidelines, the FRP system manufacturer's guidelines must also be followed. For example, the corrosion of steel reinforcing bars in concrete must be repaired (e.g., re-alkalization) before applying any FRP wrapping system. If this is not done, the large tensile force induced by corrosion could split the externally applied wrap system, resulting in potential structural failures. Surface cracks wider than 0.01 in. (0.25 mm) should be pressure grouted with epoxy in accordance with ACI 224R before the application of FRP wraps so that wrap performance can be optimized. Cracks smaller than 0.01 in. (0.25 mm) may require sealing to develop the proper bonding of FRP wraps with concrete substrate.

### 5.7.6 SURFACE PREPARATION

Surface preparation should be based on the type of applications: bond-critical applications (e.g., design controlled by flexure, shear strengthening of structural elements) or contact-critical applications (e.g., design controlled by confinement of columns or special joints).

For bond-critical applications, the concrete surface should be free of loose deposits on the concrete surface that could interfere with the bonding of the FRP-ER. Surface preparation can be accomplished using abrasive or water-blasting techniques. After removing all foreign or loose deposits, bug holes (surface voids) should be filled with putty and the concrete surface should be prepared to a minimum profile of CSP-2 as defined by the International Concrete Repair Institute (ICRI) surface profile chips. For example, localized out-of-plane variations including form lines should not exceed 0.03 in. (0.76 mm) or the tolerances recommended by the manufacturer. Such tight tolerances can be achieved by grinding the surface prior to surface blasting. The following two recommendations are made with respect to wrapping at re-entrant angles and concrete elements requiring confinement.

1. The re-entrant corners should be profiled to a minimum of a 0.5 in. (1.25 mm) radius to minimize stress concentrations in the FRP wraps and to also minimize voids between the wrap and the concrete surface. Special detailing of FRP wraps around re-entrant corners is essential to ensure a bond with the concrete substrate. All concrete surfaces should be dry, as recommended by the wrap system manufacturer. The moisture content level must be established to comply with ACI 503.3 and reduced, if necessary, to enhance resin penetration into the concrete surface and improve the mechanical interlocking of resin with concrete aggregates.

2. For concrete elements requiring confinement, uniform contact surfaces are required to develop uniform bond resistance by the FRP wraps without generating undesirable stress spikes. Large voids in the surface should be patched in both bond- and contact-critical applications. Large voids or spalls in the concrete substrate should be filled and properly bonded to substrate with high compressive-strength material to enhance the confinement effects of a concrete structural element.

### 5.7.7 APPLICATION OF CONSTITUENTS

Putty or bulk adhesive should be used to fill large voids and smooth surface discontinuities as recommended by the FRP manufacturer. Primer should be applied to all areas of the concrete surface to be covered with FRP wraps. Primer and putty should be cured fully before applying the FRP wrap system. The two types of commonly used composite wrap systems are the wet lay-up system with hand or machine application and the pre-cured system.

Wet lay-up systems typically involve the hand-laying of dry fiber sheets or fabrics, which are saturated with appropriate resins. FRP sheets or wraps can be saturated using a resin-impregnating machine. The fiber wraps should be pressed into the uncured resin as recommended by the FRP system manufacturer. To remove any trapped air, the wet wrap system should be rolled out exhaustively while the resin is still wet. When more than one layer of fiber wrap is required, laying the additional layers of fibers/fabrics before the complete cure of the previous layer of resin occurs is preferable to develop a good bond between the successive layers.

Pre-cured systems, such as FRP shells (Figure 5.1) or strips, are bonded to a clear concrete surface with an adhesive. The surface should be prepared in accordance with the manufacturer's recommendations. Adhesive should be applied uniformly to the properly prepared concrete surface before placing the pre-cured FRP systems. Any trapped air between the concrete surface and the pre-cured system should be rolled out before the adhesive is cured. Adhesive thickness should be maintained as recommended by the FRP manufacturer. Excessive adhesive thickness may not provide adequate interlaminar shear transfer capability and can even lead to poor structural compositeness between the concrete substrate and the FRP pre-cured system.

Special coatings should be provided to protect the FRP systems from UV degradation or excessive moisture ingress. Coatings should be compatible with FRP systems to minimize surface blistering or micro-cracking. Temporary protection, such as loose plastic sheathing around the FRP system, is essential for a proper cure of the resins. Such protection would help minimize the direct contact of the resin surface with rain, dust, sunlight, or even vandalism.

In wet lay-up systems, proper fiber/fabric alignment is critical. Even small variations (up to 5°) in fiber orientation from the design specifications can cause substantial strength variations. Fabric kinks or waviness should be minimized to avoid local stress concentration. Up to four plies of wraps can be used to efficiently transfer the shearing load between plies [Kshirasagar et al. 1998]. All resin systems should be cured according to the manufacturer's recommendations, and field modification of resin chemistry should not be permitted [ACI 440.2R-02]. The bond strength between a properly prepared concrete substrate and a FRP system is adequate to transfer interlaminar shear through the cured resin [GangaRao et al. 2000].

## 5.8  DESIGN PROPERTIES OF FRP-ER AND CONSTITUENTS

### 5.8.1  Design Properties of FRP-ER

The in-service thermomechanical properties of FRP composite materials deteriorate due to aging and other factors (e.g., exposure to UV rays, chemical environment)

**TABLE 5.1**
**Environmental-Reduction Factor ($C_E$) for Different FRP Systems and Exposure Conditions**

| Exposure Condition | Fiber and Resin Type | Environmental-Reduction Factor ($C_E$) |
|---|---|---|
| Interior exposure | Carbon/epoxy | 0.95 |
| | Glass/epoxy | 0.75 |
| | Aramid/epoxy | 0.85 |
| Exterior exposure (bridges, piers, and unenclosed parking garages) | Carbon/epoxy | 0.85 |
| | Glass/epoxy | 0.65 |
| | Aramid/epoxy | 0.75 |
| Aggressive environment (chemical plants and waste water treatment plants) | Carbon/epoxy | 0.85 |
| | Glass/epoxy | 0.50 |
| | Aramid/epoxy | 0.70 |

*Source*: ACI 440.2R-02.

[Arockiasamy and Zhuang 1995; Barger 2000; Green and Bisby 1998; Homam and Sheikh 2000; Javed 1996; Soudki and Green 1997; Vijay and GangaRao 1998; Vijay et al. 2002]. Therefore, FRP properties obtained through tests or those provided by manufacturers should be adjusted using reduction (or adjustment) factors for satisfactory structural performance over a service life of a minimum of 50 years. These reduction factors depend on fiber type, application, and nature of environmental exposure (see Equation 5.1 and Table 5.1). The design ultimate strength ($f_{fu}$) and rupture strain ($\varepsilon_{fu}$) are obtained by the product of the corresponding ultimate tensile strength ($f_{fu}^*$) and strain ($\varepsilon_{fu}^*$) of the FRP material as reported by the manufacturer with appropriate environmental reduction factors ($C_E$) suggested by ACI 440.2R-02.

$$f_{fu} = C_E f_{fu}^* \tag{5.1}$$

$$\varepsilon_{fu} = C_E \varepsilon_{fu}^* \tag{5.2}$$

$$E_f = \frac{f_{fu}}{\varepsilon_{fu}} \tag{5.3}$$

Environmental reduction factors ($C_E$) for different environmental exposure conditions are shown in Table 5.1.

FRP-ER obeys Hooke's law (a linear stress-strain relationship, Equation 5.3). The design ultimate strength and rupture strain (elongation) are lower than the manufacturer-reported values ($f_{fu}^*$ and $\varepsilon_{fu}$, respectively) because of the use of environmental exposure based reduction factor ($C_E$). However, for all practical purposes the modulus of elasticity (Equation 5.3) is unaffected by the environmental exposure conditions and remains the same as given by the manufacturer.

The environmental-reduction factors given in Table 5.1 are estimates based on the durability of fiber type, exposure conditions, and location. Three broad categories of exposure conditions are identified for concrete structures with FRP-ER in Table 5.1. The environmental reduction factors corresponding to the three categories are interior exposure, exterior exposure, and aggressive environment. These categories represent temperature variations associated with their location and exposure. Table 5.1 lists commonly used glass, carbon, and aramid fiber systems used specifically with epoxy resins because they are the most widely used for bonding FRP-ER to concrete structures. If other resin systems are selected, the $C_E$ values should be obtained from their manufacturers.

A perusal of Table 5.1 shows that penalties for FRP-ER are more severe with aggressive exposure conditions such as those located in chemical and wastewater treatment plants that may contain acid or alkali solutions, grease, purifiers, and other chemicals. FRP-ER properties are relatively less penalized for interior exposure conditions, where members are located within an enclosed environment such as a building. FRP-ER properties are penalized more for exterior exposure conditions than interior conditions. Examples of exterior or outdoor structures are bridges, piers, and unenclosed parking garages. These structures are subjected to deicing salts, high humidity and temperature variations, freeze-thaw cycles, and so on. Compared to interior and exterior exposure conditions, ACI 440.2R-02 cautions that future revisions of these reduction factors are possible as additional long-term performance data becomes available. Note that additional factors such as better resin systems, sizings, and protective coatings improve the durability of FRP systems.

Among the three types of fibers listed in Table 5.1, carbon fiber is the most durable under all types of exposure conditions, followed by aramid and glass fibers. For example, environmental reduction factors for carbon/epoxy, aramid/epoxy, and glass/epoxy systems subjected to interior exposure are 0.95, 0.85, and 0.75, respectively. However, when the same carbon/epoxy, aramid/epoxy, and glass/epoxy systems are subjected to aggressive environments, the reduction factors are more severe and change to 0.85, 0.70, and 0.50, respectively.

## 5.8.2 STRENGTH REDUCTION FACTORS

In designing steel-reinforced concrete beams, uncertainties of material strength, approximations in analysis, variations in dimensions of concrete sections, and variable field conditions are taken into account by applying a strength reduction factor to the nominal strength of a concrete member.

The ACI Building Code specifies the following strength reduction factors ($\phi$-factors) for steel-reinforced concrete beams under various load conditions [ACI 318-02] (see Figure 5.17):

1. Flexure (without axial loads): 0.9
2. Shear and torsion: 0.75
3. Bearing on concrete: 0.65

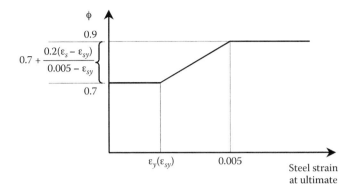

**FIGURE 5.17** Strength-reduction factor as a function of ductility. (Adapted from ACI 318-02).

For FRP-reinforced beams, a set of different strength reduction factors have been suggested to account for following parameters:

1. Basis of derivation of material properties ($\phi_{mat}$)
2. Processing methods ($\phi_{proc}$)
3. Manufacturing location ($\phi_{loc}$)
4. Long-term degradation of FRP properties ($\phi_{degr}$)
5. Cure conditions ($\phi_{cure}$)

The aforestated factors highlight several conditions that affect strength reduction factors. When applied in combination, these factors should be such that the reduction in material properties is not excessive. According to ACI 440.2R-02, the above five factors are not directly used in the current design practice suggested because sufficient data on their validity is not available; use of only the $C_E$ factor is recommended for the design of FRP-ER.

### 5.8.3 FRP Reinforcement for Flexural Strengthening

To increase flexural strength, FRP fabrics are bonded as an external reinforcement on the tension side of steel-reinforced concrete beams with fiber orientation along the member length [Hota et al. 1995; Ichimasu et al. 1993; Lee et al. 1998; Saadatmanesh and Ehsani 1991]. Depending on the ratio of FRP reinforcement area to the beam's cross-sectional area and the area of internal steel reinforcement, the increase in flexural strength can be more than 100%. However, a flexural strength increase up to 50% would be more realistic, which depends on practical considerations such as the concrete member dimensions, serviceability limits, ductility, and effective thickness of FRP fabric reinforcement [Arduini and Nanni 1997; Hota et al. 1995]. Although this chapter presents the design philosophy of strengthening rectangular RC beams, note that it is equally applicable to other shapes such as T- and I-sections having non-prestressed reinforcement.

### 5.8.4 EFFECT OF BENDING STRENGTH INCREASE ON SHEAR STRENGTH

When the flexural strength of a concrete beam is increased by using FRP-ER, verifying that the member has adequate shear strength to support the increased loads is important. If necessary, the shear strength of a reinforced concrete member can be increased by using externally bonded FRP reinforcement. Shear strengthening using FRP-ER can be carried out by orienting fibers at an angle (typically 45°) or in a transverse direction (90° orientation) with respect to the longitudinal axis of the beam (additional details are provided in Section 5.16).

### 5.8.5 INITIAL MEMBER STRAIN PRIOR TO BONDING

FRP-ER design should account for the initial strain existing in a concrete member prior to attaching the FRP-ER. FRP wraps or fabrics (FRP-ER) can be installed after partially or fully relieving the existing loads and corresponding strains on the considered member. If the strains corresponding to the existing loads including self-weight are not relieved prior to FRP-ER application, then those strains (also called *initial member strains* at the typical bonding location in tension, $\varepsilon_{bi}$) should be accounted for in the subsequent analysis and design. Initial member strain in tension at a bonding location can be determined based on elastic analysis of the cracked member (see Section 5.11, Examples 5.1 and 5.6).

### 5.8.6 NOMINAL AND DESIGN STRENGTH

The nominal flexural strength ($M_n$) of a FRP-bonded concrete member is determined based on the governing failure mode, strain compatibility, and internal force equilibrium. In the strength-design approach, the design flexural strength ($\phi M_n$) should be equal to or greater than the required flexural strength ($M_u$) as given by Equation 5.4.

$$\phi M_n \geq M_u \qquad (5.4)$$

In addition to the strength reduction factor ($\phi$) required by ACI 318 (discussed in Section 5.8.2), an additional strength reduction factor ($\Psi_f = 0.85$) is applied to the flexural strength provided by the FRP-ER to account for the higher (uncertain) reliability of FRP-ER strength properties [ACI 440.2R-02]. Use of this additional strength reduction factor is explained in the design examples on flexure.

### 5.8.7 FLEXURAL STRENGTHENING LIMITS

ACI 440.2R-02 suggests a conservative approach when using FRP-ER for structural strengthening to prevent the FRP debonding-related collapse of FRP-ER strengthened structures. FRP-ER debonding from concrete can occur due to any of the following reasons:

1. Inadequate design
2. Improper or inadequate bonding during strengthening with FRP-ER

3.  Loss of bond between adherend (concrete substrate) and adherent (FRP-ER) during the service life of the structure, wherein it is subjected to stresses and deformations, creep, fatigue, and environmental exposure
4.  Fire
5.  Vandalism and other causes

According to ACI 440.2R-02, the structural member without FRP reinforcement should not collapse and resist a certain level of future (new) loads along with the existing dead loads, prior to bonding FRP-ER and strengthening. This conservative approach of avoiding structural collapse even in the absence of FRP-ER under the action of a major portion of the new live load imposed on the member (85% of $S_{LL}$ corresponding to a FRP-ER strengthened condition) is expressed by Equation 5.5:

$$(\phi R_n)_{existing} \geq (1.2 S_{DL} + 0.85 S_{LL})_{new} \tag{5.5}$$

where

   $\phi$ = strength reduction factor
   $R_n$ = nominal member resistance prior to strengthening with FRP-ER
   $S_{DL}$ = dead load imposed on a member strengthened with FRP-ER
   $S_L$ = live load on the member strengthened with FRP-ER

## 5.8.8  FAILURE MODES

A key factor influencing the flexural strength of a steel bar reinforced concrete beam strengthened with FRP-ER is the failure mode. Some of the failure modes observed in the tests of FRP-ER strengthened concrete beams subjected to flexure are [Ganga-Rao and Vijay 1998]:

1.  Rupture of FRP-ER after yielding of tension steel reinforcement
2.  Secondary concrete crushing after yielding of tension steel reinforcement
3.  Primary concrete crushing in compression before yielding of the reinforcing steel
4.  Shear/tension delamination of the concrete cover (cover delamination)
5.  Debonding of FRP-ER from the concrete substrate (FRP-ER debonding)

Rupture of FRP-ER, secondary concrete crushing, and primary concrete crushing failure modes are generally used for designing FRP-ER strengthened concrete beams. The design is typically based on the strain values induced in FRP-ER, steel, and concrete. Before 2002, ACI 318 emphasized a failure mode of reinforced concrete beams based on the yielding of steel reinforcement in tension. However, ACI 318-02 code emphasizes "adequate ductility" (a concept defined as a stage when strain in tension steel reinforcement, $\varepsilon_s$, is greater than or equal to 0.005) in concrete beams. ACI 440.2R-02 defines the rupture of FRP-ER and concrete crushing (primary and secondary) failure modes in FRP-ER strengthened concrete beams as follows:

1. Rupture of FRP-ER occurs when the tensile strain in the FRP-ER reaches its design rupture strain ($\varepsilon_f = \varepsilon_{fu}$), before the concrete reaches its maximum usable strain ($\varepsilon_c = \varepsilon_{cu} = 0.003$) or exceeds it.
2. Concrete crushing (primary and secondary) occurs when the compressive strain in concrete reaches a maximum usable strain (i.e., $\varepsilon_c = \varepsilon_{cu} = 0.003$). Primary and secondary concrete crushing refer to the conditions where concrete crushing occurs prior to or after tension steel reinforcement yielding, respectively.

In some cases, the force in FRP-ER can be too large to be transferred to the bonded concrete substrate and can result in the delamination of the concrete cover or the debonding of FRP-ER [Arduini and Nanni 1997; Barger 2000; Maeda et al. 1997; Teng et al. 2001; Wan et al. 2004]. To prevent the debonding of FRP-ER, ACI 440.2R-02 recommends limiting the strain developed in the FRP-ER per unit bond width to a maximum value obtained by multiplying the rupture strain of the FRP fabric ($\varepsilon_{fu}$) with a bond-dependent factor, $\kappa_m$, as given by Equation 5.6:

$$
\kappa_m = \begin{cases} \dfrac{1}{60\varepsilon_{fu}}\left(1 - \dfrac{nE_f t_f}{2,000,000}\right) \leq 0.90 & \text{for } nE_f t_f \leq 1,000,000 \\[4mm] \dfrac{1}{60\varepsilon_{fu}}\left(\dfrac{500,000}{nE_f t_f}\right) \leq 0.90 & \text{for } nE_f t_f > 1,000,000 \end{cases} \tag{5.6}
$$

Equation 5.6 shows that the value of $\kappa_m$ decreases with increasing number of FRP-ER layers ($n$). An excessive number of FRP-ER plies or laminates can lead to a greater stiffness mismatch between concrete and FRP-ER than the cases with fewer (two or three) plies or laminates, resulting in the debonding of FRP-ER from the concrete substrate. For FRP-ER having a unit stiffness, $nE_f t_f > 1,000,000$ lb/in., $\kappa_m$ is designed to provide an upper limit on the force rather than strain in FRP-ER laminates.

### 5.8.9 STRAIN AND STRESS IN FRP-ER

FRP-ER is assumed to be linearly elastic up to failure and the stress in FRP-ER is proportional to its strain (Hooke's law). The maximum strain that can be achieved in the FRP-ER will be governed by either the stress developed in the FRP when concrete crushes or the point at which the FRP-ER ruptures. In addition, other failure modes such as FRP-ER debonding from the substrate can also dictate FRP-ER strain [Sharif et al. 1994; Swamy et al. 1987; Teng 2006]. This maximum strain or the effective strain ($\varepsilon_{fe}$) in the FRP-ER at the ultimate (corresponding to strength limit) can be determined from Equation 5.7:

$$
\varepsilon_{fe} = \varepsilon_{cu}\left(\frac{h-c}{c}\right) - \varepsilon_{bi} \leq \kappa_m \varepsilon_{fu} \tag{5.7}
$$

Using Hooke's law, the effective stress ($f_{fe}$) in FRP-ER can be calculated from Equation 5.8 as the product of effective strain in FRP-ER ($\varepsilon_{fe}$) and its stiffness ($E_f$):

$$f_{fe} = E_f \varepsilon_{fe} \tag{5.8}$$

## 5.8.10 DUCTILITY

The use of externally bonded FRP reinforcement for flexural strengthening can limit or reduce the ductility of the original member [Razaqpur and Ali 1996]. The reduction in ductility depends on the failure mode and strain values in concrete, steel, and FRP-ER, and may be negligible in a properly designed concrete member with FRP-ER [ACI 440.2R-02; Ahmad and Shah 1982; Iyer et al. 1996; Mander et al. 1988; Priestly et al. 1992; Razaqpur and Ali 1996; Rodriguez and Park 1994]. In conformity with ACI 318-02, ACI 440.2R-02 considers ductility as "adequate" if strain in the tension steel is at least 0.005 at the point of concrete crushing or failure of the FRP. The failure mode satisfying this strain state ($\varepsilon_s \geq 0.005$) is referred to as a tension-controlled failure mode [ACI 318-02]. ACI 318 premises that unless an unusual amount of ductility is required, the 0.005 limit will provide adequate ductile behavior for most designs. If moment redistribution is required in the design of continuous members and frames, then a net tensile strain of at least 0.0075 is recommended in the concrete member sections for "adequate ductility" in hinge regions, as recommended by ACI 318-02.

To achieve better ductility in FRP-ER strengthened beams, hybridization techniques are used where at least two types of fibers — such as carbon and glass — that have different values of strength, stiffness, and elongation at failure are combined to obtain a FRP-ER system. The hybridization of FRP-ER by using carbon fibers (which have a high stiffness and low elongation, about 1.2% strain at rupture) with glass fibers having low stiffness and high elongation (about 2.5% strain at rupture) results in pseudo ductility. Reasonable ductility at a plastic hinge location of a concrete beam strengthened with triaxially braided hybrid FRP-ER is reported [Grace et al. 2005]. According to ACI 318 and ACI 440.2R-02, a section with low ductility should be compensated with a higher reserve strength. To accomplish this, a strength reduction factor ($\phi$) of 0.70 is applied to brittle sections (where $\varepsilon_s \leq \varepsilon_{sy}$), which is much more conservative than the $\phi$ factor of 0.90 applied to ductile sections (where $\varepsilon_s \geq 0.005$). Strength-reduction factors recommended by ACI 440.2R-02 for ductile and brittle sections are expressed by Equation 5.9:

$$\phi = \begin{cases} 0.90 & \text{for } \varepsilon_s \geq 0.005 \\ 0.70 + \dfrac{0.20(\varepsilon_s - \varepsilon_{sy})}{0.005 - \varepsilon_{sy}} & \text{for } \varepsilon_{sy} < \varepsilon_s < 0.005 \\ 0.70 & \text{for } \varepsilon_s \leq \varepsilon_{sy} \end{cases} \tag{5.9}$$

where

$\varepsilon_s$ = strain in the tension steel at the ultimate (strength limit) state
$\varepsilon_{sy}$ = yield strain in steel

Equation 5.9 gives a strength reduction factor of 0.90 for ductile beam sections with "adequate" steel yield strain ($\varepsilon_s \geq 0.005$) and 0.70 for brittle sections where tension steel does not yield or barely yields ($\varepsilon_s \leq \varepsilon_{sy}$). Strength reduction factors are linearly interpolated between 0.70 and 0.90 for tension steel yield strain of 0.002 (corresponding to $f_y = 60$ ksi) and 0.005 (corresponding to $f_y = 150$ ksi, $\varepsilon_{sy} < \varepsilon_s < 0.005$).

## 5.8.11 SERVICEABILITY

From a structural standpoint, the *serviceability* of a concrete beam refers to its performance under the service load condition. Deflection and crack width are the two most widely considered serviceability criteria in the design of reinforced concrete beams. A concrete beam must be designed to satisfy the serviceability criteria of the applicable design code. Transformed section analysis can be used to determine the deflection and crack width of a concrete beam strengthened with FRP-ER, similar to the analysis of conventional steel-reinforced concrete beams. To prevent inelastic deformations of the FRP-ER strengthened reinforced concrete members under service load conditions, the yielding of the existing internal steel reinforcement is not recommended by ACI 440.2R-02. Internal steel reinforcement stress under service load condition is limited to 80% of the yield strength as given by Equation 5.10:

$$f_{s,s} \leq 0.80 f_y \tag{5.10}$$

Limiting strain in tension steel reinforcement to 80% of its yield value ($0.8 \times 0.002 = 0.0016$, corresponding to Grade 60 steel) can also limit the serviceability strain of FRP reinforcement to a level of about 0.002 depending on its distance from the neutral axis. This may lead to either a tension-controlled or tension- and compression-controlled failure mode without FRP-ER rupture in strengthened concrete sections (discussed in later sections). The failure mode of FRP-ER strengthened concrete sections depends on the amount of steel and FRP reinforcement. If the amounts of steel and FRP reinforcement in a beam were to be low, then a tension-controlled failure mode with FRP-ER rupture would be expected. Similarly, if the amount of steel and FRP reinforcement would be high, then tension- and compression-controlled failure modes without FRP rupture would likely occur. Alternatively, if the amounts of steel and FRP reinforcement would be significantly high, then a compression-controlled failure mode without FRP rupture should be expected (this failure mode is not recommended for practical designs).

## 5.8.12 CREEP-RUPTURE AND FATIGUE STRESS LIMITS

To prevent creep-rupture of the FRP reinforcement under sustained loads (dead loads and the sustained portion of the live load) or to prevent failure due to cyclic stresses

**TABLE 5.2**
**Creep-Rupture and Fatigue Load Stress Limits in FRP**
**Reinforcement**

| Creep-rupture/Fatigue load | Fiber Type | | |
|---|---|---|---|
| | Glass FRP | Aramid FRP | Carbon FRP |
| Stress limit, $F_{f,s}$ | $0.20f_{fu}$ | $0.30f_{fu}$ | $0.55f_{fu}$ |

*Source*: ACI 440.2R-02.

and fatigue of the FRP reinforcement, stress in the FRP reinforcement should not exceed the stress that corresponds to the elastic stress range of the beams.

Elastic analysis is applicable to concrete members strengthened with FRP-ER because stresses induced in concrete, steel, and FRP are designed to be within an elastic range under service loads. The sustained load required for computing creep-rupture stress is a fraction of the total service load. Therefore, elastic analysis using transformed section is used for computing creep-rupture stresses. To prevent the creep-rupture of reinforced concrete beams strengthened with different types of FRP-ER, ACI 440.2R-02 recommends that the stresses due to sustained loads should not exceed creep-rupture stress limits ($F_{f,s}$) as given by Equation 5.11:

$$F_{f,s} \geq f_{f,s} \tag{5.11}$$

For FRP-ER-strengthened members subjected to fatigue loads, ACI 440.2R-02 recommends limiting stress in the FRP-ER due to moment caused by the combination of sustained and fatigue loads to the values listed in Table 5.2. The stresses induced in FRP-ER by an applied moment due to sustained loads plus the maximum moment induced in a fatigue loading cycle can be determined based on elastic analysis. Service load stresses in FRP and steel can be calculated by using Equation 5.30h and Equation 5.31d, respectively (discussed in Section 5.11).

## 5.9  FAILURE MODES

Conceptually, all beams fail when loaded to their ultimate load capacity. After a beam reaches its ultimate load capacity, any increase in the load will cause a continuous increase in deformations (i.e., deflection, rotation, curvature) associated with a gradual or quick drop in the load capacity until the beam ruptures [Karbhari and Seible 1999; Teng 2006; Vijay and GangaRao 1999]. Different failure modes for steel-reinforced concrete beams with and without FRP-ER are discussed in this section.

In an attempt to develop a clear understanding of the failure modes of FRP-ER strengthened beams, the failure mode terminologies used for conventional steel-reinforced concrete beams that were used in the years prior to the advent of ACI 318-02, i.e., year 2002, which are summarized in this section. Based on the failure modes, steel-reinforced concrete beams were called "underreinorced," "overreinforced," or "balanced," although these terms are no longer used based in ACI 318-02. Balanced, underreinforced, and overreinforced sections were characterized as follows:

1. Balanced failure in a steel reinforced concrete beam is characterized by the simultaneous yielding of steel and crushing of concrete. This type of failure is difficult to achieve in practice.
2. The amount of steel reinforcement provided in underreinforced beams is less than what is required for a balanced failure.
3. Underreinforced beams are characterized by the yielding of tension steel reinforcement prior to the crushing of concrete in the compression zone, a failure mode now referred to as *tension failure*.
4. Overreinforced beams are characterized by concrete crushing, a failure mode now referred to as a *compression failure*. The maximum amount of compression force resultant that can be developed in the compression zone of concrete is smaller than the force required to cause design yield stress value in tensile steel whenever a beam contains an excessive amount of tensile steel. The result is that upon reaching an assumed compressive strain of 0.003, the concrete begins to crush without the onset of yielding of steel reinforcement.

Beams strengthened with FRP-ER contain an additional load carrying element — the FRP reinforcement, primarily on the tension side. Beams strengthened with FRP-ER exhibit additional failure modes of FRP-ER rupture or debonding in addition to the failure modes as those of the conventional steel-reinforced concrete beams. Note that concrete crushing during secondary compression failure accompanied by the yielding of steel in the tension zone involves a significant amount of energy absorption in FRP-ER strengthened concrete beams, particularly when concrete confinement effects (due to steel shear stirrups and transverse wrapping) are present [GangaRao and Vijay 1998]. For analysis and design purposes, the following failure modes of FRP-ER strengthened beams, classified using essentially the same approach as used in ACI 318-02 for conventional steel-reinforced concrete beams, are discussed:

1. Tension-controlled failure with FRP rupture
2. Tension-controlled failure without FRP rupture
3. Tension-and compression-controlled failure
4. Compression-controlled failure
5. Balanced failure mode

Other failure modes may include the debonding of FRP reinforcement near or at the bonded interface with concrete, which could lead to any of the failure modes 2, 3, and 4 listed above depending on the steel reinforcement strain at debonding.

ACI 440.1R-03 refers to tension-controlled failure as the one with an "adequate" amount of yielding in steel reinforcement as expressed by the condition: $\varepsilon_y \geq 0.005$. In a FRP-ER-reinforced beam, the failure strain of FRP reinforcement can vary from 0.01 to 0.015 for carbon, and from 0.015 to 0.03 for glass and aramid, two to six times greater than the steel yield strain value suggested for adequate ductility ($\varepsilon_y \geq 0.005$). A *tension-controlled* failure in a FRP-ER strengthened beam is defined to occur when the steel strain reaches 0.005. Therefore, FRP-ER strengthened beams can fail in a tension-controlled failure mode with or without rupture of the FRP reinforcement.

When the steel strain value in a FRP-ER strengthened beam at failure is between its yield strain and 0.005 (i.e., $\varepsilon_{sy} \leq \varepsilon_s < 0.005$), the failure mode is referred to as *tension-* and *compression-controlled*. Note that unless FRP debonding or other failure modes influence beam failure, tension-controlled or tension- and compression-controlled failure modes will eventually lead to secondary compression failure characterized by the crushing of concrete ($\varepsilon_{cu} = 0.003$) in the compression zone. If the strain-in-tension steel at beam failure is less than its yield strain ($\varepsilon_s < \varepsilon_y$), the corresponding failure is referred to as *compression-controlled*.

In summary, flexural failure modes in FRP-ER-strengthened beams can be classified as follows:

1. Tension-controlled failure with FRP rupture: This failure is characterized by an "adequate" amount of tension steel yielding such that $\varepsilon_s \geq 0.005 > \varepsilon_{sy}$ and eventual FRP-ER rupture ($\varepsilon_{frp} = \varepsilon_{frpu}$).
2. Tension-controlled failure without FRP rupture: This failure is characterized by an "adequate" tension steel yielding ($\varepsilon_s \geq 0.005 > \varepsilon_{sy}$) without FRP-ER rupture ($\varepsilon_{frp} < \varepsilon_{frpu}$) that eventually leads to secondary compression failure ($\varepsilon_{cu} = 0.003$).
3. Tension- and compression-controlled failure: This failure is characterized by tension steel yielding such that $\varepsilon_{sy} \leq \varepsilon_s < 0.005$ without FRP rupture ($\varepsilon_{frp} < \varepsilon_{frpu}$) that eventually leads to secondary compression failure ($\varepsilon_{cu} = 0.003$).
4. Compression-controlled failure: This failure is characterized by the absence of tension steel yielding ($\varepsilon_s \leq \varepsilon_{sy}$) and FRP rupture ($\varepsilon_{frp} < \varepsilon_{frpy}$) and manifested by concrete crushing in the compression zone.
5. Balanced failure: This is a hypothetical failure mode that is assumed to occur when strains in extreme tension and compression fibers have reached their limit values simultaneously ($\varepsilon_s = \varepsilon_u = 0.003$, $\varepsilon_{frp} = \varepsilon_{frpu}$, and as a consequence, $\varepsilon_s \geq \varepsilon_{sy}$).

## 5.10  FLEXURAL FORCES IN FRP-ER STRENGTHENED BEAMS

Idealized strain and stress distributions in a FRP-ER strengthened concrete beam and the force equilibrium at ultimate load conditions are shown in Figure 5.18. Forces in various elements of the beam cross-section are expressed in terms of stresses in concrete, steel reinforcement, and FRP reinforcement. Expressions for the nominal strength of a FRP-ER strengthened beam are derived for different failure modes as follows, based on the following assumptions:

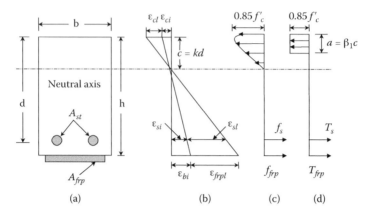

**FIGURE 5.18** Strain distribution and force equilibrium conditions at the ultimate load conditions in a reinforced concrete beam with FRP-ER: (a) beam cross-section, (b) strain distribution, (c) parabolic stress distribution, and (d) equivalent rectangular stress distribution.

1. A plane section before loading remains plane after loading.
2. The strain in concrete and the reinforcement (steel bars as well as FRP-ER) are proportional to the distance from the neutral axis.
3. The maximum usable strain in concrete ($\varepsilon_{cu}$) is 0.003.
4. The tensile strength of concrete is neglected.
5. The FRP reinforcement has a linear (elastic) relationship to failure.
6. No relative slip exists between FRP-ER and the concrete substrate.
7. The shear deformation within the adhesive layer is neglected based on the premise that the adhesive layer is very thin with slight variations in thickness.
8. Design calculations are based on the actual dimensions of the beam, internal steel reinforcing steel arrangements, and the material properties of the concrete member under consideration.

Note that the aforestated assumptions do not accurately reflect the fundamental behavior of FRP external reinforcement. For example, Assumption 6 is not quite accurate in the sense that some shear deformation will be present in the adhesive layer, which will cause distortion (no slip) between the FRP-ER and concrete substrate. However, these assumptions are necessary for computational convenience, and the degree of inaccuracy involved does not significantly affect the calculated flexural strength of the FRP-ER strengthened member. Furthermore, any discrepancy arising out of these assumptions is compensated by using an additional strength reduction factor, $\psi_f$ (discussed later).

$$\text{Uniform compressive stress in concrete} = 0.85 f_c' \qquad (5.12)$$

Tensile stress in steel:

$$\text{before yielding} \qquad f_s(\varepsilon_s < \varepsilon_{sy}) \qquad (5.13a)$$

at and after yielding        $f_y(\varepsilon_s \geq \varepsilon_{sy})$                      (5.13b)

Tensile stress in FRP-ER:

before rupture          $f_{frp}(\varepsilon_{frp} < \varepsilon_{frpu})$                   (5.14a)

at ultimate load        $f_{frpu}(\varepsilon_{frp} = \varepsilon_{frpu})$                 (5.14b)

after rupture           $f_{frpu} = 0(\varepsilon_{frp} \geq \varepsilon_{frpu})$          (5.14c)

where $f_{frpu}$ is the stress in the FRP-ER at the point of incipient rupture.

Most likely, FRP-ER strengthened beams would fail in tension-controlled mode (with or without FRP reinforcement rupture) or tension-and compression-controlled failure mode.

## 5.11  FLEXURAL STRAINS AND STRESSES IN FRP-ER STRENGTHENED BEAMS

The fundamental assumptions made in deriving the nominal strength of FRP-ER strengthened beams are the same as those for conventional concrete beams [MacGregor 1998; Nawy 2002; Park and Paulay 1975; Wang 1992] and specified in ACI 318, Section 10.2. The strain distribution is assumed linear and the stress distribution is assumed uniform over the compression zone of concrete (see Figure 5.18). However, due to the addition of FRP-ER, the following assumptions are made:

1.  No relative slip or deformation exists between the FRP-ER and concrete substrate to which it is bonded.
2.  FRP-ER follows a linear stress-strain relation (Hooke's law) up to failure (rupture).

The derivation of moment capacities of FRP-ER beams is based on satisfying two basic conditions: the static equilibrium of forces in concrete, steel, and FRP-ER; and the compatibility of strains in concrete, steel, and FRP-ER. Referring to Figure 5.18, the maximum usable strain in concrete is assumed to be 0.003 ($\varepsilon_c = \varepsilon_{cu}$ = 0.003). The strain in steel reinforcement at yield is based on Hooke's law, i.e., the yield stress $f_y$ of the reinforcing steel is equal to $E_s$ times the steel yield strain, where $E_s$ is the modulus of elasticity of steel. Assuming $E_s$ = 29,000 ksi, the following values of yield strains are obtained for Grade 40 and Grade 60 reinforcement:

Grade 40 reinforcement:

$$\varepsilon_y = \frac{f_y}{E_s} = \frac{40,000}{29,000,000} = 0.00138 \cong 0.0014 \qquad (5.15a)$$

Grade 60 reinforcement:

$$\varepsilon_y = \frac{f_y}{E_s} = \frac{60,000}{29,000,000} = 0.00207 \cong 0.002 \qquad (5.15b)$$

The failure strain of FRP composites varies depending on their type and composition. For design purposes, note that the maximum (ultimate) tensile stress in a conventional reinforced beam corresponds to the steel yield strain value, whereas for a FRP-ER strengthened beam, the maximum tensile stress corresponds to FRP rupture strain. A discussion on the mechanical properties of FRP-ER composites (e.g., carbon and glass fibers) is provided in Chapter 2. The following strain values will be used to establish strain compatibility between various elements of the beam:

| | |
|---|---|
| Concrete (all strengths) ($\varepsilon_c = \varepsilon_{cu}$) | 0.003 |
| Steel reinforcement (yield) ($\varepsilon_s = \varepsilon_{sy}$) Grade 40 | 0.0014 |
| Grade 60 | 0.002 |
| Carbon fiber fabric (rupture) ($\varepsilon_{frp} = \varepsilon_{frpu}$) | 0.01–0.015 |
| Glass fiber fabric (rupture) ($\varepsilon_{frp} = \varepsilon_{frpu}$) | 0.015–0.025 |

To establish the strain compatibility relationships, refer to Figure 5.18b, which shows strains in extreme compression fibers, steel reinforcement, and extreme tension fibers prior to and after strengthening of the beam with the fiber FRP-ER. The strain in the FRP-ER is assumed to be the same as in the extreme tension fiber of the beam, an assumption justified in view of the small thicknesses of the FRP-ER laminate and the interfacial adhesive layer. Figure 5.18b shows various strains as follows:

Before strengthening of beam:

$\varepsilon_{ci}$ = initial strain in extreme compression fibers
$\varepsilon_{si}$ = initial strain in steel reinforcement
$\varepsilon_{bi}$ = initial strain (i.e., due to dead loads) in extreme bottom tension fibers
    prior to FRP-ER strengthening

where the second subscript $i$ refers to the initial conditions corresponding to the nominal strength of the beam without external reinforcement. In parameter $\varepsilon_{bi}$, the subscript $b$ denotes the "bonded side" of the beam. In a simply supported beam, the bonded side is its tension side. In a cantilever beam, the top side (not the bottom side) of the beam will be in tension, which will be bonded with FRP reinforcement for strengthening purposes.

Upon installing the external FRP reinforcement (strengthening) with fibers oriented in the longitudinal direction, the beam develops additional flexural strength. An FRP-ER strengthened concrete beam under live loads, sustained loads, temperature, and other loads will be subjected to additional strains. These additional strains (Figure 5.18b) are as follows:

$\varepsilon_{cl}$ = additional strain in extreme compression fibers after strengthening and loading

$\varepsilon_{sl}$ = additional strain in steel reinforcement after strengthening and loading

$\varepsilon_{ci} = \varepsilon_{frp} = \varepsilon_{frpl}$ = additional strain in extreme tension fibers after strengthening and loading (note that due to the very small thickness of FRP-ER, the strain in the extreme bottom tension fiber in concrete and the bonded FRP is assumed to be the same)

The subscript $l$ refers to values resulting after beam strengthening and subsequent loading. The total strain values in concrete ($\varepsilon_c$), steel ($\varepsilon_s$), and at the extreme tension fibers where FRP-ER is bonded ($\varepsilon_b$) are defined as follows:

$$\varepsilon_c = \varepsilon_{ci} + \varepsilon_{cl} \tag{5.16}$$

$$\varepsilon_s = \varepsilon_{si} + \varepsilon_{sl} \tag{5.17}$$

$$\varepsilon_b = \varepsilon_{bi} + \varepsilon_{bl} = \varepsilon_{bi} + \varepsilon_{frpl} \tag{5.18}$$

*Strain Compatibility.* Referring to similar triangles in Figure 5.18b, the following relations are established. Before installing the FRP-ER (initial condition, typically under the effect of dead loads):

$$\frac{\varepsilon_{ci} + \varepsilon_{bi}}{h} = \frac{\varepsilon_{ci} + \varepsilon_{si}}{d} \tag{5.19}$$

After installing the FRP-ER:

$$\frac{\varepsilon_c}{c} = \frac{\varepsilon_s}{d - c} = \frac{\varepsilon_b}{h - c} \tag{5.20a}$$

Noting that the additional strain near the extreme tension fiber (i.e., at the bonded location $\varepsilon_{bl}$, which develops after installing the FRP reinforcement and subsequent loading) is equal to the strain in the FRP-ER, $\varepsilon_{frp}$ (i.e., $\varepsilon_{bl} = \varepsilon_{frp}$), Equation 5.20a can be expressed as:

$$\frac{\varepsilon_c}{c} = \frac{\varepsilon_s}{d - c} = \frac{\varepsilon_b}{h - c} = \frac{\varepsilon_{bl} + \varepsilon_{bi}}{h - c} = \frac{\varepsilon_{frp} + \varepsilon_{bi}}{h - c} \tag{5.20b}$$

From Equation 5.19, $\varepsilon_{bi}$ can be expressed as

$$\varepsilon_{bi} = \left( \frac{\varepsilon_{ci} + \varepsilon_{si}}{d} \right) h - \varepsilon_{ci} \tag{5.21}$$

*Force Equilibrium*: The compressive force in concrete ($C_c$) is

$$C_c = 0.85 f'_c ab \qquad (5.22)$$

The tensile force in steel ($T_s$) is

$$T_s = A_s f_s = A_s E_s \varepsilon_s \text{ (before steel yielding, } \varepsilon_s < \varepsilon_y) \qquad (5.23a)$$

$$T_s = A_s f_y = A_s E_s \varepsilon_y \text{ (at and after steel yielding, } \varepsilon_s \geq \varepsilon_y) \qquad (5.23b)$$

The tensile force in FRP-ER ($T_{frp}$) before or at rupture ($\varepsilon_{frp} < \varepsilon_{frpu}$) is

$$T_{frp} = A_{frp} f_{frp} = A_{frp} (E_{frp} \varepsilon_{frp}) \qquad (5.24a)$$

The tensile force in FRP-ER ($T_{frp}$) after rupture is

$$T_{frp} = 0 \qquad (5.24b)$$

## 5.11.1 Depth of Neutral Axis ($c = kd$) with and without FRP-ER

The depth of neutral axis in a beam, $kd$ (Figure 5.19), under service loads with and without FRP-ER is given by the following equations based on elastic analysis similar to conventional steel-reinforced concrete beams. The factor $k$ is sometimes referred to as the *neutral axis factor*.

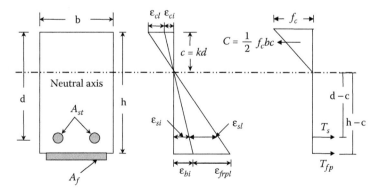

**FIGURE 5.19** Stress and strain distribution in FRP-ER reinforced concrete beam under service loads.

## 5.11.2 Value of Neutral Axis Factor ($k$) with FRP-ER

Based on the linear stress-strain relationship (Hookes' Law),

$$f_s = E_s \varepsilon_s, \quad f_f = E_f \varepsilon_f, \quad f_c = E_c \varepsilon_c \qquad (5.25)$$

Considering the force equilibrium in a cracked beam with FRP-ER and substituting the above stress values,

$$A_s f_s + A_f f_f = \frac{bc}{2} f_c \qquad (5.26a)$$

$$A_s(E_s \varepsilon_s) + A_f(E_f \varepsilon_f) = \frac{bc}{2}(E_c \varepsilon_c) \qquad (5.26b)$$

From Equation 5.20a, we obtain,

$$\varepsilon_s = \varepsilon_c \frac{d-c}{c} = \varepsilon_c\left(\frac{d}{c}-1\right), \quad \varepsilon_f = \varepsilon_c \frac{h-c}{c} = \varepsilon_c\left(\frac{h}{c}-1\right)$$

Substituting the above values into Equation 5.26b,

$$A_s E_s\left(\varepsilon_c \frac{d-c}{c}\right) + A_f E_f\left(\varepsilon_c \frac{h-c}{c}\right) = \frac{bc}{2}(E_c \varepsilon_c) \qquad (5.26c)$$

Dividing Equation 5.26c throughout by $\varepsilon_c$ and simplifying,

$$A_s E_s\left(\frac{d-c}{c}\right) + A_f E_f\left(\frac{h-c}{c}\right) = \frac{bc}{2}(E_c) \qquad (5.26d)$$

$$A_s E_s(d-c) + A_f E_f(h-c) = \frac{bc^2}{2}(E_c) \qquad (5.26e)$$

$$A_s \frac{E_s}{E_c}(d-c) + A_f \frac{E_f}{E_c}(h-c) = \frac{bc^2}{2} \qquad (5.26f)$$

Quantities ($E_s/E_c = n_s$) and ($E_f/E_c = n_f$) can be expressed as modular ratios for steel and FRP reinforcement, respectively, in Equation 5.26f, which is quadratic in $c$. With the above substitutions, Equation 5.26f can be rewritten as:

$$A_s n_s (d-c) + A_f n_f (h-c) = \frac{bc^2}{2} \tag{5.26g}$$

Dividing Equation 5.26g throughout by $bd$ yields

$$2\frac{A_s}{bd} n_s (d-c) + 2\frac{A_f}{bd} n_f (h-c) = \frac{c^2}{d} \tag{5.26h}$$

Equation 5.26h can be expressed in terms of the reinforcement ratios, defined as $(A_s/bd) = \rho_s$ and $(A_f/b_f) = \rho_f$:

$$2\rho_s n_s (d-c) + 2\rho_f n_f (h-c) = \frac{c^2}{d} \tag{5.26i}$$

Rearranging various terms in Equation 5.26i and writing it as a quadratic in $c$, we obtain

$$\frac{c^2}{d} + 2(\rho_s n_s + \rho_f n_f)c - 2(\rho_s n_s d + \rho_f n_f h) = 0 \tag{5.26j}$$

Solving Equation 5.26j for $c$,

$$c = \frac{-2(\rho_s n_s + \rho_f n_f) + \sqrt{4(\rho_s n_s + \rho_f n_f)^2 + (4)(2/d)(\rho_s n_s d + \rho_f n_f h)}}{2/d} \tag{5.27a}$$

which, when simplified, yields

$$c = \left( \sqrt{(\rho_s n_s + \rho_f n_f)^2 + 2\left(\rho_s n_s + \rho_f n_f \frac{h}{d}\right)} - (\rho_s n_s + \rho_f n_f) \right) d \tag{5.27b}$$

Noting that $c = kd$, Equation 5.27b gives the value of the neutral axis factor $k$ for a concrete beam with internal steel reinforcement and external FRP reinforcement under service load conditions:

$$k = \sqrt{(\rho_s n_s + \rho_f n_f)^2 + 2\left(\rho_s n_s + \rho_f n_f \frac{h}{d}\right)} - (\rho_s n_s + \rho_f n_f) \tag{5.28}$$

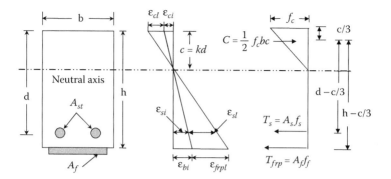

**FIGURE 5.20** Tensile and compressive forces and lever arms at service loads.

### 5.11.3 VALUE OF $k$ WITHOUT FRP-ER

The value of neutral axis factor $k$ for a beam without the external FRP reinforcement can be obtained from Equation 5.28 by setting the terms related to FRP-ER, i.e., $\rho_f$ to zero. Thus,

$$k = \sqrt{(\rho_s n_s)^2 + 2\rho_s n_s} - (\rho_s n_s) \qquad (5.29)$$

### 5.11.4 SERVICE LOAD STRESSES (POST-CRACKING) IN STEEL

Strain and stress conditions in a FRP-ER strengthened beam during the post-cracking stage due to service loads are shown in Figure 5.20.

Taking moments about the centroid of compression force resultant (Figure 5.20), and by noting that $f_s = \varepsilon_s E_s$ and $f_f = \varepsilon_f E_f$, we obtain:

$$M_s = A_s(\varepsilon_s E_s)\left(d - \frac{c}{3}\right) + A_f(\varepsilon_f E_f)\left(h - \frac{c}{3}\right) \qquad (5.30a)$$

Substituting $\varepsilon_f = \varepsilon_b - \varepsilon_{bi}$ into Equation 5.30a,

$$M_s = A_s(\varepsilon_s E_s)\left(d - \frac{c}{3}\right) + A_f(\varepsilon_b - \varepsilon_{bi})(E_f)\left(h - \frac{c}{3}\right) \qquad (5.30b)$$

$$M_s = A_s(\varepsilon_s E_s)\left(d - \frac{c}{3}\right) + A_f(\varepsilon_b)(E_f)\left(h - \frac{c}{3}\right) - A_f(\varepsilon_{bi})(E_f)\left(h - \frac{c}{3}\right) \qquad (5.30c)$$

Substituting $\varepsilon_s E_s = f_{s,s}$ and $\varepsilon_b = \varepsilon_s \dfrac{h - c}{d - c}$ from Equation 5.20a into Equation 5.30c gives,

$$M_s = A_s(f_{s,s})\left(d - \frac{c}{3}\right) + A_f\left(\varepsilon_s \frac{h-c}{d-c}\right)(E_f)\left(h - \frac{c}{3}\right) - A_f(\varepsilon_{bi})(E_f)\left(h - \frac{c}{3}\right) \quad (5.30\text{d})$$

Rearranging various terms of Equation 5.30d, we obtain:

$$\left[M_s + A_f(\varepsilon_{bi})(E_f)\left(h - \frac{c}{3}\right)\right](d-c) =$$

$$A_s(f_{s,s})\left(d - \frac{c}{3}\right)(d-c) + A_f E_f(\varepsilon_s)(h-c)\left(h - \frac{c}{3}\right) \quad (5.30\text{e})$$

Multiply both sides of Equation 5.30e by $E_s$ (note that $\varepsilon_s E_s = f_{s,s}$):

$$\left[M_s + A_f E_f(\varepsilon_{bi})\left(h - \frac{c}{3}\right)\right](d-c)(E_s) =$$

$$A_s E_s(f_{s,s})\left(d - \frac{c}{3}\right)(d-c) + A_f E_f(f_{s,s})\left(h - \frac{c}{3}\right)(h-c) \quad (5.30\text{f})$$

Rearranging and solving Equation 5.30f for $f_{s,s}$,

$$f_{s,s} = \frac{\left[M_s + A_f E_f(\varepsilon_{bi})\left(h - \frac{c}{3}\right)\right](d-c)(E_s)}{A_s E_s\left(d - \frac{c}{3}\right)(d-c) + A_f E_f\left(h - \frac{c}{3}\right)(h-c)} \quad (5.30\text{g})$$

Substituting $c = kd$ into Equation 5.30g yields the following equation, which is same as Equation 9-12 of ACI440.2R-02:

$$f_{s,s} = \frac{\left[M_s + A_f E_f(\varepsilon_{bi})\left(h - \frac{kd}{3}\right)\right](d-kd)(E_s)}{A_s E_s\left(d - \frac{kd}{3}\right)(d-kd) + A_f E_f\left(h - \frac{kd}{3}\right)(h-kd)} \quad (5.30\text{h})$$

### 5.11.5 SERVICE LOAD STRESSES IN FRP-ER

The stresses in the FRP-ER due to loads applied subsequent to its installation can be determined from Hooke's law. The strain in the FRP-ER at any given loading equals the difference between the final strain in the bonded surface of the beam ($\varepsilon_b$) that occurs after the load is applied and at the same surface before the loads were applied ($\varepsilon_{bi}$). Thus, $\varepsilon_{bl} = \varepsilon_b - \varepsilon_{bi}$ (see Equation 5.18) so that

$$f_f(\varepsilon_{bl})E_f; \quad f_f(\varepsilon_b - \varepsilon_{bi})E_f; \quad f_f = \varepsilon_b E_f - \varepsilon_{bi}E_f \tag{5.31a}$$

Substitution for $\varepsilon_b$ from Equation 5.20a and $c = kd$ yields

$$f_f = \varepsilon_s\left(\frac{h-kd}{d-kd}\right)E_f - \varepsilon_{bi}E_f \tag{5.31b}$$

or, expressing $\varepsilon_s$ as $\dfrac{f_s}{E_s}$, Equation 5.31b can be expressed as

$$f_f = \left(\frac{f_s}{E_s}\right)\left(\frac{h-kd}{d-kd}\right)E_f - \varepsilon_{bi}E_f \tag{5.31c}$$

Under the full service load condition ($f_s = f_{s,s}$) in Equation 5.31c and denoting the stress in FRP-ER ($f_f$) at service load as $f_{f,s}$:

$$f_{f,s} = f_{s,s}\left(\frac{E_f}{E_s}\right)\left(\frac{h-kd}{d-kd}\right) - \varepsilon_{bi}E_f \tag{5.31d}$$

## 5.12 NOMINAL FLEXURAL STRENGTH OF A SINGLY REINFORCED BEAM

Reinforced concrete beams encountered in practice can be singly or doubly reinforced. Five failure modes of FRP-ER strengthened beams were discussed earlier in Section 5.9. Analyses of these five different failure modes in terms of stresses and strains in concrete, steel, and FRP-ER are presented in the following sections.

### 5.12.1 TENSION-CONTROLLED FAILURE WITH FRP-ER RUPTURE

As shown in Figure 5.21, steel is assumed to yield ($\varepsilon_s \geq 0.005$) and the FRP-ER is at the point of incipient rupture ($\varepsilon_{frp} = \varepsilon_{frpu}$). The strain in concrete corresponding to the nominal strength of the strengthened beam is obtained from Equation 5.20b:

$$\frac{\varepsilon_c}{c} = \left(\frac{\varepsilon_{frpu} + \varepsilon_{bi}}{h-c}\right)a \tag{5.32a}$$

or

$$\varepsilon_c = \left(\frac{\varepsilon_{frpu} + \varepsilon_{bi}}{h-c}\right)c \tag{5.32b}$$

Note that $\varepsilon_{frpu}$ has been substituted for $\varepsilon_{frp}$ in Equation 5.20b, where $\varepsilon_{frpu}$ is the strain in the fiber reinforcement at incipient rupture.

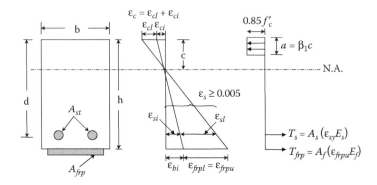

**FIGURE 5.21** Force distributon in a tension-controlled failure with FRP rupture.

Substitution of $c = a/\beta_1$ in Equation (5.32b) yields

$$\varepsilon_c = \left(\frac{\varepsilon_{frpu} + \varepsilon_{bi}}{\beta_1 h - a}\right) a \qquad (5.32c)$$

where $\beta_1$ is defined by ACI 318-02, Section 10.2.7.3.

Likewise, the strain in steel is also obtained from Equation 5.20b and Equation 5.32a:

$$\frac{\varepsilon_s}{d - c} = \left(\frac{\varepsilon_{frpu} + \varepsilon_{bi}}{h - c}\right) \qquad (5.33a)$$

so that

$$\varepsilon_s = \left(\frac{\varepsilon_{frpu} + \varepsilon_{bi}}{h - c}\right)(d - c) = \left(\frac{\varepsilon_{frpu} + \varepsilon_{bi}}{\beta_1 h - a}\right)(\beta_1 d - a) \qquad (5.33b)$$

The maximum strain in FRP-ER just before rupture:

$$\varepsilon_{frp} = \varepsilon_{frpu} \qquad (5.33c)$$

The tension and compression force equilibrium:

$$0.85 f'_c a b = A_s f_y + A_{frp}(E_{frp}\varepsilon_{frpu}) \qquad (5.34)$$

Solving for the depth of the equivalent rectangular stress block $a$,

$$a = \frac{A_s f_y + A_{frp}(E_{frp}\varepsilon_{frpu})}{0.85 f_c' b} \tag{5.35}$$

The nominal strength is obtained by taking moments of $T_s$ and $T_{frp}$ about the compression resultant $C$:

$$M_n = A_s f_y\left(d - \frac{a}{2}\right) + A_{frp}(E_{frp}\varepsilon_{frpu})\left(h - \frac{a}{2}\right) \tag{5.36a}$$

The nominal flexural strength of a FRP-ER-strengthened beam is determined as the sum of flexural strength provided by steel and the FRP (Equation 5.36a). The strength reduction factor for tension-controlled failure with steel yield and FRP rupture is $\phi = 0.9$. In addition to the use of the strength reduction factor ($\phi$) required by ACI 318, an additional strength reduction factor ($\psi_f = 0.85$) is applied to the flexural strength provided by the FRP-ER only. From Equation 5.4, the design strength $\phi M_n \geq M_u$; accordingly, the design strength of the FRP-ER strengthened beam is obtained by modifying Equation 5.36a with the appropriate strength reduction factors as follows:

$$\phi M_n = \phi\left[A_s f_y\left(d - \frac{a}{2}\right) + \psi_f A_{frp}(E_{frp}\varepsilon_{frpu})\left(h - \frac{a}{2}\right)\right] \tag{5.36b}$$

### 5.12.2 Tension-Controlled Failure without FRP Rupture

Tension-controlled failure without FRP rupture is characterized by an "adequate" amount of steel yielding ($\varepsilon_s \geq 0.005$) without FRP-ER rupture ($\varepsilon_{frp} < \varepsilon_{frpu}$, Figure 5.22). This failure mode eventually leads to secondary compression (crushing of the concrete beam in the compression zone) or shear compression failure. Due to

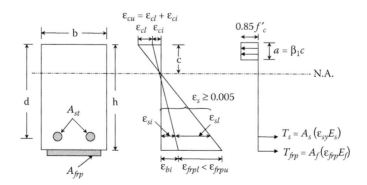

**FIGURE 5.22** Force distribution in a tension-controlled failure without FRP rupture.

secondary compression failure, assuming that the full usable strain in concrete has occurred is reasonable (i.e., concrete has crushed) so that $\varepsilon_c = \varepsilon_{cu} = 0.003$.

Strains in steel reinforcement and FRP-ER corresponding to the nominal strength of the strengthened beam are obtained from Equation 5.20b by substituting $\varepsilon_{cu}$ for $\varepsilon_c$. Thus, the strain in steel is obtained from similar triangles shown in Figure 5.18:

$$\frac{\varepsilon_s}{d-c} = \left(\frac{\varepsilon_{cu}}{c}\right) \tag{5.37a}$$

The depth of the neutral axis from the extreme compression fiber, $c$, and the depth of compression block $a$ are related by $a = \beta_1 c$:

$$\varepsilon_s = \left(\frac{\varepsilon_{cu}}{c}\right)(d-c) = \left(\frac{\varepsilon_{cu}}{a}\right)(\beta_1 d - a) \tag{5.37b}$$

The strain in FRP is

$$\frac{\varepsilon_{frp} + \varepsilon_{bi}}{h-c} = \left(\frac{\varepsilon_{cu}}{c}\right) \tag{5.38a}$$

which, upon substituting $c = a/\beta_1$, yields

$$\varepsilon_{frp} = \left(\frac{\varepsilon_{cu}}{c}\right)(h-c) - \varepsilon_{bi} = \left(\frac{\varepsilon_{cu}}{a}\right)(\beta_1 h - a) - \varepsilon_{bi} \tag{5.38b}$$

*Force Equilibrium.* In the tension-controlled failure mode with steel yield but without FRP rupture, concrete is eventually crushed ($\varepsilon_c = \varepsilon_{cu} = 0.003$). When tension steel has yielded with $\varepsilon_s \geq 0.005$, the stress in the steel is given by Equation 5.13b. Equating tensile and compressive forces:

$$0.85 f_c' ab = A_s f_y + A_{frp} f_{frp} \tag{5.39a}$$

or, expressing $f_{frp}$ in terms of strain in FRP-ER,

$$0.85 f_c' ab = A_s f_y + A_{frp} E_{frp} \varepsilon_{frp} \tag{5.39b}$$

Substituting the value of $\varepsilon_{frp}$ from Equation 5.38b into Equation 5.39b yields,

$$0.85 f_c' ba^2 + [A_{frp}E_{frp}(\varepsilon_{cu} + \varepsilon_{bi}) - A_s f_y]a - \beta_1 A_{frp}E_{frp}\varepsilon_{cu}h = 0 \qquad (5.40)$$

Equation 5.40 is a quadratic, which can be solved for $a$.

*Nominal Strength.* The nominal strength of the beam can be expressed in terms of $a$:

$$M_n = A_s f_y \left(d - \frac{a}{2}\right) + A_{frp}(E_{frp}\varepsilon_{frp})\left(h - \frac{a}{2}\right) \qquad (5.41a)$$

where

$$\varepsilon_{frp} = \left(\frac{\varepsilon_{cu}}{c}\right)(h - c) - \varepsilon_{bi} = \left(\frac{\varepsilon_{cu}}{a}\right)(\beta_1 h - a) - \varepsilon_{bi}$$

The strength reduction factor for this failure remains the same as the tension-controlled failure with FRP rupture, i.e., $\phi = 0.9$. In addition to the use of the strength reduction factor $\phi$ required by ACI 318, an additional strength reduction factor, $\psi_f = 0.85$, is applied to the flexural strength provided by the FRP-ER only (ACI 440.2R-02). Noting from Equation 5.4 that $\phi M_n \geq M_u$, Equation 5.41a can be expressed as:

$$\phi M_n = \phi \left[ A_s f_y \left(d - \frac{a}{2}\right) + \psi_f A_{frp}(E_{frp}\varepsilon_{frp})\left(h - \frac{a}{2}\right) \right] \qquad (5.41b)$$

where

$$\varepsilon_{frp} = \left(\frac{\varepsilon_{cu}}{c}\right)(h - c) - \varepsilon_{bi} = \left(\frac{\varepsilon_{cu}}{a}\right)(\beta_1 h - a) - \varepsilon_{bi}$$

### 5.12.3 Tension- and Compression-Controlled Failure with Steel Yielding and without FRP Rupture

The tension- and compression failure mode is characterized by steel yielding ($\varepsilon_{sy} \leq \varepsilon_s < 0.005$) without FRP rupture ($\varepsilon_{frp} < \varepsilon_{frpu}$, Figure 5.23) that eventually leads to secondary compression or shear compression failure ($\varepsilon_c = \varepsilon_{cu} = 0.003$). Analysis of beams with tension- and compression-controlled failure with tension steel yielding and without FRP rupture remains the same as explained in Section 5.12.2. However, the strength reduction factor for this failure mode is $\phi = 0.7 + \dfrac{0.2(\varepsilon_s - \varepsilon_{sy})}{0.005 - \varepsilon_{sy}}$ (from Figure 5.17 and Section 5.8.10).

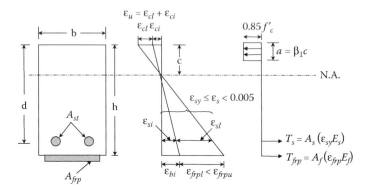

**FIGURE 5.23** Force distribution in a tension- and-compression-controlled failure with steel yielding and without FRP rupture.

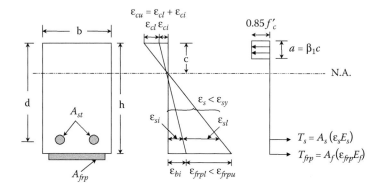

**FIGURE 5.24** Force distribution in a compression-controlled failure without steel yielding and without FRP rupture.

### 5.12.4 Compression-Controlled Failure without Steel Yielding and without FRP Rupture

This mode of failure is characterized by no tension steel yielding ($\varepsilon_{sy} \leq \varepsilon_s$), no FRP rupture ($\varepsilon_{frp} < \varepsilon_{frpu}$), and the crushing of concrete ($\varepsilon_c = \varepsilon_{cu} = 0.003$, Figure 5.24). The strength reduction factor for the compression-controlled failure mode is 0.7. However, designing beams for compression-controlled failure without steel yielding and without FRP rupture is not recommended by ACI 318-02. In addition, such compression-controlled failures may not be economical.

### 5.12.5 Balanced Failure

Balanced failure mode is a hypothetical failure mode that is assumed to occur when strains in extreme tension and compression fibers have reached their limit values simultaneously as follows (also see Figure 5.25):

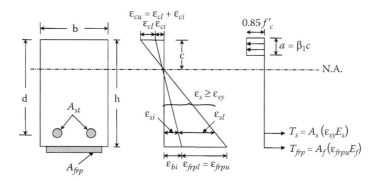

**FIGURE 5.25** Force distribution in a balanced failure with concrete crushing, steel yielding, and FRP rupture.

Strain in concrete: $\varepsilon_c = \varepsilon_{cu} = 0.003$

Strain in steel: $\varepsilon_s \geq \varepsilon_{sy}$ (this is as a consequence of strain in extreme FRP tension fiber reaching its ultimate value)

Strain in FRP-ER: $\varepsilon_{frp} = \varepsilon_{frpu}$

The strain conditions at balanced failure can be expressed as follows from the similar triangles principle:

$$\frac{c_b}{h} = \frac{\varepsilon_{cu}}{\varepsilon_{cu} + \varepsilon_{bi} + \varepsilon_{frpu}} \quad \text{or} \quad a_b = \beta_1 h \left( \frac{\varepsilon_{cu}}{\varepsilon_{cu} + \varepsilon_{bi} + \varepsilon_{frpu}} \right) \qquad (5.42)$$

where

$$\varepsilon_b = \varepsilon_{bi} + \varepsilon_{frpu}$$

$$\varepsilon_{bi} = \left( \frac{\varepsilon_{ci} + \varepsilon_{si}}{d} \right) h - \varepsilon_{ci}$$

$\varepsilon_{frpu} = 0.010$ to $0.015$ for carbon FRP

$\phantom{\varepsilon_{frpu}} = 0.015$ to $0.025$ for glass FRP

*Force Equilibrium.* Forces in concrete, steel, and the FRP-ER are assumed given, respectively, by Equation 5.22, Equation 5.23a and b, and Equation 5.24a. Substitution of these values in force equilibrium Equation 5.34 gives:

$$0.85 f'_c a_b b = A_s f_y + A_{frp} E_{frp} \varepsilon_{frpu} \qquad (5.43)$$

The above equation can be solved for $a_b$, where $a_b$ is the depth of the rectangular stress block corresponding to the balanced condition.

By knowing all other parameters in the force equilibrium equation, the area of FRP reinforcement for a balanced failure ($A_{frp,b}$) is:

$$A_{frp,b} = \frac{0.85 f_c' b a_b - A_s f_y}{E_{frp} \varepsilon_{frpu}} \qquad (5.44)$$

The nominal flexural strength of the beam is obtained by taking the moments of $T_s$ and $T_{frp}$ forces about $C_c$:

$$M_n = A_s f_y \left( d - \frac{a_b}{2} \right) + A_{frp} E_{frp} \varepsilon_{frp} \left( h - \frac{a_b}{2} \right) \qquad (5.45a)$$

The strength reduction factor for this failure mode remains the same as tension-controlled failure with FRP rupture, i.e., $\phi = 0.9$ ($\varepsilon_s \geq 0.005$). In addition to the use of the strength reduction factor ($\phi$) required by ACI 318, an additional strength reduction factor ($\psi_f = 0.85$) is applied to the flexural strength provided by the FRP-ER. Again, noting that $\phi M_n \geq M_u$, Equation 5.45a can be expressed as:

$$\phi M_n = \phi \left[ A_s f_y \left( d - \frac{a_b}{2} \right) + \psi_f A_{frp} E_{frp} \varepsilon_{frp} \left( h - \frac{a_b}{2} \right) \right] \qquad (5.45b)$$

## 5.13  TRIAL-AND-ERROR PROCEDURE FOR ANALYSIS AND DESIGN

The following trial-and-error procedure can be used for analysis and design of FRP-ER strengthened concrete beams.

1. For given beam dimensions and material properties (concrete, steel, and FRP), consider $n$ number of layers with a preferable width $b$ equal to that of the beam, unless the field conditions require a smaller width.
2. Analyze the beam as if it were a tension-controlled failure with FRP rupture and find the depth of the neutral axis (NA) $c$ or the depth of the equivalent rectangular stress block $a$ from Equation 5.35.
3. Check if $c$ or $a$ obtained in Step 2 satisfies the "tension-controlled failure criteria with FRP rupture," i.e., steel yields ($\varepsilon_s \geq 0.005$) and the FRP-ER is at the point of incipient rupture ($\varepsilon_{frp} = {}_{frpu}$) as in Section 5.9.1.1.
4. If Step 3 is satisfied, determine the moment resistance of the beam from Equation 5.36b and no further analysis is necessary.
5. If Step 3 is not satisfied, analyze the beam as if it were a tension-controlled failure without FRP rupture, and find $c$ or $a$.
6. Check if $c$ or $a$ obtained in Step 2 satisfies the tension-controlled failure criteria without FRP rupture, i.e., an "adequate" amount of steel yielding ($\varepsilon_s \geq 0.005$) is present without FRP fabric rupture ($\varepsilon_{frp} < \varepsilon_{frpu}$) as described in Section 5.12.2.
7. If Step 6 is true, find the moment resistance of the beam from Equation 5.41b and no further analysis is necessary.

8. If Step 6 is not true, verify if the beam is a tension-and compression-controlled failure characterized by steel yielding ($\varepsilon_{sy} \leq \varepsilon_s < 0.005$) without FRP rupture ($\varepsilon_{frp} < \varepsilon_{frpu}$) and find the moment resistance of the beam as described in Section 5.12.2 (i.e., make use of Equation 5.41b) and no further analysis is necessary.

9. If Step 8 indicates that the failure is characterized by no tension steel yielding ($\varepsilon_s < \varepsilon_{sy}$) and no FRP rupture ($\varepsilon_{frp} < \varepsilon_{frpu}$) but involves concrete crushing ($\varepsilon_c = \varepsilon_{cu} = 0.003$), then it establishes as a case of compression failure. This failure mode is not acceptable and the beam should be redesigned.

## 5.14 COMPUTATION OF DEFLECTION AND CRACK WIDTH

The use of FRP-ER will increase the beam stiffness and reduce deflection and crack width for a given load. Deflections and crack widths of FRP-ER can be calculated similar to traditional steel reinforced concrete beams by suitably including the stiffness contribution of FRP-ER to the structural member. The deflection computation of a FRP-ER strengthened beam is illustrated in Example 5.5 and Example 5.10.

## 5.15 DESIGN EXAMPLES ON FLEXURE

The design examples on flexural capacity of FRP-ER strengthened concrete beams provided in this chapter are based on spreadsheet values that show a higher accuracy than could be measured in the field through commonly used equipment. Numbers with higher accuracy (i.e., several decimal places) are not truncated in the following examples. In many cases, designers could proceed with their design using two decimal places for accuracy and rounding of the final values as necessary. However, truncation of values such as those of strains to a few decimal places could lead to errors in the design and analysis.

### 5.15.1 Flexural Strengthening with FRP-ER (U.S. Standard Units)

*Example 5.1*: A simply supported concrete beam in an interior location (Figure 5.26) is reinforced with four No. 8 bars and is required to carry a 100% increase in its original design live load. Assuming that the beam is safe in shear for such a strength increase, and also assuming the deflection and crack width under 100% increased loads are acceptable, design a CFRP wrap system (low grade) to carry flexural loads for future live loads. Use the beam and CFRP properties given in Table 5.3 and Table 5.4.

*Proposed wrapping scheme:* FRP system consisting of three 15-in. wide × 21.0-ft long plies will be bonded at the bottom (soffit) of the beam. Existing and new loadings and associated midspan moments for the beam are summarized in Table 5.5.

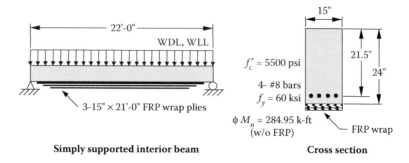

**FIGURE 5.26** Simply supported beam with FRP external reinforcement.

**TABLE 5.3**
**Dimensions of Beam and**
**Properties of Materials**

| | |
|---|---|
| Length of the beam, $l$ | 22 ft |
| Width of the beam, $w$ | 15 in. |
| $d$ | 21.5 in. |
| $h$ | 24 in. |
| $f'_c$ | 5500 psi |
| $f_y$ | 60 ksi |
| $\phi(M_n)$ w/o FRP | 284.95 k-ft |
| Bars | Four #8 |

**TABLE 5.4**
**Manufacturer's Reported FRP-System**
**Properties**

| | |
|---|---|
| Thickness per ply, $t_f$ | 0.035 in. |
| Ultimate tensile strength, $f_{fu}{}^*$ | 95,000 psi |
| Rupture strain, $\varepsilon_{fu}{}^*$ | 0.017 in./in. |
| Modulus of elasticity of FRP laminates, $E_f$ | 5600 ksi |

*Design for flexural strengthening using CFRP wrap system*: The level of strengthening is acceptable because the criterion of the strength limit as described by Equation 5.5 is satisfied. The existing moment capacity without wrap, $(\phi M_n)_{w/oFRP}$ = 284.95 k-ft, is greater than the unstrengthened moment limit, $(1.2M_{DL} + 0.85M_{LL})_{new}$ = 208.42 k-ft.

*Step 1*. Analyze the current beam. Properties of the concrete, $\beta_1$ from ACI 318-02, Section 10.2.7.3:

$$\beta_1 = 1.05 - 0.05\frac{f'_c}{1000} \text{ for } 4000 \le f'_c \le 8000 \text{ psi}$$

## TABLE 5.5
## Loadings and Corresponding Moments

| | Existing Loads | Future Loads |
|---|---|---|
| Dead loads, $w_{DL}$ | 1.10 k/ft | 1.10 k/ft |
| Live load, $w_{LL}$ | 1.25 k/ft | 2.50 k/ft |
| Unfactored loads, $w_{DL} + w_{LL}$ | 2.35 k/ft | 3.60 k/ft |
| Unstrengthened load limit, $1.2w_{DL} + 0.85w_{LL}$ | n/a | 3.445 k/ft |
| Factored loads, $1.2w_{DL} + 1.6w_{LL}$ | 3.32 k/ft | 5.32 k/ft |
| Dead-load moment, $M_{DL}$ | 66.5 k-ft | 66.5 k-ft |
| Live-load moment, $M_{LL}$ | 75.625 k-ft | 151.25 k-ft |
| Service-load moment, $M_s$ | 142.175 k-ft (1706.1 k-in.) | 217.80 k-ft (2613.6 k-in.) |
| Unstrengthened moment limit, $1.2M_{DL} + 0.85M_{LL}$ | n/a | 208.42 k-ft |
| Factored moment, $M_u$ | 200.86 k-ft | 321.86 k-ft |

$$\beta_1 = 1.05 - 0.05 \frac{5500}{1000} = 0.775$$

$$E_c = 57,000\sqrt{f_c'}$$

$$E_c = 57,000\sqrt{5500 \text{ psi}} = 4,227,233 \text{ psi}$$

The properties of the existing reinforcing steel:

$$\rho_s \equiv \frac{A_s}{bd}$$

$$A_s = \frac{(4)(\pi)(1)^2}{4} \text{ in.}^2 = 3.1416 \text{ in.}^2$$

$$\rho_s = \frac{3.1416 \text{ in.}^2}{(15 \text{ in.})(21.5 \text{ in.})} = 0.0097$$

$$n_s \equiv \frac{E_s}{E_c}$$

$$n_s = \frac{29,000,000 \text{ psi}}{4,227,233 \text{ ksi}} = 6.86$$

$$\rho_s n_s = (0.0097)(6.86) = 0.0668$$

The minimum reinforcement ratio:

$$\rho_{min} = 3\frac{\sqrt{f_c'}}{f_y} \quad \text{or} \quad \frac{200}{f_y}$$

$$\rho_{min} = 3\frac{\sqrt{5,500}}{60,000} = 0.0037$$

$$\rho_{provided}(= 0.0097) > \rho_{min}(= 0.0037)$$

Note that $200/f_y$ governs only when $f_c' \leq 4444$ psi and $f_y = 60$ ksi.

Depth of neutral axis (NA):

$$a = \frac{A_s f_y}{0.85 f_c' b}$$

$$a = \frac{(3.1416)(60,000)}{0.85(5500)(15)} = 2.69 \text{ in.}$$

$$c = a/\beta_1$$

$$c = 2.69/0.775 = 3.47 \text{ in.}$$

The ratio of the depth of the NA to the effective depth:

$$\frac{c}{d_t} = \frac{3.47}{21.5} = 0.1613$$

$$\frac{c}{d_t} = 0.1616 \leq 0.375$$

The strain in steel:

$$\varepsilon_s = 0.003\left(\frac{d-c}{c}\right)$$

$$\varepsilon_s = 0.003\left(\frac{21.5 - 3.47}{3.47}\right) = 0.0156 > 0.005$$

$$\varepsilon_y = \varepsilon_{sy} = \left(\frac{60,000}{29 \times 10^6}\right) = 0.002 \text{ in./in.}$$

The beam is ductile with a steel strain exceeding the yield value of 0.002 and the minimum limit of tension-controlled failure mode strain value of 0.005. A nominal strength reduction factor $\phi = 0.9$ will be used:

$$M_n = A_s f_y (d - a/2)$$

$$M_n = (3.1416)(60 \text{ ksi})(21.5 \text{ in.} - 2.69 \text{ in.}/2)$$

$$= 3799.3 \text{k-in.} = 316.61 \text{ k-ft.}$$

$$\phi M_n = 0.9 M_n$$

$$\phi M_n = 0.9(316.61) = 284.95 \text{ k-ft}$$

*Step 2.* Compute the FRP-system design material properties. The beam is in an interior location and CFRP material will be used. Therefore, an environmental-reduction factor $C_E$ of 0.95 is used as per Table 5.1:

$$f_{fu} = C_E f_{fu}*$$

$$f_{fu} = (0.95)(95,000 \text{ psi}) = 90,250 \text{ psi}$$

$$\varepsilon_{fu} = C_E \varepsilon_{fu}*$$

$$\varepsilon_{fu} = (0.95)(0.017 \text{ in./in.}) = 0.0161 \text{ in./in.}$$

*Step 3.* Preliminary calculations. The properties of the concrete, $\beta_1$ from ACI 318, Section 10.2.7.3:

$$\beta_1 = 1.05 - 0.05 \frac{f_c'}{1000}$$

$$\beta_1 = 1.05 - 0.05\frac{5500}{1000} = 0.775$$

$$E_c = 57,000\sqrt{f_c'}$$

$$E_c = 57,000\sqrt{5500 \text{ psi}} = 4,227,233 \text{ psi}$$

The properties of the existing reinforcing steel, calculated in Step 1:

$$A_s = 3.1416 \text{ in.}^2$$

$$\rho_s = 0.0097$$

$$n_s = 6.86$$

$$\rho_s n_s = 0.0668$$

The properties of the externally bonded CFRP reinforcement:

$$A_f = nt_f w_f$$

$$A_f = (3 \text{ plies})(0.035 \text{ in./ply})(15 \text{ in.}) = 1.575 \text{ in.}^2$$

$$\rho_f \equiv \frac{A_f}{bd}$$

$$\rho_f = \frac{1.575 \text{ in.}^2}{(15 \text{ in.})(21.5 \text{ in.})} = 0.0049$$

$$n_f \equiv \frac{E_f}{E_c}$$

$$n_f = \frac{5,600,000}{4,227,233} = 1.3247$$

$$\rho_f n_f = (0.0049)(1.3247) = 0.0065$$

*Step 4.* Determine the existing state of strain on the soffit. The existing state of strain is calculated assuming the beam is cracked and the only loads acting on the beam at the time of the FRP installation are dead loads. A cracked section analysis of the existing beam shown below gives $k = 0.2984$ and $I_{cr} = 6224$ in$^4$.

$$k = \sqrt{(\rho_s n_s)^2 + 2(\rho_s n_s)} - (\rho_s n_s) \text{ (with steel bars)}$$

$$k = \sqrt{(0.0668)^2 + 2(0.0668)} - (0.0668) = 0.2984$$

$$c = kd$$

$$c = (0.2984)(21.5) = 6.416 \text{ in.}$$

$$I_{cr} = \frac{bc^3}{3} + n_s A_s (d-c)^2 \text{ (with steel bars)}$$

$$I_{cr} = \frac{(15)(6.416)^3}{3} + (6.86)(3.14)(21.5 - 6.416)^2 = 6222 \text{ in.}^4$$

$$\varepsilon_{bi} = \frac{M_{DL}(h - kd)}{I_{cr} E_c}$$

$$\varepsilon_{bi} = \frac{(66.55 \times 12,000 \text{ lb-in.})[24 \text{ in.} - (0.2984)(21.5 \text{ in.})]}{(6222 \text{ in.}^4)(4,227,233 \text{ psi})}$$

$$\varepsilon_{bi} = 0.00053$$

*Step 5.* Determine the bond-dependent coefficient of the FRP system. The dimensionless bond-dependent coefficient for flexure, $\kappa_m$, is calculated using Equation 5.6. Compare $nE_f t_f$ to 1,000,000:

$$(3)(5,600,000 \text{ psi})(0.035 \text{ in.}) = 588,000 < 1,0000,000$$

Therefore,

$$\kappa_m = \frac{1}{60\varepsilon_{fu}} \left( 1 - \frac{nE_f t_f}{2,000,000} \right) \leq 0.90$$

$$\kappa_m = \frac{1}{60(0.0161)} \left[1 - \frac{(3)(5,600,000)(0.035)}{2,000,000}\right] \le 0.90$$

$$\kappa_m = 0.73 \le 0.90$$

*Step 6a.* Estimate $c$, the depth to the neutral axis. A reasonable initial estimate of $c$ is $c = \left(1.33 \dfrac{\varepsilon_{cu}}{\varepsilon_{cu} + \varepsilon_{fu}}\right) d$ or $c = 0.2d$, as suggested by ACI 440.R1-03. The value of $c$ is adjusted after checking equilibrium:

$$c = \left(1.33 \frac{\varepsilon_{cu}}{\varepsilon_{cu} + \varepsilon_{fu}}\right) d$$

$$c = \left(1.33 \frac{0.003}{0.003 + 0.0161}\right)(21.5 \text{ in.}) = 4.49 \text{ in.}$$

$$\frac{c}{d} = \frac{4.49}{21.5} = 0.209$$

*Step 6b.* Determine the effective level of strain in the FRP reinforcement. The effective strain level in the FRP is calculated from Equation 5.7:

$$\varepsilon_{fe} = 0.003\left(\frac{h - c}{c}\right) - \varepsilon_{bi} \le \kappa_m \varepsilon_{fu}$$

Note that for the neutral axis depth selected, we can expect a tension- and compression-controlled failure mode with steel reinforcement strain (0.0114 > 0.005 from Step 5c) without FRP rupture leading to secondary compression failure in the form of concrete crushing.

$$\varepsilon_{fe} = 0.003\left(\frac{24 - 4.49}{4.49}\right) - 0.00053 \le 0.73(0.0125)$$

$$\varepsilon_{fe} = 0.0125 \le 0.0118 \text{ (not satisfied)}$$

Note that the values are close but the condition is not satisfied; check again after obtaining final value of $c$.

*Step 6c.* Calculate the strain in the existing reinforcing steel. The strain in the reinforcing steel can be calculated using similar triangles according to Equation 5.33b:

$$\varepsilon_s = (\varepsilon_{fe} + \varepsilon_{bi})\left(\frac{d-c}{h-c}\right)$$

$$\varepsilon_s = (0.0125 + 0.00053)\left(\frac{21.5 - 4.49}{24 - 4.49}\right) = 0.0114$$

*Step 6d.* Calculate the stress level in the reinforcing steel and FRP. The stresses are calculated using Equation 5.13a and b, Equation 5.14a, b, and c, and Equation 5.25:

$$f_s = E_s\varepsilon_s \le f_y$$

$$f_s = (29,000 \text{ ksi})(0.0114) \le 60 \text{ ksi}$$

$$f_s = 329.82 \text{ ksi} \le 60 \text{ ksi}$$

Therefore, $f_s = 60$ ksi.

$$f_{fe} = E_f\varepsilon_{fe}$$

$$f_{fe} = (5600 \text{ ksi})(0.0125) = 70.06 \text{ ksi}$$

*Step 6e.* Calculate the internal force resultants and check equilibrium. Force equilibrium is verified by checking the initial estimate of *c* with Equation 5.35 by noting that $c = a/\beta_1$:

$$c = \frac{A_s f_s + A_f f_{fe}}{0.85\beta_1 f'_c b}$$

$$c = \frac{(3.14 \text{ in.}^2)(60 \text{ ksi}) + (1.575 \text{ in.}^2)(70.06 \text{ ksi})}{(0.85)(0.775)(5.5 \text{ ksi})(15 \text{ in.})} = 5.50 \text{ in.}$$

$$c = 5.50 \text{ in.} \ne 4.49 \text{ in. (assumed)}$$

Therefore, revise the estimate of *c* and repeat Step 6a through Step 6e until equilibrium is achieved. Using spreadsheet programming for iterations is suggested to obtain a hassle-free solution within a few seconds. Results of the final iteration are:

$$c = \frac{(3.14 \text{ in.}^2)(60 \text{ ksi}) + (1.575 \text{ in.}^2)(58.35 \text{ ksi})}{(0.85)(0.775)(5.5 \text{ ksi})(15 \text{ in.})} = 5.16 \text{ in.}$$

$$c = 5.16 \text{ in. (as assumed)}$$

Therefore, the value of $c$ selected for the final iteration is correct.

$$a = \beta_1 c = 0.775(5.16 \text{ in.}) = 4 \text{ in.}$$

$$c = 5.16 \text{ in.}$$

$$\varepsilon_s = 0.0095$$

$$f_s = f_y = 60 \text{ ksi}$$

$$\varepsilon_{fe} = 0.0104 \text{ (see note below)}$$

$$f_{fe} = 58.35 \text{ ksi}$$

Note:

$$\varepsilon_{fe} = 0.003\left(\frac{h-c}{c}\right) - \varepsilon_{bi} \le \kappa_m \varepsilon_{fu}$$

$$\varepsilon_{fe} = 0.003\left(\frac{24-5.16}{5.16}\right) - 0.00053 \le 0.73(0.0125)$$

$$\varepsilon_{fe} = 0.0104 \le 0.0118 \text{ (satisfied)}$$

Note that an alternate approach for Step 6a to Step 6e is shown in Example 5.3 and Example 5.4.

*Step 7.* Calculate the design flexural strength of the section. The design flexural strength is calculated using Equation 5.4 and Equation 5.36 with appropriate substitutions. An additional reduction factor, $\psi_f = 0.85$, is applied to the bending strength contributed by the FRP system. Because $\varepsilon_s = 0.0095 > 0.005$, a strength-reduction factor of $\phi = 0.90$ is used as per Equation 5.9:

$$\phi M_n = \phi\left[ A_s f_s\left(d - \frac{a}{2}\right) + \psi_f A_f f_{fe}\left(h - \frac{a}{2}\right)\right]$$

$$\phi M_n = 0.90 \left[ \begin{array}{l} (1)(3.14 \text{ in.}^2)(60 \text{ ksi})\left(21.5 \text{ in.} - \dfrac{4.0}{2}\right) + \\[2ex] (0.85)(1)(1.575 \text{ in.}^2)(58.35 \text{ ksi})\left(24 \text{ in.} - \dfrac{4}{2}\right) \end{array} \right]$$

$$\phi M_n = 4{,}854.9 \text{ k-in.} = 404.6 \text{ k-ft} \geq M_u = 321.86 \text{ k-ft}$$

Therefore, the strengthened section is capable of sustaining the new required moment strength.

*Step 8.* Check service stresses in the reinforcing steel and the FRP. Calculate the depth of the cracked neutral axis by summing the first moment of the areas of the elastic transformed section without accounting for the compression reinforcement as per Equation 5.28:

$$k = \sqrt{(\rho_s n_s + \rho_f n_f)^2 + 2\left(\rho_s n_s + \rho_f n_f\left(\frac{h}{d}\right)\right)} - (\rho_s n_s + \rho_f n_f)$$

$$k = \sqrt{(0.0668 + 0.0065)^2 + 2\left(0.0668 + 0.0065\left(\frac{24 \text{ in.}}{21.5 \text{ in.}}\right)\right)} - (0.0668 + 0.0065)$$

$$k = 0.3185$$

$$kd = (0.3185)(21.5 \text{ in.}) = 6.85 \text{ in.}$$

Calculate the stress level in the reinforcing steel using Equation 5.30h:

$$f_{s,s} = \frac{\left[M_s + \varepsilon_{bi} A_f E_f\left(h - \frac{kd}{3}\right)\right](d - kd)E_s}{A_s E_s\left(d - \frac{kd}{3}\right)(d - kd) + A_f E_f\left(h - \frac{kd}{3}\right)(h - kd)}$$

$$f_{s,s} = \frac{\left[2{,}613.60 \text{ k-in.} + (0.00053)(1.575 \text{ in.}^2)(5600 \text{ ksi})\left(24 \text{ in.} - \frac{6.85 \text{ in.}}{3}\right)\right]}{\left[\begin{array}{l}(3.14 \text{ in.}^2)(29{,}000 \text{ ksi})\left(21.5 \text{ in.} - \frac{6.85 \text{ in.}}{3}\right)(21.5 \text{ in.} - 6.85 \text{ in.}) \\ + (1.575 \text{ in.}^2)(5600 \text{ ksi})\left(24 \text{ in.} - \frac{6.85 \text{ in.}}{3}\right)(24 \text{ in.} - 6.85 \text{ in.})\end{array}\right]}$$

Verify that the stress in steel is less than the recommended limit as per Equation 5.10:

$$f_{s,s} \leq 0.80 f_y$$

$$f_{s,s} = 39.88 \text{ ksi} \leq (0.80)(60 \text{ ksi}) = 48 \text{ ksi}$$

Therefore, the stress level in the reinforcing steel is within the recommended limit.

Calculate the stress level in the FRP due to maximum service loads (assume sustained) using Equation 5.31d and verify that it is less than the creep-rupture stress limit:

$$f_{f,s} = f_{s,s}\left(\frac{E_f}{E_s}\right)\left(\frac{h-kd}{d-kd}\right) - \varepsilon_{bi}E_f$$

$$f_{f,s} = 39.88 \text{ ksi}\left(\frac{5600 \text{ ksi}}{29,000 \text{ ksi}}\right)\left(\frac{24 \text{ in.} - 6.85 \text{ in.}}{21.5 \text{ in.} - 6.85 \text{ in.}}\right) - (0.00053)(5600 \text{ ksi})$$

$$= 6.05 \text{ ksi}$$

For a carbon FRP system, the creep-rupture stress limit is listed as per Table 5.2:

$$F_{f,s} = 0.55f_{fu}$$

$$= 0.55 \ (90.25 \text{ ksi}) = 49.64 \text{ ksi}$$

Therefore, the stress level in the carbon FRP system is within the recommended limit ($f_{f,s} < F_{f,s}$).

*Example 5.2*: If the beam in Example 5.1 was to be designed with a GFRP wrap system (normal grade) to carry the same increase in flexural loads due to future live loads, what changes in design are expected? Assume that the proposed GFRP wrap system (normal grade) has same properties (hypothetically) as those of the CFRP wrap system provided in Table 5.4 of Example 5.1.

*Design*: Although the properties of GFRP are hypothetically stated to be same as those of CFRP, design values with the environmental reduction factor and creep-rupture stress limit values are lower for GFRP. Because it is tension-controlled failure without FRP rupture as noted in Example 5.1, we will verify the values of GFRP stress and strain at the ultimate and compare these with the design values (similar to Example 5.1). If stresses are within the design values, then we will proceed to check the creep-rupture strength, which is different for GFRP and CFRP systems.

Note that the procedures in Step 6 must be carried out with the new design values of the proposed GFRP because the design values of stress and strain are lower for the proposed GFRP as compared to CFRP in Example 5.1 due to lower value of $C_E$.

*Analysis and calculation details for GFRP wrap system.* FRP properties:

$$f_{fu} = C_E f_{fu}{}^*$$

$$f_{fu} = (0.75)(95,000 \text{ psi}) = 71,250 \text{ psi}$$

$$\varepsilon_{fu} = C_E \varepsilon_{fu}*$$

$$\varepsilon_{fu} = (0.75)(0.017 \text{ in./in.}) = 0.0127 \text{ in./in.}$$

From Step 6f of Example 5.1, the force equilibrium, depth of the neutral axis, stress, and strain in steel and FRP are unchanged.

$$c = \frac{(3.14 \text{ in.}^2)(60 \text{ ksi}) + (1.575 \text{ in.}^2)(58.35 \text{ ksi})}{(0.85)(0.775)(5.5 \text{ ksi})(15 \text{ in.})} = 5.16 \text{ in.}$$

$$c = 5.16 \text{ in. (as assumed)}$$

$$a = \beta_1 c = 0.775(5.16 \text{ in.}) = 4 \text{ in.}$$

$$c = 5.16 \text{ in.}$$

$$\varepsilon_s = 0.0095$$

$$f_s = f_y = 60 \text{ ksi}$$

$$\varepsilon_{fe}(0.0104 \text{ in./in.}) < \varepsilon_{fu}(0.0127 \text{ in./in.})$$

$$f_{fe} = 58.35 \text{ ksi} < 71.25 \text{ ksi}$$

Check service stresses in the FRP (similar to Step 8 of Example 5.1):

$$f_{f,s} = f_{s,s}\left(\frac{E_f}{E_s}\right)\left(\frac{h - kd}{d - kd}\right) - \varepsilon_{bi}E_f$$

$$f_{f,s} = 39.88 \text{ ksi}\left(\frac{5600 \text{ ksi}}{29,000 \text{ ksi}}\right)\left(\frac{24 \text{ in.} - 6.85 \text{ in.}}{21.5 \text{ in.} - 6.85 \text{ in.}}\right) - (0.00053)(5600 \text{ ksi}) = 6.05 \text{ ksi}$$

For a GFRP system, the creep-rupture stress limit is listed as per Table 5.2:

$$F_{f,s} = 0.20 f_{fu}$$

$$f_{f,s} = 6.05 \text{ ksi} \le (0.20)(71.25 \text{ ksi})$$

$$f_{f,s} = 6.05 \text{ ksi} \le (14.25 \text{ ksi})$$

Therefore, the stress level in the glass FRP system is within the recommended creep-rupture stress limit ($f_{f,s} < F_{f,s}$).

164

Reinforced Concrete Design with FRP Composites

*Example 5.3*: Analyze the design in Example 5.1 to determine the depth of the neutral axis for a balanced failure and the amount of FRP reinforcement needed for a balanced failure. Compare the calculated amount of FRP reinforcement for a balanced failure and compare it with the amount of FRP reinforcement provided to determine the possible failure mode for the FRP strengthened beam.

*Analysis*: Hypothetical balanced failure mode is assumed to occur when strains in extreme tension and compression fibers have reached their limit values simultaneously.

Strain in concrete: $\varepsilon_c = \varepsilon_{cu} = 0.003$
Strain in steel: $\varepsilon_s = \varepsilon_y$ (this is as a consequence of strain in extreme FRP tension fiber reaching its ultimate value)
Strain in FRP: $\varepsilon_{frp} = \varepsilon_{frpu}$

These strain conditions at balanced failure can be expressed as follows from similar triangles principle based on Equation 5.42:

$$\frac{c_b}{h} = \frac{\varepsilon_{cu}}{\varepsilon_{cu} + \varepsilon_{bi} + \varepsilon_{frpu}} \quad \text{or} \quad a_b = \beta_1 h\left(\frac{\varepsilon_{cu}}{\varepsilon_{cu} + \varepsilon_{bi} + \varepsilon_{frpu}}\right)$$

where

$$\varepsilon_b = \varepsilon_{bi} + \varepsilon_{frpu}$$

$$\varepsilon_{bi} = \left(\frac{\varepsilon_{ci} + \varepsilon_{si}}{d}\right)h - \varepsilon_{ci} \quad \text{or} \quad \varepsilon_{bi} = \frac{M_{DL}(h - kd)}{I_{cr}E_c}$$

$$\varepsilon_{bi} = 0.00053 \text{ (from Step 4, Example 5.1)}$$

$$a_b = (0.775)(24)\left(\frac{0.003}{0.003 + 0.00053 + 0.0161}\right) = 2.84 \text{ in.}$$

*Force Equilibrium.* From Equation 5.43:

$$0.85 f'_c a_b b = A_s f_y + A_{frp} E_{frp} \varepsilon_{frpu}$$

The above equation is solved for $a_b$, where $a_b$ is the depth of the rectangular stress block corresponding to the balanced condition (Equation 5.44);

$$A_{frp,b} = \frac{0.85\beta_1 f'_c b a_b - A_s f_y}{E_{frp}\varepsilon_{frpu}}$$

$$A_{frp,b} = \frac{0.85(5500)(15)(2.84) - (3.1416)(60,000)}{(5,600,000)(0.0161)} = 0.118 \text{ in.}^2 < 1.575 \text{ in.}^2 \text{ (provided)}$$

The area of FRP needed for a balanced failure is 0.118 in², which is much less than the provided area of 1.575 in². Hence, the failure mode is not tension-controlled with FRP rupture. This leaves the possible failure modes as:

- Tension-controlled without FRP rupture
- Tension-and-compression-controlled without FRP rupture
- Compression-controlled (no FRP rupture and no steel yield)

Actual failure mode can be verified as shown in Example 5.4.

*Example 5.4*: Analyze Example 5.1 to directly determine (i.e., avoid a method of iterations) the depth of the neutral axis for the same specifications.

*Analysis*: Assuming the failure mode to be tension-controlled without FRP rupture, use Equation 5.40 and calculate the depth of neutral axis $c$ where $c = a/\beta_1$. After obtaining $c$, verify if the conditions for tension-controlled failure mode without FRP rupture are satisfied, i.e., $\varepsilon_s \geq 0.005$ and $\varepsilon_{frp} < \varepsilon_{frpu}$. From Equation 5.40,

$$0.85 f_c' b a^2 + [A_{frp} E_{frp}(\varepsilon_{cu} + \varepsilon_{bi}) - A_s f_y] a - \beta_1 A_{frp} E_{frp} \varepsilon_{cu} h = 0$$

Noting that the above equation is quadratic in $a$ (the idealized depth of the neutral axis), the following coefficients are computed:

$$a = \frac{-B \pm \sqrt{B^2 - 4AC}}{2A}$$

where

$A = 0.85 f_c' b$

$B = A_{frp} E_{frp}(\varepsilon_{cu} + \varepsilon_{bi}) - A_s f_y$

$C = (-\beta_1 A_{frp} E_{frp} \varepsilon_{cu} h)$

$A = 0.85(5500)(15) = 70,125$

$B = (1.575)(5,600,000)(0.003 + 0.00053) - (3.1416)(60,000) = -157,327.9$

$C = (-0.775)(1.575)(5,600,000)(0.003)(24) = -492,156$

$$a = \frac{-(-157,327.9) + \sqrt{(-157,327.9)^2 - 4(70,125)(-492,156)}}{2(70,125)} = 4 \text{ in.}$$

$c = a/\beta_1 = 4/0.775 = 5.16$ in. (as obtained in Example 5.1)

*Verification.* Verify the strains in steel and FRP.

$$\varepsilon_s = \left(\frac{\varepsilon_{cu}}{c}\right)(d-c) = \left(\frac{0.003}{5.16}\right)(21.5 - 5.16) = 0.0095$$

$$\varepsilon_{fe} = \varepsilon_{frp} = \left(\frac{\varepsilon_{cu}}{c}\right)(h-c) - \varepsilon_{bi} = \left(\frac{0.003}{5.16}\right)(24 - 5.16) - 0.00053 = 0.01042$$

Based on strain values,

$$\varepsilon_s = 0.0095 \geq 0.005 \text{(OK)}$$

$$\varepsilon_{fe} = \varepsilon_{frp} = 0.01042 \leq \varepsilon_{frpu}(0.017)\text{(OK for CFRP)}$$

$$\varepsilon_{fe} = \varepsilon_{frp} = 0.01042 \leq \varepsilon_{frpu}(0.0127)\text{(OK for GFRP)}$$

Hence, the assumed conditions are satisfied and the failure mode is tension-controlled without FRP rupture. Other calculations for stress and moment resistance remain same as shown in Example 5.1.

*Example 5.5.* Compute deflections for the CFRP strengthened beam in Example 5.1 for a combination of dead load ($\delta_{DL}$), live load ($\delta_{LL}$), and 20% of live load sustained with dead load ($\delta_{DL+0.2LL}$). Compare the values obtained using the transformed gross moment of inertia and gross moment of inertia. What difference is expected in deflections if it were to be a GFRP strengthened beam described in Example 5.2?

$$b \text{ or } w = 15 \text{ in.}$$

$$d = 21.5 \text{ in.}$$

$$h = 24 \text{ in.}$$

$$l = 22 \text{ ft} = 262 \text{ in.}$$

$$n_s = 6.86$$

$$A_s = 3.1416 \text{ in.}^2$$

$$n_f = 1.32$$

$$A_f = 1.575 \text{ in.}^2$$

$$n = 3 \text{ (no. of wraps)}$$

$$t_f = 0.035 \text{ in./ply}$$

$$k \text{ (with wrap)} = 0.3185$$

$$c = kd \text{ (with wrap)} = 6.86 \text{ in.}$$

*Solution*: Because the GFRP and CFRP properties in terms of area and stiffness are the same (hypothetically) in Example 5.1 and Example 5.2, the short-term deflection values computed in this example will remain the same. However, part of the long-term deflections will be influenced by the type of fibers (glass or carbon) based on their creep coefficient values.

Gross section:

$$y = \frac{h}{2}$$

$$y = \frac{24}{2} = 12 \text{ in.}$$

$$I_g = \frac{bh^3}{12}$$

$$I_g = \frac{(15)(24)^3}{12} = 17,280 \text{ in.}^4$$

$$f_r = 7.5\sqrt{f_c'}$$

$$f_r = 7.5\sqrt{5500} = 556.2 \text{ psi}$$

$$M_{cr} = \frac{I_g f_r}{y}$$

$$M_{cr} = \frac{(17,280)(556.2)}{12} = 66,746 \text{ lb-ft}$$

$$I_{cr} = bc^3 / 3 + n_s A_s (d-c)^2 + n_f A_f (h + nt_f - c)^2$$

$$I_{cr} = (15)(6.86)^3 / 3 + (6.86)(3.1416)(21.5 - 6.86)^2$$
$$+ (1.32)(1.575)(24 + 3 \times 0.035 - 6.86)^2 = 6627.2 \text{ in.}^4$$

Note that typically, $nt_f$ is ignored in the cracked moment of inertia calculation due to the negligible difference in the values as shown below.

$$I_{cr} = (15)(6.86)^3/3 + (6.86)(3.1416)(21.5 - 6.86)^2$$

$$+ (1.32)(1.575)(24 - 6.86)^2 = 6622.2 \text{ in.}^4$$

Transformed section:

$$\bar{y} = \frac{(bh^2/2) + (n_s - 1)A_s d + n_f A_f h}{bh + (n_s - 1)A_s + n_f A_f}$$

$$\bar{y} = \frac{[(15)(24^2)/2] + (6.86 - 1)(3.1416)(24) + (1.32)(1.575)(24)}{(15)(24) + (6.86 - 1)(3.1416) + (1.32)(1.575)} = 12.523 \text{ in.}$$

$$y_t = (h - \bar{y})$$

$$y_t = (24 - 12.523) = 11.474 \text{ in.}$$

$$I_{gt} = \frac{bh^3}{12} + bh(\bar{y} - y)^2 + (n_s - 1)A_s(d - \bar{y})^2 + n_f A_f(h - \bar{y})^2$$

$$I_{gt} = \frac{(15)(24)^3}{12} + (15)(24)(12.523 - 12)^2 + (6.86 - 1)(3.1416)(24 - 12.523)^2$$

$$+ (1.32)(1.575)(24 - 12.523)^2 = 19,139 \text{ in.}^4$$

$$f_r = 7.5\sqrt{5500} = 556.2 \text{ psi}$$

$$M_{cr} = \frac{I_{gt} f_r}{y_t}$$

$$M_{cr} = \frac{(19,139.5)(556.2)}{11.47} = 77,316 \text{ lb-ft} = 77.316 \text{ k-ft}$$

$I_{cr} = 6622.2$ in.$^4$ (as calculated in the gross section analysis)

Note that typically, $nt_f$ is ignored in the cracked moment of inertia calculation due to the negligible difference in the values as shown below.

$$I_{cr} = (15)(6.86)^3 / 3 + (6.86)(3.1416)(21.5 - 6.86)^2$$
$$+ (1.32)(1.575)(24 - 6.86)^2 = 6622.2 \text{ in.}^4$$

Calculations based on transformed section analysis:

$$I_{cr} = 6622 \text{ in.}^4$$

$$I_{gt} = 19,139 \text{ in.}^4$$

$$M_{cr} = 77.136 \text{ k-ft (transformed section analysis)}$$

$$I_e = \left(\frac{M_{cr}}{M_a}\right)^3 I_{gt} + \left[1 - \left(\frac{M_{cr}}{M_a}\right)^3\right] I_{cr}$$

$$I_e = I_{cr} + \left(\frac{M_{cr}}{M_a}\right)^3 (I_{gt} - I_{cr})$$

Calculations based on the gross section analysis:

$$I_{cr} = 6622 \text{ in.}^4$$

$$I_g = 17,280 \text{ in.}^4$$

$$M_{cr} = 66.746 \text{ k-ft}$$

$$I_e = I_{cr} + \left(\frac{M_{cr}}{M_a}\right)^3 (I_g - I_{cr})$$

The comparison of calculations based on gross and transformed section analyses is listed in Table 5.6 and Table 5.7.

Deflection requirements within parenthesis correspond to transformed and gross sectional analyses:

$$\frac{l}{180} = 1.47 \text{ in. (> 0.3453 in. and 0.3724 in.) is satisfied for all load combinations}$$

$$\frac{l}{240} = 1.1 \text{ in. (> 0.3453 in. and 0.3724 in.) is satisfied for all load combinations}$$

**TABLE 5.6**
**Moment of Inertia through Transformed and Gross Section Analysis**

| Load Combination | $M_a$ (kip-ft) | $M_{cr}/M_a$ Transformed Section | $M_{cr}/M_a$ Gross Section | Effective Moment of Inertia $(I_e)$(in.⁴) Transformed Section | Effective Moment of Inertia $(I_e)$(in.⁴) Gross Section |
|---|---|---|---|---|---|
| $\delta_{DL}$ (before LL) | 66.55 | $M_{cr} > M_a$ | $M_{cr} > M_a$ | 19,139 | 17,280 |
| $\delta_{LL}$ | 151.25 | 0.5512 | 0.4413 | 8,294 | 8,046 |
| $\delta_{DL+LL}$ | 217.80 | 0.3550 | 0.3065 | 7,182 | 7,099 |
| $\delta_{DL}$ (after LL) | | $\delta_{DL} = \delta_{DL+LL} - \delta_{LL}$ | | | |
| $\delta_{DL+0.2LL}$ | 96.80 | 0.7987 | 0.6895 | 13,000 | 12,053 |
| $\delta_{0.2LL}$ | | $\delta_{0.2LL} = \delta_{DL+0.2LL} - \delta_{DL}$ | | | |

*Note:* $M_{cr} = 77.316$ k-ft (transformed section analysis) and $M_{cr} = 66.746$ k-ft (gross section analysis).

**TABLE 5.7**
**Deflections through Transformed and Gross Section Analysis**

| Load Combination | Effective Moment of inertia $(I_e)$ from Table 5.6 (in.⁴) Transformed Section | Effective Moment of inertia $(I_e)$ from Table 5.6 (in.⁴) Gross Section | Deflection $(\delta)$ (in.) Transformed Section | Deflection $(\delta)$ (in.) Gross Section |
|---|---|---|---|---|
| $\delta_{DL}$ (before LL) | 19,139 | 17,280 | 0.0717 | 0.0794 |
| $\delta_{LL}$ | 8,294 | 8,046 | 0.3758 | 0.3874 |
| $\delta_{DL+LL}$ | 7,182 | 7,099 | 0.6250 | 0.6323 |
| $\delta_{DL}$ (after LL) | $\delta_{DL} = \delta_{DL+LL} - \delta_{LL}$ | | 0.2492 | 0.2449 |
| $\delta_{DL+0.2LL}$ | 13,000 | 12,053 | 0.3453 | 0.3724 |
| $\delta_{0.2LL}$ | $\delta_{0.2LL} = \delta_{DL+0.2LL} - \delta_{DL}$ | | 0.0961 | 0.1275 |

$\dfrac{l}{360} = 0.73$ in. (> 0.3453 in. and 0.3724 in.) is satisfied for all load combinations

$\dfrac{l}{480} = 0.55$ in. (< 0.3453 in. and 0.3724 in.) is not satisfied for the $\delta_{DL+LL}$ combination

Note 1: The difference in deflection values from the gross section and transformed section analyses are small. Hence, gross section analysis is generally adequate for deflection calculation.

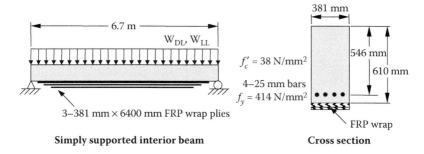

**FIGURE 5.27** Simply supported beam with FRP external reinforcement.

Note 2: Higher crack opening and lower effective moment of inertia ($I_e$) is observed when a full live load is applied as compared to a lower crack opening and relatively higher $I_e$ when only part of the live load is applied.

### 5.15.2 FLEXURAL STRENGTHENING WITH FRP-ER (SI UNITS)

*Example 5.6*: A simply supported concrete beam in an interior location (Figure 5.27) reinforced with four 25-mm bars is required to carry a 100% increase in its original design live load. Assuming that the beam is safe in shear for such a strength increase, and also assuming a deflection and crack width under 100% increased loads are acceptable, design a CFRP wrap system (low grade) to carry flexural loads for future live loads. Use the beam and CFRP properties given in Table 5.8 and Table 5.9.

Existing and new loadings and associated midspan moments for the beam are summarized in Table 5.10. Strengthening the existing reinforced concrete beam using FRP system is proposed having the properties shown in Table 5.9. Three 381-mm wide × 6.7-m long plies will be bonded at the bottom (soffit) of the beam.

*Part 1*: Design of FRP wrap system for flexural strengthening using CFRP wraps. The level of strengthening is acceptable because the criterion of strength limit as

---

**TABLE 5.8**
**Dimensions and Beam Properties**

| | |
|---|---|
| Length of the beam, $l$ | 6.7 m |
| Width of the beam, $w$ | 381 mm |
| $d$ | 546 mm |
| $h$ | 610 mm |
| $f_c'$ | 38 N/mm² |
| $f_y$ | 414 N/mm² |
| $\phi(M_n)$ w/o *FRP* | 386.34 kN-m |
| Bars | Four 25mm |

**TABLE 5.9**
**Manufacturer's Reported FRP-System Properties**

| | |
|---|---|
| Thickness per ply, $t_f$ | 0.89 mm |
| Ultimate tensile strength, $f_{fu}{}^*$ | 0.65 kN/mm² |
| Rupture strain, $\varepsilon_{fu}{}^*$ | 0.017 mm/mm |
| Modulus of elasticity of FRP laminates, $E_f$ | 38.6 kN/mm² |

**TABLE 5.10**
**Loadings and Corresponding Moments**

| | Existing Loads | Future Loads |
|---|---|---|
| Dead loads, $w_{DL}$ | 16.05 N/mm | 16.05 N/mm |
| Live load, $w_{LL}$ | 18.24 N/mm | 36.47 N/mm |
| Unfactored loads, $w_{DL} + w_{LL}$ | 34.28 N/mm | 52.52 N/mm |
| Unstrengthened load limit, $1.2w_{DL} + 0.85w_{LL}$ | n/a | 50.26 N/mm |
| Factored loads, $1.2w_{DL} + 1.6w_{LL}$ | 48.44 N/mm | 77.62 N/mm |
| Dead-load moment, $M_{DL}$ | 90.16 kN-m | 90.16 kN-m |
| Live-load moment, $M_{LL}$ | 102.53 kN-m | 205.06 kN-m |
| Service-load moment, $M_s$ | 192.76 kN-m | 295.29 kN-m |
| Unstrengthened moment limit, $1.2M_{DL} + 0.85M_{LL}$ | n/a | 282.58 kN-m |
| Factored moment, $M_u$ | 272.33 kN-m | 436.38 kN-m |

described by Equation 5.5 is satisfied. The existing moment capacity without wrap, $(\phi M_n)_{w/o\ FRP} = 386.34$ kN-m, is greater than the unstrengthened moment limit, $(1.2M_{DL} + 0.85M_{LL})_{new} = 282.58$ kN-m.

*Step 1.* Analyze the current beam. The properties of the concrete are (SI units) from ACI 318, Section 10.2.7.3:

$$\beta_1 = 1.09 - 0.008 f_c'$$

$$\beta_1 - 1.09 - 0.008(38) = 0.786$$

$$E_c = 4750\sqrt{f_c'}$$

$$E_c = 4750\sqrt{38}\ \text{N/mm}^2 = 29,280\ \text{N/mm}^2$$

The properties of the existing reinforcing steel:

$$\rho_s \equiv \frac{A_s}{bd}$$

$$A_s \equiv \frac{(4)(\pi)(25.4)^2}{4}\text{ mm}^2 = 2026.83\text{ mm}^2$$

$$\rho_s \equiv \frac{2026.83\text{ mm}^2}{(381\text{ mm})(546\text{ mm})} = 0.0097$$

$$n_s \equiv \frac{E_s}{E_c}$$

$$n_s \equiv \frac{200\text{ kN/mm}^2}{29.15\text{ kN/mm}^2} = 6.86$$

$$\rho_s n_s = (0.0097)(6.86) = 0.0668$$

The minimum reinforcement ratio:

$$\rho_{min} = 0.249\frac{\sqrt{f'_c}}{f_y}$$

$$\rho_{min} \equiv 0.249\frac{\sqrt{38}}{414} = 0.0037$$

$$\rho_{provided}(= 0.0097) > \rho_{min}(= 0.0037)$$

The depth of neutral axis (NA):

$$a = \frac{A_s f_y}{0.85 f'_c b}$$

$$a = \frac{(2026.83)(414)}{(0.85)(38)(381)} = 68.18\text{ mm}$$

$$c = a/\beta_1$$

$$c = 68.18/0.786 = 86.74 \text{ mm}$$

The ratio of the depth of the NA to effective depth:

$$\frac{c}{d_t} = \frac{86.75}{546} = 0.1588$$

$$\frac{c}{d_t} = 0.1588 \le 0.375$$

The strain in steel:

$$\varepsilon_s = 0.003\left(\frac{d-c}{c}\right)$$

$$\varepsilon_s = 0.003\left(\frac{546 - 86.74}{86.74}\right) = 0.0158 > 0.005$$

$$\varepsilon_y = \varepsilon_{sy} = \left(\frac{414}{200,000}\right) = 0.002 \text{ mm/mm}$$

The beam is ductile with the steel strain exceeding the yield value of 0.002 and the minimum limit of tension-controlled failure mode strain value of 0.005. A nominal strength reduction factor $\phi = 0.9$ will be used.

$$M_n = A_s f_y (d - a/2)$$

$$M_n = (2026.83)(414)\left(546 \text{ m} - \frac{68.18}{2} \text{ mm}\right) = 429.55 \text{ kN-m}$$

$$\phi M_n = 0.9 \, M_n$$

$$\phi M_n = 0.9(429.55) = 386.6 \text{ kN-m}$$

*Step 2.* Compute the FRP-system design material properties. The beam is in an interior location and CFRP material will be used. Therefore, an environmental-reduction factor $C_E$ of 0.95 is used as per Table 5.1:

$$f_{fu} = C_E f_{fu}^*$$

$$f_{fu} = (0.95)(650 \text{ N/mm}^2) = 617.5 \text{ N/mm}^2$$

$$\varepsilon_{fu} = C_E \varepsilon_{fu}^*$$

$$\varepsilon_{fu} = (0.95)(0.017 \text{ mm/mm}) = 0.0161 \text{ mm/mm}$$

*Step 3.* Preliminary calculations. The properties of the concrete are from ACI 318, Section 10.2.7.3:

$$\beta_1 = 1.09 - 0.008 f_c'$$

$$\beta_1 = 1.09 - 0.008(38) = 0.786$$

$$E_c = 4750\sqrt{f_c'}$$

$$E_c = 4750\sqrt{38 \text{ N/mm}^2} = 29,280 \text{ N/mm}^2$$

The properties of the existing reinforcing steel (calculated in Step 1):

$$A_s = 2026.83 \text{ mm}^2$$

$$\rho_s = 0.0097$$

$$n_s = 6.86$$

$$\rho_s n_s = 0.0668$$

The properties of the externally bonded CFRP reinforcement:

$$A_f = n t_f w_f$$

$$A_f = (3 \text{ plies})(0.89 \text{ mm/ply})(381 \text{ mm}) = 1017.27 \text{ mm}^2$$

$$\rho_f \equiv \frac{A_f}{bd}$$

$$\rho_f \equiv \frac{1017.27 \text{ mm}^2}{(381 \text{ mm})(546 \text{ mm})} = 0.0049$$

$$n_f \equiv \frac{E_f}{E_c}$$

$$n_f \equiv \frac{38.6 \text{ kN/mm}^2}{29.28 \text{ kN/mm}^2} = 1.32$$

$$\rho_f n_f = (0.0049)(1.32) = 0.0065$$

*Step 4.* Determine the existing state of strain on the soffit. The existing state of strain is calculated assuming the beam is cracked and the only loads acting on the beam at the time of the FRP installation are dead loads. A cracked section analysis of the existing beam gives $k = 0.2984$ and $I_{cr} = 2.827 \times 10^9$ mm⁴.

$$k = \sqrt{(\rho_s n_s)^2 + 2(\rho_s n_s)} - (\rho_s n_s) \text{ (with steel bars)}$$

$$k = \sqrt{(0.0668)^2 + 2(0.0668)} - (0.0668) = 0.3048$$

$$c = kd$$

$$c = (0.3048)(546) = 166.42 \text{ mm}$$

$$I_{cr} = \frac{bc^3}{3} + n_s A_s (d - c)^2 \text{ (with steel bars)}$$

$$I_{cr} = \frac{(546)(166.42)^3}{3} + (6.86)(2026.83)(564 - 163.92)^2 = 3.037 \times 10^9 \text{ mm}^4$$

$$\varepsilon_{bi} = \frac{M_{DL}(h - kd)}{I_{cr}E_c}$$

$$\varepsilon_{bi} = \frac{(90.16 \times 10^6 \text{ N-mm})[610 \text{ mm} - (0.3048)(546 \text{ mm})]}{(3.037 \times 10^9 \text{ mm}^4)(29,280 \text{ N/mm}^4)}$$

$$\varepsilon_{bi} = 0.00045$$

*Step 5.* Determine the bond-dependent coefficient of the FRP system. The dimensionless bond-dependent coefficient for flexure, $\kappa_m$, is calculated using Equation 5.6. Compare $nE_f t_f$ to 1,000,000:

$$(3)(38,600 \text{ N/mm}^2)(0.89 \text{ mm}) = 103,062 < 175,336$$

Therefore,

$$\kappa_m = \frac{1}{60\varepsilon_{fu}}\left(1 - \frac{nE_f t_f}{360,000}\right) \le 0.90nE_f t_f \le 180,000$$

$$\kappa_m = \frac{1}{60\varepsilon_{fu}}\left(\frac{90,000}{nE_f t_f}\right) \le 0.90nE_f t_f > 180,000$$

$$\kappa_m = \frac{1}{60(0.0161)}\left(1 - \frac{(3)(38,600)(0.89)}{360,000}\right) \le 0.90$$

$$\kappa_m = 0.74 \le 0.90$$

*Step 6a.* Estimate $c$, the depth to the neutral axis. A reasonable initial estimate of $c$ is $c = \left(1.33\dfrac{\varepsilon_{cu}}{\varepsilon_{cu} + \varepsilon_{fu}}\right)d$ or $c = 0.2d$, as suggested by ACI 440.2R-02. The value of $c$ is adjusted after checking equilibrium.

$$c = \left(1.33\frac{\varepsilon_{cu}}{\varepsilon_{cu} + \varepsilon_{fu}}\right)d$$

$$c = \left(1.33\frac{0.003}{0.003 + 0.0161}\right)(546 \text{ mm}) = 114.06 \text{ mm}$$

$$\frac{c}{d} = \frac{114.06}{546} = 0.209$$

*Step 6b.* Determine the effective strain in the FRP reinforcement. The effective strain in the FRP is calculated from Equation 5.7:

$$\varepsilon_{fe} = 0.003\left(\frac{h - c}{c}\right) - \varepsilon_{bi} \le \kappa_m \varepsilon_{fu}$$

Note that for the neutral axis depth selected, we can expect a tension- and compression-controlled failure mode with steel reinforcement strain (0.0114 > 0.005, from Step 5c) without FRP rupture leading to secondary compression failure in the form of concrete crushing.

$$\varepsilon_{fe} = 0.003 \left( \frac{610 - 114.06}{114.06} \right) - 0.00048 \le 0.74(0.0161)$$

$$\varepsilon_{fe} = 0.0125 \le 0.0119 \text{ (not satisfied)}$$

Note that the values are close but the condition is not satisfied; check again after obtaining final value of $c$.

*Step 6c.* Calculate the strain in the existing reinforcing steel. The strain in the reinforcing steel can be calculated using similar triangles according to Equation 5.33b:

$$\varepsilon_s = (\varepsilon_{fe} + \varepsilon_{bi}) \left( \frac{d - c}{h - c} \right)$$

$$\varepsilon_s = (0.0125 + 0.00045) \left( \frac{546 - 114.06}{610 - 114.06} \right) = 0.0113$$

*Step 6d.* Calculate the stress level in the reinforcing steel and FRP. The stresses are calculated using Equation 5.13a and b, Equation 5.14a, b, and c, and Equation 5.25.

$$f_s = E_s \varepsilon_s \le f_y$$

$$f_s = (200 \text{ kN/mm}^2)(0.0113) \le 0.14 \text{ kN/mm}^2$$

$$f_s = 2.28 \text{ kN/mm}^2 \le 0.14 \text{ kN/mm}^2$$

Therefore, $f_s = 0.14 \text{ kN/mm}^2$.

$$f_{fe} = E_f \varepsilon_{fe}$$

$$f_{fe} = (38.6 \text{ kN/mm}^2)(0.0125) = 0.4825 \text{ kN/mm}^2$$

*Step 6e.* Calculate the internal force resultants and check equilibrium. Force equilibrium is verified by checking the initial estimate of $c$ with Equation 5.35 by noting that $c = a/\beta_1$.

$$c = \frac{A_s f_s + A_f f_{fe}}{0.85\beta_1 f_c' b}$$

$$c = \frac{(2026.83 \text{ mm}^2)(414 \text{ N/mm}^2) + (1017.27 \text{ mm}^2)(482.5 \text{ N/mm}^2)}{(0.85)(0.786)(38 \text{ N/mm}^2)(381 \text{ mm})}$$

$$= 137.49 \text{ mm}$$

$$c = 137.49 \text{ mm} \neq 144.06 \text{ mm (assumed)}$$

Therefore, revise the estimate of $c$ and repeat Step 6a through Step 6e until equilibrium is achieved. Using spreadsheet programming for iterations is suggested to obtain a hassle-free solution within a few seconds. Results of the final iteration are:

$$c = \frac{(2026.83 \text{ mm}^2)(414 \text{ N/mm}^2) + (1017.27 \text{ mm}^2)(407.9 \text{ N/mm}^2)}{(0.85)(0.786)(38 \text{ N/mm}^2)(381 \text{ mm})}$$

$$= 129.84 \text{ mm}$$

$$c = 129.84 \text{ mm (as assumed)}$$

Therefore, the value of $c$ selected for the final iteration is correct.

$$a = \beta_1 c = 0.786(129.84 \text{ mm}) = 102.05 \text{ mm}$$

$$c = 129.84 \text{ mm}$$

$$\varepsilon_s = 0.0095$$

$$f_s = f_y = 414 \text{ N/mm}^2$$

$$\varepsilon_{fe} = 0.0104 \text{ (see note below)}$$

$$f_{fe} = 407.9 \text{ N/mm}^2$$

Note:

$$\varepsilon_{fe} = 0.003 \left( \frac{h-c}{c} \right) - \varepsilon_{bi} \leq \kappa_m \varepsilon_{fu}$$

$$\varepsilon_{fe} = 0.003 \left( \frac{610 - 129.84}{129.84} \right) - 0.00045 \leq 0.74(0.0161)$$

$$\varepsilon_{fe} = 0.0106 \leq 0.0119 \text{ (satisfied)}$$

Note that the alternate approach for Step 6a to Step 6e is shown in Example 5.8 and Example 5.9.

*Step 7.* Calculate the design flexural strength of the section. The design flexural strength is calculated using Equation 5.4 and Equation 5.36 with appropriate substitutions. An additional reduction factor, $\psi_f = 0.85$, is applied to the bending strength contributed by FRP system. Because $\varepsilon_s = 0.0095 > 0.005$, a strength-reduction factor of $\phi = 0.90$ is used as per Equation 5.9:

$$\phi M_n = \phi\left[ A_s f_s\left(d - \frac{a}{2}\right) + \psi_f A_f f_{fe}\left(h - \frac{a}{2}\right)\right]$$

$$\phi M_n = 0.90\left[\begin{array}{c}(2026.83 \text{ mm})(414 \text{ N/mm}^2)\left(546 \text{ mm} - \dfrac{102.05}{2}\right) + \\ \\ (0.85)(1017.27 \text{ mm}^2)(407.9 \text{ N/mm}^2)\left(610 \text{ mm} - \dfrac{102.05}{2}\right)\end{array}\right]$$

$$\phi M_n = 551.2 \text{ kN-m} \geq M_u = 436.38 \text{ kN-m}$$

Therefore, the strengthened section is capable of sustaining the new required moment strength.

*Step 8.* Check the service stresses in the reinforcing steel and the FRP. Calculate the depth of the cracked neutral axis by summing the first moment of the areas of the elastic transformed section without accounting for the compression reinforcement as per Equation 5.28:

$$k = \sqrt{(\rho_s n_s + \rho_f n_f)^2 + 2\left(\rho_s n_s + \rho_f n_f\left(\frac{h}{d}\right)\right)} - (\rho_s n_s + \rho_f n_f)$$

$$k = \sqrt{(0.0668 + 0.0065)^2 + 2\left(0.0668 + 0.0065\left(\frac{610 \text{ mm}}{546 \text{ mm}}\right)\right)} - (0.0668 + 0.0065)$$

$$k = 0.3185$$

$$kd = (0.3185)(546 \text{ mm}) = 173.9 \text{ mm}$$

Calculate the stress level in the reinforcing steel using Equation 5.30h:

$$f_{s,s} = \frac{\left[M_s + \varepsilon_{bi} A_f E_f\left(h - \dfrac{kd}{3}\right)\right](d - kd)E_s}{A_s E_s\left(d - \dfrac{kd}{3}\right)(d - kd) + A_f E_f\left(h - \dfrac{kd}{3}\right)(h - kd)}$$

$$f_{s,s} = \cfrac{\left[\begin{array}{c} 295.29\times10^3 \text{ kN-mm} + (0.00045)(1017.27 \text{ mm}^2)(38.6 \text{ kN/mm}^2) \\ \left(610 \text{ mm} - \dfrac{173.9 \text{ mm}}{3}\right) \end{array}\right]}{\left[\begin{array}{c} (546 \text{ mm} - 173.9 \text{ mm})(210 \text{ kN/mm}^2) \\ (2026.83 \text{ mm}^2)(200 \text{ kN/m}^2)\left(546 \text{ mm} - \dfrac{173.9 \text{ mm}}{3}\right)(546 \text{ mm} - 173.9 \text{ mm}) \\ + (1017.27 \text{ mm}^2)(38.6 \text{ kN/mm}^2)\left(610 \text{ mm} - \dfrac{173.9 \text{ mm}}{3}\right) \\ (610 \text{ mm} - 173.9 \text{ mm}) \end{array}\right]}$$

$$f_{s,s} = 0.287 \text{ kN/mm}^2$$

Verify that the stress in steel is less than recommended limit as per Equation 5.10:

$$f_{s,s} \le 0.80f_y$$

$$f_{s,s} = 287 \text{ N/mm}^2 \le (0.80)(414 \text{ N/mm}^2) = 331.2 \text{ N/mm}^2$$

Therefore, the stress level in the reinforcing steel is within the recommended limit.
   Calculate the stress level in the FRP due to maximum service loads (assume sustained) using Equation 5.31d and verify that it is less than the creep-rupture stress limit.

$$f_{f,s} = f_{s,s}\left(\frac{E_f}{E_s}\right)\left(\frac{h - kd}{d - kd}\right) - \varepsilon_{bi}E_f$$

$$f_{s,s} = 0.287 \text{ kN/mm}^2\left(\frac{38.6 \text{ kN/mm}^2}{200 \text{ kN/mm}^2}\right)\left(\frac{610 \text{ mm} - 173.9 \text{ mm}}{546 \text{ mm} - 173.9 \text{ mm}}\right) -$$
$$(0.00045)(38.6 \text{ kN/mm}^2)$$

For a carbon FRP system, the creep-rupture stress limit is listed as per Table 5.2:

$$F_{f,s} = 0.55f_{fu}$$

$$f_{s,s} = 47.55 \text{ N/mm}^2 \le (0.55)(617.5 \text{ N/mm}^2)$$

$$f_{s,s} = 47.55 \text{ N/mm}^2 \le (339.625 \text{ N/mm}^2)$$

Therefore, the stress level in the FRP is within the recommended creep-rupture stress limit.

*Example 5.7*: If the beam in Example 5.6 was to be designed with a GFRP wrap system (normal grade) to carry the same increase in flexural loads due to future live loads, what changes in design are expected? Assume that the proposed GFRP wrap system (normal grade) has same properties (hypothetically) as those of the CFRP wrap system provided in Table 5.9.

*Design*: Though the properties of GFRP are hypothetically stated to be same as those of CFRP, design values with environmental reduction factor and creep-rupture stress limit values are lower for GFRP. Because it is tension-controlled failure without FRP rupture as noted in Example 5.1, we will verify the values of GFRP stress and strain at the ultimate and compare these with the design values (similar to Example 5.1). If stresses are within the design values, then check the creep-rupture strength, which is different for GFRP and CFRP systems.

Note that the procedures in Step 6 of Example 5.6 must be calculated with the new design values of the proposed GFRP because the design values of stress and strain are lower for the proposed GFRP as compared to CFRP in Example 5.6 due to lower value of $C_E$.

*Analysis and calculation details for GFRP wrap system.* FRP properties:

$$f_{fu} = C_E f_{fu}^*$$

$$f_{fu} = (0.75)(0.65 \text{ kN/mm}^2) = 0.4875 \text{ kN/mm}^2$$

$$\varepsilon_{fu} = C_E \varepsilon_{fu}^*$$

$$\varepsilon_{fu} = (0.75)(0.017 \text{ mm/mm}) = 0.0127 \text{ mm/mm}$$

From Step 6e of Example 5.6, the depth of the neutral axis and the stress and strain in steel and FRP:

$$c = \frac{A_s f_s + A_f f_{fe}}{0.85 \beta_1 f_c' b}$$

$$c = \frac{(2026.83 \text{ mm}^2)(414 \text{ N/mm}^2) + (1017.27 \text{ mm}^2)(407.91 \text{ N/mm}^2)}{(0.85)(0.786)(38 \text{ N/mm}^2)(381 \text{ mm})}$$

$$= 129.84 \text{ mm}$$

$$c = 129.84 \text{ mm (as assumed)}$$

$$a = \beta_1 c = 0.786(129.84 \text{ mm}) = 102.05 \text{ mm}$$

$$c = 129.84 \text{ mm}$$

$$\varepsilon_s = 0.0095$$

$$f_s = f_y = 414 \text{ N/mm}^2$$

$$\varepsilon_{fe}(0.0104 \text{ mm/mm}) < \varepsilon_{fe}(0.0127 \text{ mm/mm})$$

$$f_{fe} = 407.9 \text{ N/mm}^2 < 487.5 \text{ N/mm}^2$$

*Step 8.* Check the service stresses in the reinforcing steel and the FRP.

$$f_{f,s} = f_s \left( \frac{E_f}{E_s} \right) \left( \frac{h - kd}{d - kd} \right) - \varepsilon_{bi} E_f$$

$$f_{f,s} = 0.273 \text{ kN/mm}^2 \left( \frac{38.6 \text{ kN/mm}^2}{200 \text{ kN/mm}^2} \right) \left( \frac{610 \text{ mm} - 173.9 \text{ mm}}{546 \text{ mm} - 173.9 \text{ mm}} \right) -$$

$$(0.00048)(38.8 \text{ kN/mm}^2)$$

For a carbon FRP system, the creep-rupture stress limit is listed as per Table 5.2:

$$F_{f,s} = 0.55 f_{fu}$$

$$f_{f,s} = 44.76 \text{ N/mm}^2 \leq (0.2)(487.5 \text{ N/mm}^2)$$

$$f_{f,s} = 44.76 \text{ N/mm}^2 \leq (97.5 \text{ N/mm}^2)$$

Therefore, the stress level in the FRP is within the recommended creep-rupture stress limit.

*Example 5.8*: Analyze the design in Example 5.6 to determine the depth of the neutral axis for a balanced failure and the amount of FRP reinforcement needed for a balanced failure. Compare the calculated amount of FRP reinforcement for a balanced failure and compare it with the amount of FRP reinforcement provided to determine the possible failure mode for the FRP strengthened beam.

*Analysis*: The hypothetical balanced failure mode is assumed to occur when strains in extreme tension and compression fibers have reached their limit values simultaneously.

Strain in concrete: $\varepsilon_c = \varepsilon_{cu} = 0.003$

Strain in steel: $\varepsilon_s = \varepsilon_y$ (this is as a consequence of the strain in extreme FRP tension fiber reaching its ultimate value)

Strain in FRP: $\varepsilon_{frp} = \varepsilon_{frpu}$

These strain conditions at balanced failure can be expressed as follows from similar triangles principle based on Equation 5.42:

$$\frac{c_b}{h} = \frac{\varepsilon_{cu}}{\varepsilon_{cu} + \varepsilon_{bi} + \varepsilon_{frpu}} \quad \text{or} \quad a_b = \beta_1 h \left( \frac{\varepsilon_{cu}}{\varepsilon_{cu} + \varepsilon_{bi} + \varepsilon_{frpu}} \right)$$

where

$$\varepsilon_b = \varepsilon_{bi} + \varepsilon_{frpu}$$

$$\varepsilon_{bi} = \left( \frac{\varepsilon_{ci} + \varepsilon_{si}}{d} \right) h - \varepsilon_{ci} \quad \text{or} \quad \varepsilon_{bi} = \frac{M_{DL}(h - kd)}{I_{cr}E_c}$$

$$\varepsilon_{bi} = 0.00045 \text{ (from Step 4, Example 5.6)}$$

$$a_b = (0.786)(610) \left( \frac{0.003}{0.003 + 0.00045 + 0.0161} \right) = 73.6 \text{ mm}$$

Force equilibrium from Equation 5.43:

$$0.85 f_c' a_b b = A_s f_y + A_{frp} E_{frp} \varepsilon_{frpu}$$

The above equation is solved for $a_b$, where $a_b$ is the depth of the rectangular stress block corresponding to the balanced condition (Equation 5.44).

$$A_{frp,b} = \frac{0.85 \beta_1 f_c' b a_b - A_s f_y}{E_{frp} \varepsilon_{frpu}}$$

$$A_{frp,b} = \frac{0.85(38)(381)(73.6) - (2026.83)(414)}{(38,600)(0.0161)}$$

$$= 104.45 \text{ mm}^2 < 1017.27 \text{ mm}^2 \text{ (provided)}$$

The area of FRP needed for a balanced failure is 104.45 mm², which is much less than the provided area of 1017.27 mm². Hence, the failure mode is not tension-controlled with FRP rupture. This leaves the possible failure modes as:

- Tension-controlled without FRP rupture
- Tension-and-compression-controlled without FRP rupture
- Compression-controlled (no FRP rupture and no steel yield)

Actual failure modes can be verified as shown in Example 5.8.

*Example 5.9*: Analyze Example 5.6 to directly determine the depth of the neutral axis for the same specifications (i.e., avoid method of iterations).

*Solution*: Assuming the failure mode is tension-controlled without FRP rupture, use Equation 5.40 and calculate the depth of neutral axis $c$ where $c = a/\beta_1$. After obtaining $c$, verify if the conditions for the tension-controlled failure mode without FRP rupture are satisfied, i.e., $\varepsilon_s \geq 0.005$ and $\varepsilon_{frp} < \varepsilon_{frpu}$. From Equation 5.40,

$$0.85 f_c'ba^2 + [A_{frp}E_{frp}(\varepsilon_{cu} + \varepsilon_{bi}) - A_s f_y]a - \beta_1 A_{frp}E_{frp}\varepsilon_{cu}h = 0$$

Noting that the above equation is quadratic in $a$ (the idealized depth of the neutral axis), the following coefficients are computed:

$$a = \frac{-B \pm \sqrt{B^2 - 4AC}}{2A}$$

where

$$A = 0.85 f_c'b$$

$$B = [A_{frp}E_{frp}(\varepsilon_{cu} + \varepsilon_{bi}) - A_s f_y]$$

$$C = (-\beta_1 A_{frp}E_{frp}\varepsilon_{cu}h)$$

$$A = 0.85(38)(381) = (12,306.3)$$

$$B = [(1017.27)(38,600)(0.003 + 0.00045) - (2026.83)(414)] = -703,637.8$$

$$C = [-(0.775)(1017.27)(38,600)(0.003)(610)] = (-55,689,886.65)$$

$$a = \frac{-(-703,637.8) + \sqrt{(-703,637.8)^2 - 4(12,306.3)(-55,689,886.65)}}{2(12,306.3)} = 101.7 \text{ mm}$$

$$c = a/\beta_1 = 101.7/0.786 = 129.4 \text{ mm}$$

as obtained in Example 5.6 (129.84 mm). Note that the minor difference in value is due to computational round-offs.

*Verification.* Verify the strains in steel and FRP:

$$\varepsilon_s = \left(\frac{\varepsilon_{cu}}{c}\right)(d-c) = \left(\frac{0.003}{129.13}\right)(546-129.13) = 0.0096$$

$$\varepsilon_{fe} = \varepsilon_{frp} = \left(\frac{\varepsilon_{cu}}{c}\right)(h-c) - \varepsilon_{bi} = \left(\frac{0.003}{129.13}\right)(610-129.13) - 0.00045 = 0.01072$$

Based on strain values,

$$\varepsilon_s = 0.0096 \geq 0.005 \text{ (OK)}$$

$$\varepsilon_{fe} = \varepsilon_{frp} = 0.01072 \leq \varepsilon_{frpu}(0.017) \text{ (OK for CFRP)}$$

$$\varepsilon_{fe} = \varepsilon_{frp} = 0.01072 \leq \varepsilon_{frpu}(0.0127) \text{ (OK for GFRP)}$$

Hence, the assumed conditions are satisfied and the failure mode is tension-controlled without FRP rupture. Other calculations for stress and moment resistance remain same as shown in Example 5.6.

*Example 5.10*: Compute deflections for the CFRP strengthened beam in Example 5.6 for a combination of dead load ($\delta_{DL}$), live load ($\delta_{LL}$), and 20% of live load sustained with dead load $\delta_{DL+0.2LL}$. Compare the values obtained using transformed gross moment of inertia and gross moment of inertia. What difference is expected in the deflections if it were to be a GFRP strengthened beam described in Example 5.7?

$$b \text{ or } w = 381 \text{ mm}$$

$$d = 546 \text{ mm}$$

$$h = 610 \text{ mm}$$

$$l = 6.7 \text{ m}$$

$$n_s = 6.86$$

$$A_s = 2026.83 \text{ mm}^2$$

$$n_f = 1.32$$

$$A_f = 1017.27 \text{ mm}^2$$

$$n = 3 \text{ (no. of wraps)}$$

$$t_f = 0.89 \text{ mm/ply}$$

$$k \text{ (with wrap)} = 0.3185$$

$$c = kd \text{ (with wrap)} = 173.9 \text{ mm}$$

*Analysis*: Because the GFRP and CFRP properties in terms of area and stiffness are the same (hypothetically) in Example 5.6 and Example 5.7, the short-term deflection values computed in this example remain the same. However, part of the long-term deflections affected by FRP will be influenced by their contribution towards the creep coefficient values.

Gross section:

$$y = \frac{h}{2}$$

$$y = \frac{610}{2} = 305 \text{ mm}$$

$$I_{gr} = I_g = \frac{bh^3}{12}$$

$$I_g = \frac{(381)(610)^3}{12} = 7.2066 \times 10^6 \text{ mm}^4$$

$$f_r = 0.62\sqrt{f_c'}$$

$$f_r = 0.62\sqrt{38} = 3.8219 \text{ N/mm}^2$$

$$M_{cr} = \frac{I_g f_r}{y}$$

$$M_{cr} = \frac{(7.2066 \times 10^6)(3.8219)}{305} = 90.305 \times 10^6 \text{ N-mm}$$

$$I_{cr} = bc^3/3 + n_s A_s (d-c)^2 + n_f A_f (h + nt_f - c)^2$$

$$I_{cr} = (381)(173.9)^3/3 + (6.86)(2026.83)(546 - 173.9)^2$$
$$+ (1.32)(1017.27)(610 + 3 \times 0.89 - 173.9)^2 = 2.8515 \times 10^9 \text{ mm}^4$$

Note that typically, $nt_f$ is ignored in the cracked moment of inertia calculation due to the negligible difference in the values, as shown below.

$$I_{cr} = (381)(173.9)^3 / 3 + (6.86)(2026.83)(546 - 173.9)^2$$

$$+ (1.32)(1017.27)(610 - 173.9)^2 = 2.8484 \times 10^9 \text{ mm}^4$$

Transformed section:

$$\bar{y} = \frac{(bh^2 / 2) + (n_s - 1)A_s d + n_f A_f h}{bh + (n_s - 1)A_s + n_f A_f}$$

$$\bar{y} = \frac{[(381)(610)^2 / 2] + (6.86 - 1)(2026.83)(546) + (1.32)(1017.27)(610)}{(381)(610) + (6.86 - 1)(2026.83) + (1.32)(1017.27)}$$

$$y_t = (h - \bar{y})$$

$$y_t = (610 - 321.16) = 288.84 \text{ mm}$$

$$I_{gt} = \frac{bh^3}{12} + bh(\bar{y} - y)^2 + (n_s - 1)A_s(d - \bar{y})^2 + n_f A_f (h - \bar{y})^2$$

$$I_{gt} = \frac{(381)(610)^3}{12} + (381)(610)(321.16 - 305)^2 + (6.86 - 1)(2026.83)(546 - 321.16)^2$$

$$+ (1.32)(1017.27)(610 - 321.16)^2 = 8.37 \times 10^9 \text{ mm}^4$$

$$f_r = 0.62\sqrt{38} = 3.8219 \text{ N/mm}^2$$

$$M_{cr} = \frac{I_{gt} f_r}{y_t}$$

$$M_{cr} = \frac{(8.37 \times 10^9)(3.8219)}{288.84} = 110.75 \times 10^6 \text{ N-mm}$$

$I_{cr} = 2.8484 \times 10^9$ mm$^4$ (as calculated in gross section analysis)

Note that typically, $nt_f$ is ignored in the cracked moment of inertia calculation due to the negligible difference in the values, as shown below.

$$I_{cr} = (381)(173.9)^3 / 3 + (6.86)(2026.83)(546 - 173.9)^2$$
$$+ (1.32)(1017.27)(610 - 173.9)^2 = 2.8484 \times 10^9 \text{ mm}^4$$

Calculations based on transformed section analysis:

$$I_{cr} = 2.8484 \times 10^9 \text{ mm}^4$$

$$I_{gt} = 8.37 \times 10^9 \text{ mm}^4$$

$M_{cr} = 110.75 \times 10^6$ N-mm (transformed section analysis)

$$I_e = \left(\frac{M_{cr}}{M_a}\right)^3 I_{gt} + \left[1 - \left(\frac{M_{cr}}{M_a}\right)^3\right] I_{cr}$$

$$I_e = I_{cr} + \left(\frac{M_{cr}}{M_a}\right)^3 (I_{gt} - I_{cr})$$

Calculations based on gross section analysis:

$$I_{cr} = 2.8484 \times 10^9 \text{ mm}^4$$

$$I_{gr} = 7.2066 \times 10^6 \text{ mm}^4$$

$$M_{cr} = 90.305 \times 10^6 \text{ N-mm}$$

$$I_e = I_{cr} + \left(\frac{M_{cr}}{M_a}\right)^3 (I_{gr} - I_{cr})$$

The comparison of calculations based on gross and transformed section analyses is listed in Table 5.11 and Table 5.12.

Deflection requirements:

$$\frac{l}{180} = 37.22 \text{ mm (satisfied for all load conditions)}$$

$$\frac{l}{240} = 27.92 \text{ mm (satisfied for all load conditions)}$$

**TABLE 5.11**
**Moment of Inertia through Transformed and Gross Section Analysis**

| Load Combination | $M_a$ (kN-m) | $M_{cr}/M_a$ Transformed Section | $M_{cr}/M_a$ Gross Section | Effective Moment of Inertia $(I_e)$ (mm⁴) Transformed Section | Effective Moment of Inertia $(I_e)$ (mm⁴) Gross Section |
|---|---|---|---|---|---|
| $\delta_{DL}$ (before LL) | 90.16 | $M_{cr} > M_a$ | $M_{cr} > M_a$ | $8.37 \times 10^9$ | $7.2066 \times 10^9$ |
| $\delta_{LL}$ | 205.067 | 0.5400 | 0.4403 | $3.7178 \times 10^9$ | $3.2204 \times 10^9$ |
| $\delta_{DL+LL}$ | 295.29 | 0.3750 | 0.3058 | $3.1395 \times 10^9$ | $2.9730 \times 10^9$ |
| $\delta_{DL}$ (after LL) | | | $\delta_{DL} = \delta_{DL+LL} - \delta_{LL}$ | | |
| $\delta_{DL+0.2LL}$ | 131.173 | 0.8443 | 0.6884 | $6.1716 \times 10^9$ | $4.2702 \times 10^9$ |
| $\delta_{0.2LL}$ | | | $\delta_{0.2LL} = \delta_{DL+0.2LL} - \delta_{DL}$ | | |

**TABLE 5.12**
**Deflections through Transformed and Gross Section Analysis**

| Load Combination | Effective Moment of Inertia $(I_e)$ (mm⁴) Transformed Section | Effective Moment of Inertia $(I_e)$ (mm⁴) Gross Section | Deflection ($\delta$) (mm) Transformed Section | Deflection ($\delta$) (mm) Gross Section |
|---|---|---|---|---|
| $\delta_{DL}$ (before LL) | $8.37 \times 10^9$ | $7.2066 \times 10^9$ | 1.7665 | 2.0517 |
| $\delta_{LL}$ | $3.7178 \times 10^9$ | $3.2204 \times 10^9$ | 8.7905 | 10.1483 |
| $\delta_{DL+LL}$ | $3.1395 \times 10^9$ | $2.9730 \times 10^9$ | 14.9910 | 15.8305 |
| $\delta_{DL}$ (after LL) | $\delta_{DL} = \delta_{DL+LL} - \delta_{LL}$ | | 6.2005 | 5.6822 |
| $\delta_{DL+0.2LL}$ | $6.1716 \times 10^9$ | $4.2702 \times 10^9$ | 9.0121 | 6.4664 |
| $\delta_{0.2LL}$ | $\delta_{0.2LL} = \delta_{DL+0.2LL} - \delta_{DL}$ | | 2.8116 | 0.7842 |

$$\frac{l}{360} = 18.61 \text{ mm (satisfied for all load conditions)}$$

$$\frac{l}{480} = 13.95 \text{ mm (not satisfied for } \delta_{DL+LL} \text{ combination)}$$

Note 1: The difference in deflection values from the gross section and transformed section analyses are small. Hence, the gross section analysis is generally adequate.

Note 2: Higher crack opening and lower effective moment of inertia $(I_e)$ is observed when a full live load is applied as compared to a lower crack opening and relatively higher $I_e$ when only part of the live load is applied.

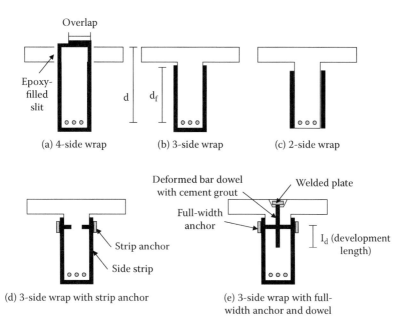

FIGURE 5.28 Wrapping configurations for increasing shear strength.

## 5.16  SHEAR BEHAVIOR OF WRAPPED CONCRETE MEMBERS

FRP wraps increase the shear strength of concrete beams and columns [Chajes et al. 1995; Fyfe et al. 1998; Khalifa et al. 1998; Sheheta et al. 1998]. The additional shear strength is obtained by orienting the fibers in a direction that is transverse to the axis of the concrete member or perpendicular to the shear cracks ACI SP-138 1993; NCHRP Report 514 2004].

### 5.16.1  WRAPPING CONFIGURATIONS FOR SHEAR STRENGTHENING

Beams and columns can be wrapped with FRP either partially (a few sides of a member) or fully (all around the member) to increase shear strength. Full wrapping of FRP fabric or tow sheet around a reinforced concrete member is most suitable for columns and piers, whereas beams cast integrally with slabs allow partial wrapping around three sides only. Wrapping on three sides is commonly referred to as a *U-wrap* (continuous) or *U-strip* (discrete) configuration. Different wrapping configurations commonly employed to increase shear strength are shown in Figure 5.28.

The FRP system can be installed continuously along the length of a member or in the form of discrete strips (Figure 5.28). Maximum shear resistance through wrapping can be achieved by wrapping as many sides of the beam as possible. Wrapping on all sides provides confinement-related benefits such as enhanced bending and shear strength for a reinforced concrete section.

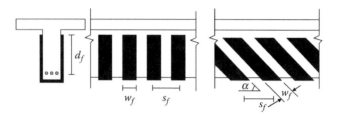

**FIGURE 5.29** Installing FRP wraps as straight or oriented strips.

## 5.16.2 ULTIMATE STRENGTH

The nominal shear strength of a concrete member strengthened with an FRP system should be greater than the required shear strength calculated by using Equation 11-1 of ACI 440.2R-02, which is:

$$\phi V_n \geq V_u \tag{5.46}$$

The strength reduction factor for shear $\phi = 0.85$ as recommended by ACI 440.2R-02 will be used in this chapter; however, the ACI 318-02 recommends a shear strength reduction factor of 0.75, in lieu of 0.85 as recommended by the ACI 318-99 code. The shear strength provided by the FRP wrap can be evaluated from forces resulting from the tensile stress in the FRP wrap, which depends on the fiber and crack orientation ($\alpha$) angle with reference to the longitudinal axis of a concrete beam. To simplify the analysis, the inclination of the crack ($\alpha$) is assumed to be 45°. The fibers are oriented at 45° (vertical strip) to the crack or at 90° (inclined strip) as shown in Figure 5.29. Calculations can be carried out using the actual crack inclination in Equation 5.50a. Tensile stress ($f_{fe}$) in the FRP shear reinforcement at ultimate is determined by calculating the effective strain in the FRP.

The nominal shear strength of a FRP strengthened concrete member can be determined using Equation 5.47. The contribution of the FRP system to shear strength $V_f$ with a reduction factor $\psi_f$ is the additional term in Equation 5.47, which is otherwise similar to that of ACI 318-02.

$$\phi V_n = \phi(V_c + V_s + \psi_f V_f) \tag{5.47}$$

Similar to Equation 11-3 of ACI 318-02,

$$V_c = \lambda \times 2.0\sqrt{f_c'} b_w d \tag{5.48}$$

where, $\lambda = 1$ for normal weight concrete.
From Equation 11-16 of ACI 318-02,

$$V_s = \frac{A_v f_y (\sin\alpha + \cos\alpha)d}{s} \tag{5.49}$$

**TABLE 5.13**
**Additional Reduction Factors ($\psi_f$) for FRP Shear Reinforcement**

$\psi_f = 0.95$     Four-side (completely) wrapped members
                   (contact critical applications)
$\psi_f = 0.85$     Three-sided U-wraps (bond critical applications)

*Source:* ACI 440.2R-02.

The shear strength contribution of the FRP wrap in the form of discrete strips or continuous fabric along the length of a beam can be calculated by using:

$$V_f = \frac{A_{fv}f_{fe}(\sin\alpha + \cos\alpha)d_f}{s_f} \qquad (5.50a)$$

where

$$A_{fv} = n(2t_f w_f) \qquad (5.50b)$$

$$f_{fe} = \varepsilon_{fe}E_f \qquad (5.50c)$$

ACI 440.2R-02 suggests using an additional reduction factor, $\psi_s$, for shear contribution from the FRP reinforcement as shown in Table 5.13.

### 5.16.3 EFFECTIVE STRAIN

The *effective strain* ($\varepsilon_{fe}$) in the FRP system used as the basis for the shear strength is defined as the maximum tensile strain that can be developed before shear failure of the section at ultimate loads. The effective strain is influenced by the failure mode of the FRP system and of the strengthened concrete member. The loss of concrete aggregate interlock is observed at fiber strain levels to be less than the fiber failure strain. In some members wrapped with FRP for shear strength enhancement, debonding was observed prior to or after the loss of aggregate interlock. To account for such observed failure mode, ACI 440.2R-02 limits the maximum effective strain used for design to 0.4% as shown by Equation 5.51a. This value is a fraction of the ultimate tensile failure strain of FRP, which is approximately 0.015 for carbon and 0.025 for glass. For FRP wrapping on two- and three-side bonding, ACI 440.2R-02 suggests an effective strain calculation by applying a bond-reduction coefficient, $\kappa_v$, applicable to shear. The bond-reduction coefficient $\kappa_v$ depends on the concrete strength, wrap strain and stiffness properties, and the thickness and number of wraps.

$$\varepsilon_{fe} = 0.004 \le 0.75\varepsilon_{fu}$$
(for completely wrapping around the member's cross section)    (5.51a)

$$\varepsilon_{fe} = \kappa_v\varepsilon_{fu} \le 0.004 \text{ (for U-wrapping or bonding to two sides)} \qquad (5.51b)$$

where

$$\kappa_v = \frac{k_1 k_2 L_e}{468\varepsilon_{fu}} \leq 0.75 \text{ (US Units)} \tag{5.52a}$$

$$\kappa_v = \frac{k_1 k_2 L_e}{11,900\varepsilon_{fu}} \leq 0.75 \text{ (SI Units)} \tag{5.52b}$$

The bond-reduction coefficient depends on two modification factors: $k_1$, that depends on concrete strength, and $k_2$, that depends on type of wrapping scheme to be employed in field. In addition, $\kappa_v$ depends on the ultimate strain of FRP and active bond length, $L_e$. The active bond length $L_e$ is the length over which the majority of the bond stress is developed.

$$k_1 = \left(\frac{f'_c}{4000}\right)^{2/3} \text{ (US Units)} \tag{5.53a}$$

$$k_1 = \left(\frac{f'_c}{27}\right)^{2/3} \text{ (SI Units)} \tag{5.53b}$$

$$k_2 = \begin{cases} \dfrac{d_f - L_e}{d_f} & \text{for 3-sided U wraps} \\[2ex] \dfrac{d_f - 2L_e}{d_f} & \text{for 2-side bonding} \end{cases} \text{ (US Units)} \tag{5.54a}$$

$$k_2 = \left(\frac{d_f - L_e}{d_f}\right) \text{ (SI Units)} \tag{5.54b}$$

$$L_e = \frac{2500}{(nt_f E_f)^{0.58}} \text{ (US Units)} \tag{5.55a}$$

$$L_e = \frac{23,300}{(nt_f E_f)^{0.58}} \text{ (SI Units)} \tag{5.55b}$$

In practice, mechanical anchorages have also been used to improve bonding and facilitate the transfer of large forces, and ACI 440.R2-02 suggests limiting effective strain values to a maximum of 0.004 even when additional anchorages are used.

### 5.16.4 Spacing

The spacing of FRP strips ($s_f$) is defined as the distance between the centerline of two consecutive strips. The height of the strips ($d_f$) should be taken as the distance between the centroid of tension reinforcement and the terminal point of the strip at the other end (Figure 5.29). The spacing of strips $s_f$ decreases with increasing ($V_n - V_c$). To ensure that a vertical FRP strip resists the potential diagonal crack, maximum spacing limitations similar to those in Section 11.5.4 of ACI 318-02 are recommended.

$$\text{If } (V_n - V_c) \le 4\sqrt{f_c'}b_w d: \text{ then } s_{max} = d/2 \le 24 \text{ in.} \tag{5.56}$$

$$\text{If } (V_n - V_c) > 4\sqrt{f_c'}b_w d: \text{ then } s_{max} = d/4 \le 24 \text{ in.} \tag{5.57}$$

If $(V_n - V_c) > 8\sqrt{f_c'}b_w d$ : then revise the section such that

$$(V_n - V_c) \le 8\sqrt{f_c'}b_w d \tag{5.58}$$

### 5.16.5 Total Shear Strength

The total shear strength of a FRP strengthened concrete member is given by Equation 5.47. Total shear reinforcement is the sum of steel and FRP reinforcement. The total shear reinforcement should be limited as per Section 11.5.6.9 of ACI 318-02, which is originally suggested for steel reinforcement alone as given by Equation 5.59:

$$V_s - V_f \le 8\sqrt{f_c'}b_w d \tag{5.59}$$

## 5.17 DESIGN EXAMPLES ON SHEAR

The design examples on the shear capacity of FRP-ER strengthened concrete beams provided in this section are based on spreadsheet values that show a higher accuracy than could be measured in the field through generally used equipment. Numbers with higher accuracy (i.e., several decimal places) are not truncated in the following examples. In many cases, designers could proceed with their design using two decimal places of accuracy and rounding of the final values as necessary. However, truncation of values such as those of strains to a few decimal places could lead to errors in the design and analysis.

### 5.17.1 Shear Strengthening with FRP-ER (U.S. Units)

*Example 5.11*: An interior reinforced concrete T-beam shown in Figure 5.30 ($f_c' = 5000$ psi) is required to carry live loads in excess of the original design. Assuming that the beam is safe in flexure due to additional live loads, design FRP wrap strips

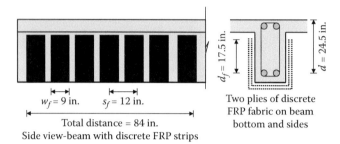

$w_f = 9$ in.     $s_f = 12$ in.

Total distance = 84 in.

Side view-beam with discrete FRP strips

$d_f = 17.5$ in.

$d = 24.5$ in.

Two plies of discrete
FRP fabric on beam
bottom and sides

**FIGURE 5.30** Shear reinforcement with discrete FRP strips.

### TABLE 5.14
### Manufacturer's Reported FRP System Properties

| | |
|---|---|
| Thickness per ply, $t_f$ | 0.075 in. |
| Ultimate tensile strength, $f_{fu}*$ | 505,000 psi |
| Rupture strain, $\varepsilon_{fu}*$ | 0.0153in./in. |
| Modulus of elasticity, $E_f$ | 33,000,000 psi |

### TABLE 5.15
### Configuration of the Supplemental FRP Shear Reinforcement

| | |
|---|---|
| $d$ | 24.5 in. |
| $d_f$ | 17.5 in. |
| Width of each sheet, $w_f$ | 9 in. |
| Span between each sheet, $s_f$ | 12 in. |

for increasing the shear strength by 25 kips. Use discrete carbon FRP strips for bonding the sides and bottom of the beam for a total distance of 84 in. from the starting point of bonding near the support to end point. The FRP system manufacturer's reported material properties are provided in Table 5.14.

*Design*: FRP shear reinforcement is designed as shown in Figure 5.30. Each FRP strip consists of two plies ($n = 2$) of carbon sheets installed by wet lay-up with dimensions as shown in Table 5.15.

*Shear design procedure and calculations for discrete FRP*:

*Step 1*. Design material properties. The beam is located in an interior location and CFRP fabrics will be used. Therefore, an environmental-reduction factor of 0.95 is used.

$$f_{fu} = C_E f_{fu}*$$

$$f_{fu} = (0.95)(505 \text{ ksi}) = 479.75 \text{ ksi}$$

$$\varepsilon_{fu} = C_E \varepsilon_{fu}^*$$

$$\varepsilon_{fu} = (0.95)(0.0153) = 0.0145$$

*Step 2.* Calculate the effective strain level in the FRP shear reinforcement. The effective strain in FRP U-wraps should be determined using the bond-reduction coefficient, $\kappa_v$. This coefficient can be computed using Equation 5.52a through Equation 5.55a.

$$L_e = \frac{2500}{(nt_f E_f)^{0.58}}$$

$$L_e = \frac{2500}{[(2)(0.0075 \text{ in./ply})(33 \times 10^6 \text{ psi})]^{0.58}} = 1.24 \text{ in.}$$

$$k_1 = \left(\frac{f_c'}{4000}\right)^{2/3}$$

$$k_1 = \left(\frac{5000 \text{ psi}}{4000}\right)^{2/3} = 1.16$$

$$k_2 = \left(\frac{d_f - L_e}{d_f}\right)$$

$$k_2 = \left(\frac{17.5 \text{ in.} - 1.16 \text{ in.}}{17.5 \text{ in.}}\right) = 0.93$$

$$k_v = \frac{k_1 k_2 L_e}{468 \varepsilon_{fu}} \leq 0.75$$

$$k_v = \frac{(1.16)(0.93)(1.24 \text{ in.})}{468(0.0145)} = \underline{0.197} \leq 0.75$$

The effective strain can then be computed using Equation 5.51b as follows:

$$\varepsilon_{fe} = k_v \varepsilon_{fu} \leq 0.004$$

$$\varepsilon_{fe} = 0.197(0.0145) = 0.0029 \leq 0.004$$

*Step 3.* Calculate the contribution of the FRP reinforcement to the shear capacity. The area of FRP shear reinforcement can be computed as follows:

$$A_{fv} = 2nt_f w_f$$

$$A_{fv} = 2(2)(0.0075 \text{ in.})(9 \text{ in.}) = 0.27 \text{ in.}^2$$

The effective stress in the FRP is:

$$f_{fe} = \varepsilon_{fe} E_f$$

$$f_{fe} = (0.0029)(33,000 \text{ ksi}) = 95.7 \text{ ksi}$$

The shear contribution of the FRP is calculated from Equation 5.50a:

$$V_f = \frac{A_{fv} f_{fe}(\sin\alpha + \cos\alpha)d_f}{s_f}$$

Note that the ply angle $\alpha = 90°$.

$$V_f = \frac{(0.27 \text{ in.}^2)(95.7 \text{ ksi})(1)(17.5 \text{ in.})}{(12 \text{ in.})}$$

$$V_f = 37.68 \text{ kips}$$

*Step 4.* Calculate the shear capacity of the section. The design shear strength is calculated from Equation 5.47 with $\psi_f = 0.85$ for FRP U-wraps.

$$\phi V_n = \phi(V_c + V_s + \psi_f V_f)$$

The wrap contribution is $\phi(\psi_f V_f)$. Note that the value of $\phi = 0.85$ is used as per ACI 440.R2-02.

$$= 0.85[(0.85)(37.68)]$$

$$= 27.22 \text{ kips} > 25 \text{ kips (required)}$$

Therefore, the strengthened section is capable of supporting an additional shear load due to the increased live load with $\phi = 0.85$.

With $\phi = 0.75$ as per ACI 318-02,

$$\phi V_n = \phi(V_c + V_s + \psi_f V_f)$$

$$= 0.75[(0.85)(37.68)]$$

$$= 24.02 \text{ kips} < 25 \text{ kips (required)}$$

Therefore, the strengthened section is not capable of supporting an additional shear load due to the increased live load with $\phi = 0.75$.

*Example 5.12*: If the CFRP fabric used in Example 5.11 is continuous instead of 9-in.-wide discrete strips (Figure 5.30), find the difference in shear capacity contributed by FRP.

*Analysis*: By substituting the width of the strip $w_f = 12$ in. in Step 3 of Example 5.11, we find that the shear contribution with continuous FRP to be 35.88 kips for $\phi = 0.85$ [ACI 440.R2-02] and 31.66 kips for $\phi = 0.75$ [ACI 318-02]. For comparison, all the steps shown in Example 5.11 are provided.

*Shear design procedure and calculations for continuous FRP fabric*:

*Step 1*. Compute the design material properties. The beam is located in an interior location and CFRP fabrics will be used. Therefore, an environmental-reduction factor of 0.95 is used.

$$f_{fu} = C_E f_{fu}*$$

$$f_{fu} = (0.95)(505 \text{ ksi}) = 479.75 \text{ ksi}$$

$$\varepsilon_{fu} = C_E \varepsilon_{fu}*$$

$$\varepsilon_{fu} = (0.95)(0.0153) = 0.0145$$

*Step 2*. Calculate the effective strain level in the FRP shear reinforcement. The effective strain in FRP U-wraps should be determined using the bond-reduction coefficient, $\kappa_v$. This coefficient can be computed using Equation 5.52a through Equation 5.55a.

$$L_e = \frac{2500}{(nt_f E_f)^{0.58}}$$

$$L_e = \frac{2500}{[(2)(0.00075 \text{ in./ply})(33 \times 10^6 \text{ psi})]^{0.58}} = 1.24 \text{ in.}$$

$$k_1 = \left( \frac{f'_c}{4000} \right)^{2/3}$$

$$k_1 = \left( \frac{5000 \text{ psi}}{4000} \right)^{2/3} = 1.16$$

$$k_2 = \left(\frac{d_f - L_e}{d_f}\right)$$

$$k_2 = \left(\frac{17.5 \text{ in.} - 1.16 \text{ in.}}{17.5 \text{ in.}}\right) = 0.929$$

$$k_v = \frac{k_1 k_2 L_e}{468 \varepsilon_{fu}} \leq 0.75$$

$$k_v = \frac{(1.16)(0.929)(1.24 \text{ in.})}{468(0.0145)} = \underline{0.197} \leq 0.75$$

The effective strain can then be computed using Equation 5.51b as follows:

$$\varepsilon_{fe} = k_v \varepsilon_{fu} \leq 0.004$$

$$\varepsilon_{fe} = 0.197(0.0145) = 0.0029 \leq 0.004$$

*Step 3.* Calculate the contribution of the FRP reinforcement to the shear capacity. The area of FRP shear reinforcement can be computed as follows:

$$A_{fv} = 2nt_f w_f$$

$$A_{fv} = 2(2)(0.0075 \text{ in.})(12 \text{ in.}) = 0.36 \text{ in.}^2$$

The effective stress in the FRP is:

$$f_{fe} = \varepsilon_{fe} E_f$$

$$f_{fe} = (0.0029)(33,000 \text{ ksi}) = 95.7 \text{ ksi}$$

The shear contribution of the FRP is calculated from Equation 5.50a:

$$V_f = \frac{A_{fv} f_{fe} (\sin \alpha + \cos \alpha) d_f}{s_f}$$

Note that the ply angle $\alpha = 90°$.

$$V_f = \frac{(0.36 \text{ in.}^2)(95.7 \text{ ksi})(1)(17.5 \text{ in.})}{(12 \text{ in.})}$$

$$V_f = 50.24 \text{ kips}$$

*Step 4.* Calculate the shear capacity of the section. The design shear strength is calculated from Equation 5.47 with $\psi_f = 0.85$ for U-wraps.

$$\phi V_n = \phi(V_c + V_s + \psi_f V_f)$$

The wrap contribution is $(\phi_f V_f)$. Note that the value of $\phi = 0.85$ is used as per ACI 440.R2-02.

$$= 0.85[(0.85)(50.24)]$$

$$= 36.29 \text{ kips} > 25 \text{ kips (required)}$$

Therefore, the strengthened section is capable of supporting an additional shear load due to the increased live load with $\phi = 0.85$.

With $\phi = 0.75$ as per ACI 318-02,

$$\phi V_n = \phi(V_c + V_s + \psi_f V_f)$$

$$= 0.75[(0.85)(50.24)]$$

$$= 32.03 \text{ kips} > 25 \text{ kips (required)}$$

Therefore, the strengthened section is capable of supporting an additional shear load due to the increased live load with $\phi = 0.75$.

## 5.17.2 SHEAR STRENGTHENING WITH FRP-ER (SI UNITS)

*Example 5.13*: An interior reinforced concrete T-beam shown in Figure 5.31 ($f_c' = 38 \text{ N/mm}^2$) is required to carry additional live loads in excess of the original design. Assuming the beam to be safe in flexure due to the additional live loads, design

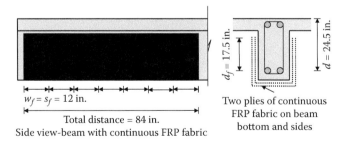

$d_f = 17.5$ in.   $d = 24.5$ in.

$w_f = s_f = 12$ in.

Total distance = 84 in.

Side view-beam with continuous FRP fabric

Two plies of continuous FRP fabric on beam bottom and sides

**FIGURE 5.31** Shear reinforcement with continuous FRP fabric.

### TABLE 5.16
### Manufacturer's Reported FRP
### System Properties

| | |
|---|---|
| Thickness per ply, $t_f$ | 0.254 mm |
| Ultimate tensile strength, $f_{fu}{}^*$ | 3481.9 N/mm² |
| Rupture strain, $\varepsilon_{fu}{}^*$ | 0.0153 mm/mm |
| Modulus of elasticity, $E_f$ | 227.53 kN/mm² |

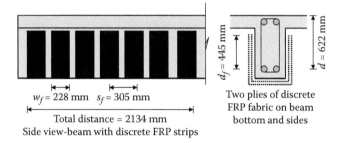

$w_f$ = 228 mm    $s_f$ = 305 mm

Total distance = 2134 mm
Side view-beam with discrete FRP strips

$d_f$ = 445 mm     $d$ = 622 mm

Two plies of discrete
FRP fabric on beam
bottom and sides

**FIGURE 5.32**  FRP shear reinforcement.

### TABLE 5.17
### Configuration of the Supplemental
### FRP Shear Reinforcement

| | |
|---|---|
| $d$ | 622 mm |
| $d_f$ | 445 mm |
| Width of each sheet, $w_f$ | 228 mm |
| Span between each sheet, $s_f$ | 305 mm |
| FRP strip length | 2134 mm |

FRP wrap strips for increasing the shear strength by 112 kN. Use discrete carbon FRP strips for bonding the sides and the bottom of the beam for a total distance of 2134 mm from the starting point of bonding near the support to the end point. The FRP system manufacturer's reported material properties are provided in Table 5.16.

*Design*: FRP shear reinforcement is designed as shown in Figure 5.32. Each FRP strip consists of two plies ($n = 2$) of carbon sheets installed by wet lay-up with dimensions as shown in Table 5.17.

*Step 1*. Compute the design material properties. The beam is located in an interior location and CFRP material will be used. Therefore, an environmental-reduction factor of 0.95 is used.

$$f_{fu} = C_E f_{fu}{}^*$$

$$f_{fu} = (0.95)(3481.9 \text{ N/mm}^2) = 3.31 \text{ kN/mm}^2$$

$$\varepsilon_{fu} = C_E \varepsilon_{fu}^*$$

$$\varepsilon_{fu} = (0.95)(0.0153) = 0.0145$$

*Step 2.* Calculate the effective strain level in the FRP shear reinforcement. The effective strain in FRP U-wraps should be determined using the bond-reduction coefficient, $\kappa_v$. This coefficient can be computed using Equation 5.52b through Equation 5.55b.

$$L_e = \frac{23,300}{(nt_f E_f)^{0.58}}$$

$$L_e = \frac{23,300}{[(2)(0.1905 \text{ mm/ply})(227,530 \text{ N/mm}^2)]^{0.58}} = 31.87 \text{ mm}$$

$$k_1 = \left( \frac{f_c'}{27} \right)^{2/3}$$

$$k_1 = \left( \frac{38 \text{ N/mm}^2}{27} \right)^{2/3} = 1.256$$

$$k_2 = \left( \frac{d_f - L_e}{d_f} \right)$$

$$k_2 = \left( \frac{445 \text{ mm} - 31.87 \text{ mm}}{445 \text{ mm}} \right) = 0.93$$

$$\kappa_v = \frac{k_1 k_2 L_e}{11,900 \varepsilon_{fu}^*} \leq 0.75$$

$$\kappa_v = \frac{(1.256)(0.93)(31.87 \text{ mm})}{11,900(0.0145)} = 0.2197 \leq 0.75$$

The effective strain can then be computed using Equation 5.51b as follows:

$$\varepsilon_{fe} = \kappa_v \varepsilon_{fu} \leq 0.004$$

$$\varepsilon_{fe} = (0.2197)(0.0145) = 0.0031 \leq 0.004 \ (OK)$$

*Step 3.* Calculate the contribution of the FRP reinforcement to the shear capacity. The area of FRP shear reinforcement can be computed as follows:

$$A_{fv} = 2nt_f w_f$$

$$A_{fv} = 2(2)(0.1905 \ \text{mm})(228 \ \text{mm}) = 173.73 \ \text{mm}^2$$

The effective stress in the FRP is:

$$f_{fe} = \varepsilon_{fe} E_f$$

$$f_{fe} = (0.0031)(227.53 \ \text{kN/mm}^2) = 0.705 \ \text{kN/mm}^2$$

The shear contribution of the FRP is calculated from Equation 5.50a:

$$V_f = \frac{A_{fv} f_{fe}(\sin\beta + \cos\beta)d_f}{s_f}$$

Note that the ply angle = 90°.

$$V_f = \frac{(173.73 \ \text{mm}^2)(0.705 \ \text{kN/mm}^2)(1)(445 \ \text{mm})}{(305 \ \text{mm})}$$

$$V_f = 178.70 \ \text{kN}$$

*Step 4.* Calculate the shear capacity of the section. The design shear strength is calculated from Equation 5.47 with $\psi_f = 0.85$ for U-wraps.

$$\phi V_n = \phi(V_c + V_s + \psi_f V_f)$$

The wrap contribution is $\phi(\psi_f V_f)$. Note that the value of $\phi = 0.85$ is used as per ACI 440.R2-02.

$$= 0.85[(0.85)(178.70)]$$

$$= 129.11 \ \text{kN} > 112 \ \text{kN} \ (\text{required})$$

Therefore, the strengthened section is capable of supporting an additional shear load due to the increased live load with $\phi = 0.85$.

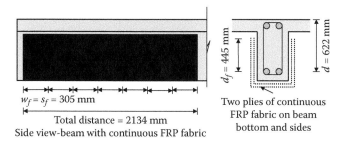

$w_f = s_f = 305$ mm

Total distance = 2134 mm

Side view-beam with continuous FRP fabric

$d_f = 445$ mm

Two plies of continuous
FRP fabric on beam
bottom and sides

$d = 622$ mm

**FIGURE 5.33** Shear reinforcement with continuous FRP fabric.

With $\phi = 0.75$ as per ACI 318-02,

$$\phi V_n = \phi(V_c + V_s + \psi_f V_f)$$

$$= 0.75[(0.85)(178.70)]$$

$$= 113.92 \text{ kN} > 112 \text{ kN (required)}$$

Therefore, the strengthened section is capable of supporting an additional shear load due to the increased live load with $\phi = 0.75$.

*Example 5.14*: If the CFRP fabric used in Example 5.13 is continuous (Figure 5.33) instead of 228.75-mm wide discrete strips, find the difference in shear capacity contributed by FRP.

*Analysis*: By substituting the width of the strip $w_f = 305$ mm in Step 3 of Example 5.13, we find that the shear contribution with continuous FRP to be 159.7 kN for $\phi = 0.85$ [ACI 440.R2-02] and 140.89kN for $\phi = 0.75$ [ACI 318-02]. For comparison, all the steps shown in Example 5.13 are provided.

*Shear design procedure and calculations for continuous FRP fabric*:

*Step 1*. Compute the design material properties. The beam is located in an interior location and CFRP fabrics will be used. Therefore, an environmental-reduction factor of 0.95 is used.

$$f_{fu} = C_E f_{fu}{}^*$$

$$f_{fu} = (0.95)(3479.45 \text{ N/mm}^2) = 3305.4 \text{ N/mm}^2$$

$$\varepsilon_{fu} = C_E \varepsilon_{fu}{}^*$$

$$\varepsilon_{fu} = (0.95)(0.0153) = 0.0145$$

*Step 2*. Calculate the effective strain level in the FRP shear reinforcement. The effective strain in FRP U-wraps should be determined using the bond-reduction coefficient, $\kappa_v$. This coefficient can be computed using Equation 5.52b through Equation 5.55b.

$$L_e = \frac{23,300}{(nt_f E_f)^{0.58}}$$

$$L_e = \frac{23,300}{[(2)(0.1905 \text{ mm/ply})(227,370 \text{ N/mm}^2)]^{0.58}} = 31.88 \text{ mm}$$

$$k_1 = \left(\frac{f_c'}{4000}\right)^{2/3}$$

$$k_1 = \left(\frac{34.45 \text{ N/mm}^2}{27}\right)^{2/3} = 1.17$$

$$k_2 = \left(\frac{d_f - L_e}{d_f}\right)$$

$$k_2 = \left(\frac{445 \text{ mm} - 31.88}{445 \text{ mm}}\right) = 0.93$$

$$k_v = \frac{k_1 k_2 L_e}{11,900\varepsilon_{fu}} \leq 0.75$$

$$k_v = \frac{(1.16)(0.93)(31.88 \text{ mm})}{11,900(0.0145)} = \underline{0.199} \leq 0.75$$

The effective strain can then be computed using Equation 5.51b as follows:

$$\varepsilon_{fe} = k_v \varepsilon_{fu} \leq 0.004$$

$$\varepsilon_{fe} = 0.199(0.0145) = 0.0029 \leq 0.004$$

*Step 3.* Calculate the contribution of the FRP reinforcement to the shear capacity. The area of FRP shear reinforcement can be computed as follows:

$$A_{fv} = 2nt_f w_f$$

$$A_{fv} = 2(2)(0.1905 \text{ mm})(305 \text{ mm}) = 232.41 \text{ mm}^2$$

The effective stress in the FRP is:

$$f_{fe} = \varepsilon_{fe}E_f$$

$$f_{fe} = (0.0029)(227{,}370 \text{ N/mm}^2) = 659.37 \text{ N/mm}^2$$

The shear contribution of the FRP is calculated from Equation 5.50a:

$$V_f = \frac{A_{fv}f_{fe}(\sin\alpha + \cos\alpha)d_f}{s_f}$$

Note that the ply angle $\alpha = 90°$.

$$V_f = \frac{(232.41 \text{ mm}^2)(659.37 \text{ N/mm}^2)(1)(445 \text{ mm})}{(305 \text{ mm})}$$

$$V_f = 223.59 \text{ kN}$$

*Step 4.* Calculate the shear capacity of the section. The design shear strength is calculated from Equation 5.47 with $\psi_f = 0.85$ for U-wraps.

$$\phi V_n = \phi(V_c + V_s + \psi_f V_f)$$

The wrap contribution is $\phi(\psi_f V_f)$. Note that the value of $\phi = 0.85$ is used as per ACI 440.R2-02.

$$= 0.85[(0.85)(223.59)]$$

$$= 161.54 \text{ kN} > 112 \text{ kN (required)}$$

Therefore, the strengthened section is capable of supporting an additional shear load due to the increased live load with $\phi = 0.85$.
  With $\phi = 0.75$ as per ACI 318-02,

$$\phi V_n = \phi(V_c + V_s + \psi_f V_f)$$

$$= 0.75[(0.85)(223.59)]$$

$$= 142.54 \text{ kN} > 112 \text{ kN (required)}$$

Therefore, the strengthened section is capable of supporting an additional shear load due to the increased live load with $\phi = 0.75$.

## NOTATION

| | |
|---|---|
| $A_f$ | Area of FRP external reinforcement (in.²) |
| | $A_f = nt_f w_f$ |
| $A_{fv}$ | Area of FRP shear reinforcement within spacing $s$ (in.²) |
| $A_g$ | Gross area of section (in.²) |
| $A_s$ | Area of nonprestressed steel reinforcement (in.²) |
| $a$ | Depth of equivalent rectangular stress block (in.) |
| $a_b$ | $a$ for balanced failure (in.) |
| $b$ | Width of a rectangular cross section (in.) |
| $c$ | Distance from extreme compression fiber to the neutral axis (in.) |
| $c_b$ | $c$ for a balanced failure (in.) |
| $C_E$ | Environmental reduction factor |
| $d$ | Distance from extreme compression fiber to the centroid of the nonprestressed steel tension reinforcement (in.) |
| $d_f$ | Depth of FRP shear reinforcement (in.) |
| $E_c$ | Modulus of elasticity of concrete (psi) |
| $E_f$ | Tensile modulus of elasticity of FRP (psi) |
| $E_s$ | Modulus of elasticity of steel (psi) |
| $f_c$ | Compressive stress in concrete (psi) |
| $f_c$ | Specified compressive strength of concrete (psi) |
| $\sqrt{f_c'}$ | Square root of specified compressive strength of concrete, psi |
| $f_f$ | Stress level in the FRP reinforcement (psi) |
| $f_{f,s}$ | Stress level in the FRP caused by a moment within the elastic range of the member (psi) |
| $F_{f,s}$ | Creep-rupture stress limit in the FRP (psi) |
| $f_{fe}$ | Effective stress in the FRP; stress level attained at section failure (psi) |
| $f_{frpu}$ | Stress in the FRP-ER at the point of incipient rupture |
| $f_{fu}^*$ | Ultimate tensile strength of the FRP material as reported by the manufacturer (psi) |
| $f_{fu}$ | Design ultimate tensile strength of FRP (psi) |
| $\overline{f_{fu}}$ | Mean ultimate tensile strength of FRP based on a population of 20 or more tensile tests per ASTM D3039 (psi) |
| $f_s$ | Stress in nonprestressed steel reinforcement (psi) |
| $f_{s,s}$ | Stress level in nonprestressed steel reinforcement at service loads (psi) |
| $f_y$ | Specified yield strength of nonprestressed steel reinforcement (psi) |
| $h$ | Overall thickness (height) of a member (in.) |
| $k$ | Ratio of the depth of the neutral axis to the reinforcement depth measured on the same side of neutral axis |
| $k_f$ | Stiffness per unit width per ply of the FRP reinforcement (lb/in.) = $E_f t_f$ |
| $k_1$ | Modification factor applied to $\kappa_v$ to account for the concrete strength |
| $k_2$ | Modification factor applied to $\kappa_v$ to account for the wrapping scheme |
| $L_e$ | Active bond length of FRP laminate (in.) |

| $M_n$ | Nominal moment strength (lb-in.) |
|---|---|
| $M_s$ | Moment within the elastic range of the member due to service loads (lb-in.) |
| $M_u$ | Factored moment at section (lb-in.) |
| $M_T$ | Moment capacity in tension-controlled failure (lb-in.) |
| $M_{TF,UL}$ | Upper limit on moment capacity in tension-controlled failure with FRP rupture (lb-in.) |
| $M_{T,UL}$ | Upper limit on moment capacity in tension-controlled failure without FRP rupture (lb-in.) |
| $M_{TC,UL}$ | Upper limit on moment capacity in tension-and-compression-controlled failure without FRP rupture (lb-in.) |
| $n$ | Number of plies of FRP reinforcement |
| $n_f$ | Modular ratio of elasticity of FRP reinforcement to concrete $= E_f/E_c$ |
| $n_s$ | Modular ratio of elasticity of reinforcing steel to concrete $= E_s/E_c$ |
| $p_{fu}*$ | Tensile strength per unit width per ply of the FRP reinforcement (lb/in.) $= f_{fu}*t_f$ |
| $t_f$ | Nominal thickness of one ply of the FRP reinforcement (in.) |
| $V_c$ | Nominal shear strength provided by concrete with steel flexural reinforcement (lb) |
| $V_n$ | Nominal shear strength (lb) |
| $V_s$ | Nominal shear strength provided by steel stirrups (lb) |
| $V_f$ | Nominal shear strength provided by FRP stirrups (lb) |
| $V_u$ | Required shear strength based on factored loads (lb) |
| $w$ | Width of a rectangular cross section (also $b$) (in.) |
| $w_f$ | Width of the FRP reinforcing plies (in.) |
| $\beta_1$ | Ratio of the depth of the equivalent rectangular stress block to the depth to the neutral axis |
| $\varepsilon_b$ | Strain level in the concrete substrate developed by a given bending moment at the FRP bonded location (tension is positive) (in./in.) |
| $\varepsilon_{bi}$ | Strain level in the concrete substrate at the FRP bonded location at the time of the FRP installation (tension is positive) (in./in.) |
| $\varepsilon_c$ | Strain level in the concrete (in./in.) |
| $\varepsilon_{cu}$ | Maximum usable compressive strain of concrete (in./in.) |
| $\varepsilon_f$ | Strain level in the FRP reinforcement (in./in.) |
| $\varepsilon_{fe}$ | Effective strain level in FRP reinforcement; strain level attained at section failure (in./in.) |
| $\varepsilon_{fu}*$ | Ultimate rupture strain of the FRP reinforcement (in./in.) |
| $\varepsilon_{fu}$ | Design rupture strain of FRP reinforcement (in./in.) |
| $\overline{\varepsilon_{fu}}$ | Mean rupture strain of FRP reinforcement based on a population of 20 or more tensile tests per ASTM D3039 (in./in.) |
| $\varepsilon_s$ | Strain level in the nonprestressed steel reinforcement (in./in.) |
| $\varepsilon_{sy}$ | Strain corresponding to the yield strength of nonprestressed steel reinforcement (in./in.) |
| $\phi$ | Strength-reduction factor |

$\kappa_m$       Bond-reduction coefficient for flexure
$\kappa_v$       Bond-reduction coefficient for shear
$\rho_f$       FRP reinforcement ratio
$\sigma$       Standard deviation
$\psi_f$       Additional FRP strength-reduction factor

## REFERENCES

ACI SP-138, Shear Capacity of RC and PC Beams Using FRP Reinforcement, Detroit, MI: American Concrete Institute, 1993.

ACI 224.1R Causes, Evaluation, and Repair of Cracks in Concrete Structures, Detroit, MI: American Concrete Institute.

ACI 224R Control of Cracking in Concrete Structures, Detroit, MI: American Concrete Institute.

ACI 318 Building Code Requirements for Reinforced Concrete and Commentary, Detroit, MI: American Concrete Institute, 2002.

ACI 318-99, American Concrete Institute Building Code Requirements for Reinforced Concrete, Detroit, MI: American Concrete Institute, 1999.

ACI 364.1R Guide for Evaluation of Concrete Structures Prior to Rehabilitation, Detroit, MI: American Concrete Institute.

ACI 437R Strength Evaluation of Existing Concrete Buildings, Detroit, MI: American Concrete Institute.

ACI 440.1R-03: Guide for the Design and Construction of Concrete Reinforced with FRP Bars, Detroit, MI: American Concrete Institute, 2003.

ACI 440.2R-02: Design and Construction of Externally Bonded FRP Systems for Strengthening Concrete Structures, Detroit, MI: American Concrete Institute, 2002.

ACI 503.1-92 - 503.4-92: Use of Epoxy Specifications (4 volumes), Detroit, MI: American Concrete Institute.

ACI 546R Concrete Repair Guide, Detroit, MI: American Concrete Institute.

ACMBS-III, Advanced Composite Materials for Bridges and Structures, Ottawa, Ontario, Canada, 2000.

Ahmad, S.H. and Shah, S.P., Stress-strain curves of concrete confined by spiral reinforcement, ACI Journal, 79(6): 484–490, 1982.

Alkhrdaji T. and Thomas J., Upgrading Parking Structures: Techniques and Design Considerations, Bulletin-The Construction Specifier, Structural Group Preservation Systems, Hanover, MD, 2005.

Arduini, M. and Nanni, A., Behavior of pre-cracked RC beams strengthened with carbon FRP sheets, *Journal of Composites in Construction*, 1(2): 63–70, 1997.

Arockiasamy, M. and Zhuang, M., Durability of Concrete Beams Pretensioned with CFRP Tendons, *Proceedings of the International Conference on Fiber Reinforced Structural Plastics in Civil Engineering at Indian Institute of Technology*, Madras, December, pp.18–20, 1995.

Baaza, I.M., Missihoun, M., and Labossiere, P., Strengthening of reinforced concrete beams with CFRP sheets, First International Conference on Composites in Infrastructures ICCI 96, Tucson, pp. 746–759, January 1996.

Barger, J.D., Effect of Aging on the Bond between FRP and Concrete, M.S. thesis, West Virginia University, Morgantown, 2000.

Benmokrane, B.and Rahman, H. (Eds.), *CDCC-1998, Proceedings of 1st International Conference on Durability of Fiber Reinforced Polymer (FRP) Composites for Construction (CDCC 98)*, Canada, August 1998.

CFR-49: Code of Federal Regulations: Chapter C Transportation.

Chajes, M.J., Januzska, T.F., Mertz, D.R., Thomson, T.A. Jr., and Flench, W.W. Jr., Shear strengthening of reinforced concrete beams using externally applied composite fabrics, *ACI Structural Journal*, 92(3): 295–303, 1995.

Crasto, A.S., Kim, R.Y., Fowler, C., and Mistretta, J.P., Rehabilitation of concrete bridge beams with externally-bonded composite plates. Part I, First International Conference on Composites in Infrastructures ICCI 96, Tucson, pp. 857–869, January 1996.

Dolan, C., FRP development in the United States, fiber-reinforced-plastic reinforcement for concrete structures: properties and applications, *Developments in Civil Engineering*, 42, 129–163, 1993.

Fyfe, E.R., Gee, D.J., and Milligan, P.B., Composite systems for seismic applications, *Concrete International*, 20(6): 31–33, 1998.

GangaRao, H.V.S. and Thippeswamy, H.K., A Conference on Polymer Composites — Infrastructural Renewal and Economic Development, April 19-21, Parkersburg, West Virginia, in *Advanced Polymer Composites for Transportation Infrastructure*, Creese, R.C. and H. GangaRao, H.V.S. (Eds.), Lancaster, PA: Technomic Publishing Company, pp. 3–20, 1999.

GangaRao, H.V.S. and Vijay, P.V., Bending behavior of concrete beams wrapped with carbon fabric, *Journal of Structural Engineering*, 124(1): 3–10, 1998.

GangaRao, H.V.S. and Vijay, P.V., Draft design guidelines for concrete beams externally strengthened with FRP, NIST Workshop on Standards Development for the Use of Fiber Reinforced Polymers for the Rehabilitation of Concrete and Masonry Structures, 1998, Tucson, Duthinh, D. (Ed.), NISTIR 6288, pp. 3/113–120, January 1999.

GangaRao, H.V.S., Vijay, P.V., and Barger, J.D., Durability of Reinforced Concrete Members Wrapped with FRP Fabrics (Phase III) — Effects of Aging on Bond between FRP and Concrete, CFC Report #00-280, West Virginia Department of Transportation (WVDOH) Report, RP #T-699-CIDDMOC, Charleston, WV, May 2000.

Grace, N.F., Ragheb, W.F., and Sayed, G.A., Ductile FRP strengthening systems, *Concrete International*, January, pp. 31–36, 2005.

Green, M. and Bisby, L.A., Effects of freeze-thaw action on the bond of FRP sheets to concrete, *Proceedings of the International Conference on Durability of Fiber Reinforced Polymer (FRP) Composite for Construction*, Montreal, Quebec, Canada, pp.179–190, May 1998.

Hollaway L., Adhesive and bolted joints, in *Polymers and Polymer Composites in Construction*, London: Thomas Telford Ltd., 1990.

Homam, S.M. and Sheikh, S.A., Durability of fibre reinforced polymers used in concrete structures, *Proceedings of the Third International Conference on Advanced Composite Materials in Bridges and Structures*, Ottawa, Ontario, Canada, pp.751–758, August 2000.

Hota, V.S., GangaRao, Faza, S.S., and Vijay, P.V., Behavior of Concrete Beams Wrapped with Carbon Tow Sheet, CFC report No.95-196, April, submitted to Tonen Corporation, Tokyo, 1995.

ICCI-1998, Fiber composites in infrastructure, *Proceedings of the Second International Conference on Composites in Infrastructure*, ICCI'98, Tucson, AZ.

ICRI 03730, Guide for Surface Preparation for the Repair of Deteriorated Concrete Resulting from Reinforcing Steel Corrosion.

ICRI 03733, Guide for Selecting and Specifying Materials for Repairs of Concrete Surfaces.

Ichimasu, H., Maruyama, M., Watanabe, H., and Hirose, T., RC slabs strengthened by bonded carbon FRP plates, Part-1 Laboratory Study and Part-2 Applications, *Fiber Reinforced-Plastic Reinforcement for Concrete Structures*, International Symposium, A. Nanni and C. W. Dolan, Eds., ACI SP-138, pp. 933–970, 1993.

Iyer S.L., Kortikere A., and Khubchnadani A., Concrete and Sand Confined with Composite Tubes, Materials for the New Millennium, *Proceedings of the Fourth Materials Engineering Conference*, Washington, D.C., November 10–14, Chong K.P. (Ed.), ASCE, pp. 1308–1319, 1996.

Javed, S., Accelerated Aging in Concrete Beams Externally Bonded with Carbon Fiber Tow sheet, CFC report No.96-239, West Virginia University, Morgantown, 1996.

JSCE, Application of Continuous Fiber Reinforcing Materials to Concrete Structures, JSCE Subcommittee on Continuous Fiber Reinforcing Materials, Concrete Library International of JSCE, No. 19, Tokyo, 1992

Karbhari, V. M. and Seible, F., Design Considerations for the Use of Fiber Reinforced Polymeric Composites in the Rehabilitation of Concrete Structures, *NIST Workshop on Standards Development for the Use of Fiber Reinforced Polymers for the Rehabilitation of Concrete and Masonry Structures*. Proceedings January 7–8, 1998, Tucson, AZ, D. Duthinh (Ed.), NISTIR 6288, pp. 3/59-72, 1999.

Khalifa, A., Gold, W., Nanni, A., and Abel-Aziz, M., Contribution of externally bonded FRP to the shear capacity of RC flexural members, *Journal of Composites in Construction*, 2(4): 195–203, 1998.

Kshirsagar, S., Lopez, A. R., and Gupta, R. K., Durability of Fiber Reinforced, Composite Wrapping for the Rehabilitation of Concrete Piers, *Proceedings of CDCC 98*, Montreal, pp. 117–128, 1998.

Lee, Y.J., Boothby, T.E., Nanni, A., and Bakis, C.E., Tension Stiffening Model for FRP Sheets Bonded to Reinforced Concrete, *Proceedings, ICCI98*, Tucson, AZ, January 5–7, 1998.

MacGregor, G. J., *Reinforced Concrete Mechanics and Design*, Upper Saddle River, NJ: Prentice Hall, 1998.

Machida, A., (Ed.), State-of-the-art report on continuous fiber reinforcing materials, *Concrete Engineering Series 3*, Research Committee on Continuous Fiber Reinforcing Materials, Japan Society of Civil Engineering, Tokyo, 164 pp.

Maeda, T., Asano, Y., Sato, Y., Ueda, T., and Kakuta, Y., A study on bond mechanisms of carbon fiber sheets, in *Non-Metallic (FRP) Reinforcement for Concrete Structures*, Proceedings of the Third Symposium, Vol. 1, pp. 279–286, 1997.

Mallick, P. K., *Fiber Reinforced Composite Materials, Manufacturing and Design*, New York: Marcel Dekker, 1993.

Malvar, L., Warren, G., and Inaba, C., Rehabilitation of Navy Pier Beams With Composite Sheets, *Second FRP International Symposium*, Non-Metallic (FRP) Reinforcements for Concrete Structures, Gent, Belgium, pp. 533–540, 1995.

Mander, J. B., Priestly, M. J. N., and Park, R., Observed stress-strain behavior of confined concrete, *Journal of Structural Engineering*, 114 (8): 1827–1849, 1988.

Marshall, O. S., Sweeney, S. C., and Trovillion, J. C., Seismic Rehabilitation of Unreinforced Masonry Walls, *Fourth International Symposium Fiber Reinforced Polymer Reinforcement for Reinforced Concrete Structures*, ACI SP-188, pp. 287–295, 1999.

Master Builders-Design Guidelines, Design Manual, Master Builders, Cleveland, Ohio, 2005.

McConnel, V. P., Infrastructure update, *High Performance Composites*, pp. 21–25, 1995.

Meier, U., Carbon fiber-reinforced polymers: modern materials in bridge engineering, Reprint from *Structural Engineering International*, International Association for Bridge and Structural Engineering, CH-8093, Zurich, pp. 7–12, 1992.

Meier, U., Deuring, M., Meier, H., and Schwegler, G., Strengthening of structures with advanced composites, in *Alternative Materials for the Reinforcement and Prestressing of Concrete*, J. L. Clarke (Ed.), Blackie Academic and Professional, 1993.

Mufti, A., Erki, M. A., and Jaeger, L. (Eds.), *Advanced Composite Materials with Application to Bridges*, Canadian Society of Civil Engineers, Montreal, 1991.

Nawy, E.G., *Reinforced Concrete: Fundamental Approach*, Upper Saddle River, NJ: Prentice Hall, 2002.

NCHRP Report 514, Bonded Repair and Retrofit of Concrete Structures Using FRP Composites: Recommended Construction Specifications and Process Control Manual, Transportation Research Board, 2004.

Neale, K. W. and Labossière, P., State-of-the-art report on retrofitting and strengthening by continuous fibre in Canada, *Non-Metallic (FRP) Reinforcement for Concrete Structures*, Japan Concrete Institute, Tokyo, Vol. 1, pp. 25–39, 1997.

Nikolaos, P., Triantafillou, T. C., and Veneziano, D., Reliability of RC members strengthened with CFRP laminates, *Journal of Structural Engineering*, 121(7): 1037–1044, 1995.

Oehlers, D. J., Reinforced concrete beams with plates glued to their soffits, *Journal of Structural Engineering*, 118(8): 2023–2038, 1992.

Pantelides, C. P., Gergely, J., Reaveley, L. D., and Volnyy, V. A., Retrofit of RC bridge pier with CFRP advanced composites, *Journal of Structural Engineering*, 125(10): 1094–1099, 1999.

Park, R. and Paulay, T., *Reinforced Concrete Structure*, New York: John Wiley & Sons, 1975.

Priestly, M. J. N., Seible, F., and Fyfe, E., Column seismic retrofit using fiberglass/epoxy jackets, in *Advanced Composite Materials in Bridges and Structures*, K. W. Neale and P. Labossiere (Eds.), Canadian Society for Civil Engineering, pp. 287–298, 1992.

Razaqpur G. and Ali, A. M., A New Concept for Achieving Ductility in FRP-Reinforced Concrete, ICCI '96, *Proceedings of First International Conference on Composites in Infrastructure*, Tucson, AZ, pp. 401–413, 1996.

Rodriguez, M. and Park, R., Seismic load tests on reinforced concrete columns strengthened by jacketing, *ACI Structural Journal*, 91(2): 150–159, 1994.

Saadatmanesh, H. and Ehsani, M. R., RC beams strengthened with GFRP plates-I: experimental study, *Journal of Structural Engineering*, 117(11): 3417–3433, 1991.

Sen R., Mullins G., Winters D., Suh K., Underwater Repair of Corroded Piles, Structural Faults and Repair, 11th International Conference and Exhibits, Edinburgh, UK, June 2006.

Sharif, A., Al-Sulaimani, G.J., Basunbul, I.A., Baluch, M.H., and Ghaleb, B.N., Strengthening of initially loaded reinforced concrete beams using FRP plates, *ACI Structural Journal*, 2, 160–168, 1994.

Sheheta, E., Morohy, R., and Rizkalla, S., Use of FRP as shear reinforcement for concrete structures, in Proceedings of ICCI '98, M. R. Eshani and H. Saadatmanesh (Eds.), pp. 300–315, 1998.

Sika Design Manual, Engineering Guidelines for the Use of Sika Carbodur (CFRP) Laminates for Structural Strengthening of Concrete Structures, Sika Corporation, 1997.

Sonobe, Y., Fukuyama, H., Okamoto, T., Kani, N., Kimura, K., Kobayashi, K., Masuda, Y., Matsuzaki, Y., Nochizuki, S., Nagasaka, T., Shimizu, A., Tanano, H., Tanigaki, M., and Teshigawara, M., Design guidelines of FRP reinforced concrete building structures, *Journal of Composites for Construction*, 90–115, 1997.

Soudki, K. A., and Green, M. F., Freeze-thaw response of CFRP wrapped concrete, *Concrete International*, 19(8): 64–67, 1997.

Swamy, R.N., Jones, R., and Bloxham, J.W., Structural behavior of rReinforced concrete beams strengthened by epoxy-bonded steel plates, *The Structural Engineer*, 65A(2): 1987.

Tarricone P., Composite sketch, ASCE, 65(5): 52–55, May 1995.

Tavakkolizadeh M. and Saadatmanesh H., Galavanic corrosion of carbon and steel in aggressive environments, *Journal of Composites for Construction*, ASCE, 5(3): 200–210, August 2001.

Teng J.G., Debonding Failures of RC Beams Flexurally Strengthended with Externally Bonded FRP Reinforcement, Structural Faults and Repair, 11th International Conference and Exhibits, Edinburgh, UK 13–15, June 2006.

Teng, J.G., Smith, S.T., Yao, J., and Chen, J.F., Intermediate crack induced debonding in RC beams and slabs, *Construction and Building Materials*, 17(6–7): 447–462, 2001.

Tonen-Design Guidelines, FORCA TOW SHEET Technical Manual, Tonen Corporation, Japan, 1996.

Vijay, P.V., Aging Behavior of Concrete Beams Reinforced with GFRP Bars, Ph.D. dissertation, West Virginia University, Morgantown, 1999.

Vijay, P.V. and GangaRao, H.V.S., , Development of Fiber Reinforced Plastics for Highway Application: Aging Behavior of Concrete Beams Reinforced with GFRP bars, *CFC-WVU Report No. 99-265 (WVDOH RP #T-699-FRP1)*, 202 pp., 1999

Vijay, P.V. and GangRao, H.V.S., Accelerated aging and durability of GFRP bars, *Proceedings of the First International Conference on Durability of Fiber Reinforced Polymer (FRP) Composite for Construction*, Montreal, Quebec, Canada, July, 1998.

Vijay, P.V. GangRao, H.V.S., Webster, K., and Prachasaree, W., Durability of concrete beams with fiber wraps, *Proceedings of the Second International Conference on Durability of Fiber Reinforced Polymer (FRP) Composite for Construction*, Montreal, Quebec, Canada, pp.51–62, May 2002.

Wan, B., Sutton, M., Petrou, M.F., Harries, K.A., and Li, N., Investigation of bond between FRP and concrete undergoing global mixed mode I/II loading, *ASCE Journal of Engineering Mechanics*, 130(12): 1467–1475, 2004.

Wang, C.K., *Reinforced Concrete Design*, 5d ed., New York: Harper Collins, 1992.

Wheat, H.G., Karpate H., Jirsa J.O., Fowler D.W., and Whitney D.P., Examination of FRP Wrapped Columns and Beams in a Corrosive Environment, Structural Faults and Repair, 11th International Conference and Exhibits, Edinburgh, UK, 13–15, June 2006.

# 6 Design and Behavior of Internally FRP-Reinforced Beams

## 6.1 INTRODUCTION

In Chapter 5, we discussed design and analysis for the strengthening of conventionally reinforced concrete beams by wrapping them with FRP fabrics. We shall refer to them as *externally reinforced beams*; those beams were (internally) reinforced with steel reinforcement. In this chapter, we shall discuss the design and analysis of concrete beams internally reinforced with FRP bars instead of steel reinforcing bars. Such beams are often referred to as *internally FRP-reinforced beams*.

Strength, stiffness, and bond characteristics of FRP bars (as a substitute for steel reinforcement) in concrete beams have been extensively researched for understanding their flexural behavior and reported in the literature [ACMBS-II 1996; ACMBS-III 2000; FRPRCS-IV 1999; ICCI 1996]. FRP reinforcement is highly suitable for structures subjected to corrosive environments. These include concrete bridge decks and other superstructures, concrete pavements treated with deicing salts, waste water and chemical treatment plants, and structures built in or close to sea water (e.g., seawalls and offshore structures). The lightweight nature of FRPs is a distinct advantage for weight sensitive structures. In facilities and structures supporting magnetic resonance imaging (MRI) units or other equipment sensitive to electromagnetic fields, which require low electric conductivity or electromagnetic neutrality, the nonmagnetic characteristics of FRP reinforcement are very useful.

When using FRP reinforcement that exhibits nonductile behavior, its use should be limited to structures that will benefit from properties such as its noncorrosive or nonconductive characteristics. Note that FRP reinforcement is not recommended at this stage for continuous structures, such as moment frames or zones where moment redistribution is required, because of the lack of experience and research data on its use for such structures.

This chapter presents a comprehensive discussion on design and behavior of *FRP-reinforced* concrete beams, i.e., concrete beams that are reinforced with FRP bars (GFRP, CFRP, or AFRP) to resist tension rather than steel bars that are used in conventionally reinforced concrete beams (a topic discussed in texts on reinforced concrete design with which readers might be familiar). Typically, any one of the three types of commercially available FRP bars can be used as tensile reinforcement for concrete beams: glass fiber reinforced polymer (GFRP) bars, carbon fiber reinforced polymer (CFRP) bars, and aramid fiber reinforced polymer (AFRP) bars. The terms "FRP-reinforced" and "steel-reinforced" beams are used throughout this chapter to distinguish them from each other. A parallel exists in the analysis and design of these two types of beams; to that extent, a knowledge of the analysis and design of steel-reinforced concrete beams and familiarity with *Building Code Requirements for Structural Concrete (ACI 318) and Commentary (ACI 318R)* (referred to as ACI 318 in this chapter) will be very helpful in understanding the discussion that follows.

Recognizing that FRP reinforcing bars differ from steel reinforcing bars in many respects including mechanical properties is very important. FRP bars consist of continuous longitudinally aligned fibers embedded in a matrix and are characterized by a high tensile strength in the longitudinal direction only (i.e., in the direction of fibers). As discussed in Chapter 2, FRP materials are inherently anisotropic, and this anisotropy affects the shear strength, dowel action, and bond properties of FRP bars. Because fibers buckle easily when loaded in compression, FRP reinforcement is not recommended as reinforcing material in designing columns or other compression members. For the same reasons, FRP reinforcement bars are unsuitable for use as compression reinforcement in flexural members. Therefore, the discussion in this chapter is limited to singly reinforced concrete members only. Doubly reinforced FRP beams are not discussed here because the compression strength of reinforcing bars is not well established yet.

## 6.2 ADVANTAGES AND LIMITATIONS OF STEEL AND FRP REINFORCEMENTS

FRP reinforcement offers many advantages because of their many desirable characteristics as compared to those of conventional steel reinforcement. These advantages are as follows:

1. Corrosion resistance
2. High longitudinal (unidirectional) strength
3. High fatigue endurance (varies with type of reinforcing fiber and bar)
4. Magnetic transparency
5. Lightweight (about one-fifth to one-fourth the density of steel)
6. Low thermal and electric conductivity (for glass and aramid fibers)

However, the FRP reinforcements do have limitations on their applications because of the following undesirable characteristics:

1. Many types of FRP bars do not exhibit yielding before rupture (i.e., they are brittle)
2. Low modulus of elasticity (varies with fiber type)

3. Low transverse strength
4. Low shear strength
5. Reduced durability in moist, acid/salt, and alkaline environments
6. High coefficient of thermal expansion perpendicular to the fibers relative to concrete
7. Fire resistance can be less than adequate, depending on the type of matrix used for producing FRP bars

## 6.3 DESIGN PHILOSOPHY

FRP-reinforced concrete structures can be designed using both strength and working stress design approaches. The emphasis in this chapter is placed on the strength method as suggested in the ACI Guide [ACI 440.1R-03]. Provisions of strength design method as per *ACI 318 Building Code Requirements for Structural Concrete and Commentary* (1995, 1999, and 2002 editions) are followed in this chapter.

Typically, steel-reinforced concrete sections are designed as tension-controlled, which ensures the yielding of steel reinforcement. Such beams, if loaded beyond a certain curvature limit, will lead to secondary compression failure, manifested by the crushing of concrete. The yielding behavior of steel reinforcement is referred to as *ductile*, which provides a warning of impending failure. Because most of the FRP reinforcements do not exhibit a yield plateau, the traditional design approaches used for designing steel-reinforced concrete beams must be suitably modified to account for the linear stress-strain behavior of the FRP bars up to bar rupture. Note that some types of FRP reinforcement bars, such as those with helical ribs or hybrid fibers, e.g., a combination of carbon and glass fibers, exhibit quasi- or pseudo-ductility [Bakis et al. 1996; Belarbi et al. 1996; Harris et al. 1998; Huesgen 1997; Nanni et al. 1994a, 1994b; Razaqpur and Ali 1996; Somboonsong 1997; Tamuzs et al. 1996]. However, these types of bars have not been used so far for large-scale field applications, and as such they are not discussed in this book.

Characteristically, in the case of a tension failure of a FRP-reinforced concrete member, the FRP reinforcement ruptures followed by sudden failure of the member. Large crack-widths and deflections can be observed prior to the rupture of FRP bars. A more gradual and less catastrophic failure with a higher deformability factor (discussed in Section 6.9) has been observed in compression failures of FRP-reinforced concrete members. Recognizing that as compared to tension failure, the compression failure of FRP-reinforced concrete sections have exhibited improved ductile and plastic behavior is important.

To account for the lack of traditional ductility in a FRP-reinforced concrete member and to safeguard against the long-term degradation of strength under different exposure conditions, the chosen failure mode should satisfy both strength and serviceability criteria in addition to providing for a higher reserve strength to carry loads even beyond the design loads. Although a compression failure mode may be preferable in some cases such as T-beams, achieving it could require a disproportionately high area of reinforcement with redundant reserve strength and higher cost. Hence, both tension and compression failure modes are acceptable for design as long as the serviceability criteria and long-term performance criteria are satisfied.

### 6.3.1 DESIGN ASSUMPTIONS

The following assumptions are made for design and analysis of FRP-reinforced concrete beams. These assumptions are very similar to those used for designing concrete members with conventional steel reinforcement, except that the stress-strain behavior of FRP reinforcement is linear and that it fails in a brittle mode:

1. A plane section remains plane before and after loading.
2. Strains in concrete and FRP reinforcement are proportional to the distance from the neutral axis.
3. Concrete and FRP reinforcement have a good interfacial bond with no relative slip.
4. FRP reinforcement exhibits a linear stress-strain behavior up to failure (i.e., the reinforcement response follows Hooke's law).
5. The maximum usable compressive strain in the concrete is assumed to be 0.003.
6. The tensile strength of concrete is very small and is ignored.

### 6.3.2 STRENGTH LIMIT STATES AND STRENGTH REDUCTION FACTORS

The load factors given in ACI 318 are also used to determine the required strength of a FRP-reinforced concrete member. Appendix B of ACI 318 recommends a strength reduction factor $\phi$ of 0.70 for a steel-reinforced concrete member with failure controlled by concrete crushing. The same strength reduction factor ($\phi = 0.7$) is recommended for concrete crushing failures in FRP-reinforced concrete members. However, a reduction factor of 0.5 is recommended for tension failures manifested by FRP reinforcement rupture because tension-controlled failures exhibit less ductility than those exhibiting concrete crushing.

Although theoretical delineation of the concrete crushing failure mode of concrete beams is possible, the actual member may not fail as predicted. For example, if the concrete strength is higher than specified, the member can fail due to FRP rupture. Similarly, if the strength of the FRP reinforcement degrades as a result of long-term exposure to a harsh environment, the member can fail due to FRP rupture. If concrete crushing is the desired failure mode, then to ensure concrete crushing, a reinforcement ratio of $1.4\rho_{fb}$ is recommended by the ACI Committee 440, where $\rho_{fb}$ is the balanced steel ratio for FRP reinforcement as discussed in Section 6.5.2.1. Experimental and statistical analyses of FRP-reinforced concrete beams failing in compression mode suggest that such beams should have $1.33\rho_{fb}$ as the minimum reinforcement ratio to ensure compression failure [Vijay and GangaRao 1999]. ACI 440 recommends the use of a linearly interpolated transition value of $\phi$ between $\rho_f$ and $1.4\rho_{fb}$ such that $0.5 \leq \phi \leq 0.7$. For this reason, ACI 440 defines a section controlled by concrete crushing as a section in which $\rho_f \geq 1.4\rho_{fb}$, and a section controlled by

**TABLE 6.1**
**Environmental Reduction Factors for the Tensile Strength,**
**$f_{fu}^*$, of Various Fiber Types and Exposure Conditions**

| Exposure Condition | Fiber Type | Environmental Reduction Factor ($C_E$) |
|---|---|---|
| Concrete not exposed to earth | Carbon | 1.0 |
| and weather | Glass | 0.8 |
| | Aramid | 0.9 |
| Concrete exposed to earth | Carbon | 0.9 |
| and weather | Glass | 0.7 |
| | Aramid | 0.8 |

*Source*: ACI 440.1R-03.

FRP rupture as one in which $\rho_f < \rho_{fb}$. The relationships between the strength reduction factor and the FRP reinforcement ratio is expressed by Equation 6.1:

$$\phi = \begin{cases} 0.50 & \text{for } \rho_f \leq \rho_{fb} \\ 0.5\left(\dfrac{\rho_f}{\rho_{fb}}\right) & \text{for } \rho_{fb} < \rho_f < 1.4\rho_{fb} \\ 0.70 & \text{for } \rho_f \geq 1.4\rho_{fb} \end{cases} \tag{6.1}$$

### 6.3.3 MATERIAL PROPERTIES FOR DESIGN

#### 6.3.3.1 Strength of Straight FRP Bars

In contrast to the properties of conventional construction materials (e.g., steel, concrete, wood), certain mechanical properties of FRP bars — such as the tensile strength, creep-rupture strength, and fatigue endurance — degrade owing to the long-term exposure to environmental conditions. Therefore, when designing FRP-reinforced concrete members, using environmental reduction factors, $C_E$, on the manufacturer-specified tensile strength, $f_{fu}^*$ (see Table 6.1) is recommended (ACI 440). These reduction factors for different FRP types and exposure conditions are assumed to be conservative.

The design tensile strengths and rupture strain of FRP reinforcement recommended by ACI 440 are given, respectively, by Equation 6.2 and Equation 6.3:

$$f_{fu} = C_E f_{fu}^* \tag{6.2}$$

where

$f_{fu}$ = the design tensile strength of FRP, considering reductions for service environment (psi)

$C_E$ = the environmental reduction factor, given in Table 6.1 (ACI 440) for various fiber types and exposure conditions

$f_{fu}^*$ = the guaranteed tensile strength of an FRP bar, defined as the mean tensile strength of a sample of test specimens minus three times the standard deviation, $f_{fu,avg} - 3\sigma$ (psi)

$$\varepsilon_{fu} = C_E \varepsilon_{fu}^* \qquad (6.3)$$

where

$\varepsilon_{fu}$ = the design rupture strain of FRP reinforcement

$\varepsilon_{fu}^*$ = the guaranteed rupture strain of FRP reinforcement, defined as the mean tensile strain at failure of a sample of test specimens minus three times the standard deviation, $\varepsilon_{u,avg} - 3\sigma$

Because the same environmental reduction factor is applied to both the stress and strain, the stiffness (i.e., modulus of elasticity, $E_f$) of FRP reinforcement under different exposure conditions remains unaffected. The stiffness of FRP reinforcement has been found to be relatively unaffected by environmental exposure as compared to the strength properties [Vijay and GangaRao 1999]. Therefore, applying the environmental reduction factor on the specified stiffness of the FRP reinforcement is not necessary.

It is recommended that the FRP reinforcement be used at temperature levels at least 25°F below the glass transition temperature, $T_g$.

### 6.3.3.2  Strength of FRP Bars at Bends

According to JSCE [1997], the design tensile strength of FRP bars in the bent portion is given by:

$$f_{fb} = \left(0.05\frac{r_b}{d_b} + 0.3\right)f_{fu} \leq f_{fu} \qquad (6.4)$$

where

$f_{fb}$ = the design tensile strength of the bend of FRP bar (psi)

$r_b$ = the radius of the bend (in)

$d_b$ = the diameter of reinforcing bar (in)

$f_{fu}$ = the design tensile strength of FRP

### 6.3.3.3  Shear

The Canadian Highway Bridge Design Code [CSA 1996] limits the tensile strain in FRP shear reinforcement to 0.002 in/in. The Eurocrete Project provisions limit the value of the shear strain in FRP reinforcement to 0.0025 in./in. [Dowden and Dolan 1997].

## 6.4 FLEXURAL BEHAVIOR AND FAILURE MODES OF RECTANGULAR FRP-REINFORCED BEAMS

### 6.4.1 CONCEPT OF TENSION AND COMPRESSION FAILURE MODES

Theoretically, reinforced concrete beams and slabs can be designed for tension, balanced, or compression failure modes. Traditionally, steel-reinforced concrete beams and slabs are designed for tension failure to take advantage of the elastic-plastic behavior of steel. But unlike steel, the FRP reinforcement does not exhibit this elasto-plastic behavior; instead, it exhibits linear stress-strain behavior (i.e., the plot of stress-strain relationship is a straight line) to failure. Therefore, when designing FRP-reinforced concrete members, one needs to consider: (a) the mechanics of failure or the failure mode, (b) the magnitude and nature of energy absorption, (c) and the physical and chemical interaction between the FRP bars and concrete, e.g., bond and alkalinity effects. Delineation of failure modes and the corresponding analytical expressions are presented in the subsequent sections.

### 6.4.2 BALANCED FAILURE: THE CONCEPT

A balanced failure mode represents an idealized condition that assumes that strains in concrete and GFRP bars reach their predefined limiting strain values simultaneously, i.e., $\varepsilon_c = \varepsilon_{cu} = 0.003$ and $\varepsilon_f = \varepsilon_{fu} = f_{fu}/E_f$, respectively, in concrete and FRP bars. Although this condition is difficult to achieve in practice, it does represent a limit state in delineating the tension and compression failure modes.

#### 6.4.2.1 (c/d) Ratio Approach for Balanced Strain Condition in Rectangular and Nonrectangular Concrete Beams

Concrete sections are said to be in a balanced strain condition, manifested by the simultaneous attainment of predefined ultimate compressive strain in concrete and tensile strain in FRP reinforcement. The ratio of neutral axis depth, $c_b$, to the effective depth of beam, $d$ (i.e., the $c_b/d$ ratio) for a balanced strain condition is determined based on the assumption of a linear strain distribution along the depth of the beam between the extreme compression fibers and the tensile FRP reinforcement (see Figure 6.1).

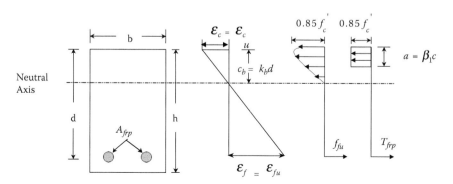

**FIGURE 6.1** Stress and strain distribution on a balanced beam section.

Referring to the strain distribution shown in Figure 6.1, we obtain

$$\left(\frac{c_b}{d}\right) = \left(\frac{\varepsilon_{cu}}{\varepsilon_{cu} + \varepsilon_f}\right) \tag{6.5}$$

where the strain in the FRP can be obtained from Hooke's law:

$$\varepsilon_f = \frac{f_{fu}}{E_f} \tag{6.6}$$

where
    $f_{fu}$ = the ultimate tensile strength of FRP reinforcement (ksi)
    $E_f$ = the modulus of elasticity of FRP tensile reinforcement (ksi)

The ratio $c_b/d$ in Equation 6.5 plays an important role in the analysis of FRP-reinforced members as it defines the transition from a tension-controlled failure mode to a compression-controlled failure mode:

1. The tension failure of FRP bars in concrete members, which is manifested by their sudden rupture, is expected when the ratio $(c/d) < (c_b/d)$, and
2. By concrete crushing (i.e., compression failure) when $(c/d) > (c_b/d)$.

Note that this approach is valid for both rectangular and nonrectangular sections. The substitution of $\varepsilon_{cu} = 0.003$ for concrete in Equation 6.5 yields

$$\left(\frac{c_b}{d}\right) = \frac{0.003}{0.003 + (f_{fu}/E_f)} \tag{6.7}$$

## 6.4.2.2  Balanced (Percentage) Reinforcement Ratio Approach for a Rectangular Concrete Beam

The balanced reinforcement ratio, $\rho_b$, in a concrete rectangular beam is defined as the ratio of tension reinforcement, $A_f$ (corresponding to the balanced failure condition), to the cross-sectional area of beam, $bd$. Referring to the equivalent stress distribution diagram shown in Figure 6.1, the force in compression block of concrete, $C$, and tensile force in the FRP reinforcement, $T$, can be expressed as

$$C = 0.85f_c'ab \tag{6.8}$$

$$T = A_f f_{fu}$$

Equating the tensile and compressive forces for horizontal force equilibrium yields

$$0.85f'_c ab = A_f f_{fu}$$

Rearranging and dividing throughout by $d$,

$$\frac{A_f}{bd} = \frac{0.85f'_c}{f_{fu}}\left(\frac{a}{d}\right) \tag{6.9}$$

In Equation 6.9, the depth of compressions block, $a$, is given by

$$a = \beta_1 c \tag{6.10}$$

The value of coefficient $\beta_1$ depends on the compressive strength of concrete and is defined by ACI 318 as follows:

$$\beta_1 = 0.85 \text{ for } f'_c \leq 4000 \text{ psi}$$

$$\beta_1 = 0.85 - 0.05\left(\frac{f'_c - 4000}{1000}\right) \text{ for } f'_c \geq 4000 \text{ psi} \tag{6.11}$$

$$\beta_1 \text{ not } < 0.65$$

Substitution of $a = \beta_1 c$ in Equation 6.9 yields

$$\frac{A_f}{bd} = \frac{0.85f'_c\beta_1}{f_{fu}}\left(\frac{c}{d}\right) \tag{6.12}$$

Substituting $c = c_b$ and $\rho = \rho_{fb} = \frac{A_f}{bd}$ for the balanced condition, Equation 6.12 yields

$$\rho_{fb} = 0.85\beta_1 \frac{f'_c}{f_{fu}}\left(\frac{c_b}{d}\right) \tag{6.13}$$

Substitution for $c_b/d$ from Equation 6.5 in Equation 6.13 yields

$$\rho_{fb} = 0.85\beta_1 \frac{f'_c}{f_{fu}}\left(\frac{\varepsilon_{cu}}{\varepsilon_{cu} + \varepsilon_{fu}}\right) \tag{6.14}$$

where, $\rho_{fb}$ is the balanced reinforcement ratio of a rectangular concrete beam. For a tension failure mode, $\rho < \rho_{fb}$, and for a compression failure mode, $\rho > \rho_{fb}$. Substituting the compression failure limiting strain value of concrete $\varepsilon_{cu} = 0.003$, Equation 6.14 can be written as

$$\rho_{fb} = \frac{0.85\beta_1 f_c'}{f_{fu}} \frac{0.003E_f}{f_{fu} + 0.003E_f} \qquad (6.15)$$

### 6.4.3  TENSION FAILURE MODE

Tension failure in FRP-reinforced concrete beams is manifested by the rupture of bars in the tensile zone of the beam. In the case of steel-reinforced concrete beams, the tension failure mode is indicated by steel yielding, which leads to either the secondary compression failure of concrete or the rupture of tension in steel. Note that the rupture of steel bars in tension is highly unlikely, except in cases of highly under-reinforced beam sections. In FRP-reinforced concrete sections, the primary failure modes are either tension failure manifested by bar rupture or primary compression failure manifested by concrete crushing. The depth of neutral axis in FRP-reinforced beams designed for tension failure can be as low as 13% and 16.67% of the effective beam depth, respectively, corresponding to failure strains of 2% (i.e., $\varepsilon_f = 0.02$) and 1.5% (i.e., $\varepsilon_f = 0.015$) in FRP bars, which can be verified from Equation 6.7. In addition to the tension or compression failure modes, FRP beams can also fail in other modes, such as shear failure, concrete cover failure, debonding, and creep-rupture and fatigue of FRP bars.

A tension failure of a FRP members is initiated by the rupture of FRP bars, which occurs when $\rho_f < \rho_{fb}$. The nominal strength of such a member can be determined by either linear or rectangular stress-distribution approaches. A discussion on these two approaches follows.

### 6.4.3.1  Linear Stress Distribution ($f_c < f_c'$)

The assumption that the stress in concrete is less that its compressive strength and linear stress distribution is valid. This condition represents *allowable stress design* (ASD) or the working stress design approach. The moment resistance of a beam under this condition can be determined as the moment due to couple formed by tensile and compressive forces (they are both equal) acting on the beam cross section. Referring to Figure 6.2,

$$\text{Tensile stress resultant} = A_f f_{fu}$$

$$\text{Compressive stress resultant} = 0.5 f_c bkd$$

For the horizontal equilibrium of forces acting on the beam cross section,

$$\text{Tensile stress resultant} = \text{Compressive stress resultant}$$

$$A_f f_{fu} = 0.5 f_c bkd \qquad (6.16)$$

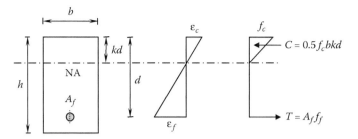

**FIGURE 6.2** Linear stress distribution in a reinforced concrete beam.

The compressive stress in concrete due to service loads can be expressed from Equation 6.16 as

$$f_c = \frac{f_{fu}A_f}{0.5bkd} = \frac{f_{fu}\rho_f}{0.5k} \tag{6.17}$$

In Equation 6.16 and Equation 6.17, $k$ is the ratio of the depth of the neutral axis to the centroid of tensile reinforcement. Its value can be determined from statics as given by Equation 6.18:

$$k = \sqrt{(\rho_f n_f)^2 + 2\rho_f n_f} - \rho_f n_f \tag{6.18}$$

Discussion on determination of $k$ is presented in Chapter 5 and not repeated here.

Referring to Figure 6.2, the lever arm between the tensile and compressive stress resultants is $d - kd/3$. The moment resistance of the beam can be determined as a couple formed by the compressive and tensile force resultants. Thus, using tensile force,

$$M_{R,ASD} = A_f f_{fu}(d - kd/3) \tag{6.19}$$

Using compressive force,

$$M_{R,ASD} = 0.5 f_c bkd^2 \left(1 - \frac{k}{3}\right) \tag{6.20}$$

where $M_{R,ASD}$ = ASD-level moment resistance.

### 6.4.3.2 Rectangular Stress Distribution ($f_c = f_c'$)

When using the rectangular stress distribution approach, the conventional ACI rectangular stress block (as used in ACI 318 strength design approach, where concrete strength is equal to $f_c'$) cannot be used because the maximum concrete strain may not reach the specified value of 0.003 as in the case of beams having balanced or compression failure. However, based on the strength design approach of ACI 318, the nominal strength of a beam can be determined in a manner similar to that of steel-reinforced beams with some modification.

ACI 440 suggests that for a given section, the quantity $\beta_1 c$ in Equation 6.8 and Equation 6.9 varies with the material properties and the FRP reinforcement ratio (for steel-reinforced concrete beams, $a = \beta_1 c$). The maximum value of this quantity corresponding to a balanced failure (which corresponds to a concrete strain of 0.003) is equal to $\beta_1 c_b$. Therefore, the nominal moment strength of the beam can be determined by taking the moment of tensile stress resultant (in the FRP bars) about the compression stress resultant in concrete:

$$M_n = A_f f_{fu} \left( d - \frac{\beta_1 c_b}{2} \right) \tag{6.21}$$

In Equation 6.21, the depth of compression block ($a = \beta_1 c_b$) is uncertain as the "rectangular" shape is merely an approximation of the parabolic stress distribution. The analysis of a FRP beam failing in tension involves two unknowns: the strain in concrete, $\varepsilon_c$, and the depth of neutral axis, $c$. Also, the depth of the rectangular compression stress block (ACI 440 refers to this in terms of an unknown factor, $\alpha_1$, the ratio of the average concrete stress, $f_c$, to concrete strength, $f_c'$) and the ratio of the depth of the rectangular compression stress block and the depth of neutral axis (i.e., factor $\beta_1$) are also unknown. In view of this uncertainty, ACI 440 recommends using a multiplier of 0.8 to the value given by Equation 6.21 as a conservative approach for determining the nominal strength of a FRP-reinforced member, resulting in Equation 6.22:

$$M_n = 0.8 A_f f_{fu} \left( d - \frac{\beta_1 c_b}{2} \right) \tag{6.22}$$

where the depth of a neutral axis at the balanced failure, $c_b$, can be determined from Equation 6.23, based on the strain diagram corresponding to the balanced failure condition (Figure 6.1):

$$c_b = \left( \frac{\varepsilon_{cu}}{\varepsilon_{cu} + \varepsilon_{fu}} \right) d \tag{6.23}$$

### 6.4.4 COMPRESSION FAILURE MODE

Compression failure in a FRP-reinforced beam is defined as one in which the FRP bars do not fail. The beam failure is manifested by concrete crushing, typically under the load points and spalling in the midsection. Theoretically, the compression failure of an FRP-reinforced beam can be ensured by providing a reinforcement ratio higher than that required for the balanced failure case (so that tension failure of FRP bars is precluded).

The minimum reinforcement required to ensure a compression failure has been determined to be equal to $\rho_{fb}/(1 - 3\sigma)$, where $\sigma = 8.88\%$ and represents the standard deviation [Vijay and GangaRao 2001]. Based on the statistical analysis, a minimum reinforcement equal to $\rho_{fb}/(1 - 3\sigma)$ is recommended for determining the nominal moment strength of FRP-reinforced concrete beams failing in compression. This standard deviation value ($\sigma = 8.88\%$) was determined from published test results of 64 GFRP-reinforced concrete beams exhibiting compression mode failures as reported in Vijay and Gangarao [2001] and presented in Appendix B [Al-Salloum et al. 1996; Benmokrane and Masoudi 1996; Brown and Bartholomew 1993; Cozensa et al. 1997; GangaRao and Faza 1991; Nawy and Neuwerth 1977; Nawy et al. 1971; Sonobe et al. 1977; Theriault and Benmokrane 1998; Vijay and GangaRao 1996; Zhao et al. 1997]. Note that ACI 318 suggests a value of $3/4\rho_b$ as the upper limit of the reinforcement ratio to ensure a steel-reinforced concrete beam to fail in tension. Hence, for ensuring a compression failure of GFRP-reinforced beam, the minimum reinforcement should be $1.33\rho_{fb}$. Compliance with deformability criteria (discussed later) also ensures the minimum reinforcement criteria for a cracked section in compression failure.

Referring to Figure 6.1, the nominal strength of a concrete section failing in compression can be determined by taking the moment of the compression force resultant about the tensile force resultant as expressed by Equation 6.24,

$$M_n = 0.85 f_c'ab(d - a/2) \tag{6.24}$$

The value of $a$ in Equation 6.24 can be determined from the equilibrium of horizontal forces acting on the beam cross section:

$$\text{Compression stress resultant} = 0.85 f_c'ab \tag{6.25}$$

$$\text{Tensile stress resultant} = A_f f_f \tag{6.26}$$

$$0.85 f_c'ab = A_f f_f \tag{6.27}$$

The stress in the FRP bars can be expressed in terms of strain from Hookes's law (Equation 6.28):

$$\frac{f_f}{\varepsilon_f} = E_f \tag{6.28}$$

so that

$$f_f = \varepsilon_f E_f \tag{6.29}$$

Substitution of Equation 6.28 in Equation 6.27 yields

$$0.85 f_c'ab = A_f E_f \varepsilon_f \tag{6.30}$$

From similar triangles based on the linear strain distribution (Figure 6.1), the relationship between the concrete strain $\varepsilon_{cu}$ and the strain in FRP bars can be expressed as given by Equation 6.31:

$$\frac{\varepsilon_{cu}}{\varepsilon_f} = \frac{c}{d-c} = \frac{a}{\beta_1 d - a} \qquad (6.31)$$

where $c = a/\beta_1$. From Equation 6.31, the strain in FRP can be expressed as

$$\varepsilon_f = \frac{(\beta_1 d - a)\varepsilon_{cu}}{a} \qquad (6.32)$$

Substitution of $\varepsilon_f$ from Equation 6.32 in Equation 6.30 yields

$$0.85 f'_c ab = A_f E_f \frac{(\beta_1 d - a)\varepsilon_{cu}}{a} \qquad (6.33)$$

Rearranging various terms in Equation 6.33 in terms of $a$ yields

$$\left(\frac{0.85 f'_c b}{A_f E_f \varepsilon_{cu}}\right) a^2 + a - \beta_1 d = 0 \qquad (6.34)$$

Note that Equation 6.34 is a quadratic in $a$, and knowing the properties of concrete and FRP bars, the depth of compression block $a$ can be determined.

### 6.4.5 EXAMPLES ON NOMINAL STRENGTH OF FRP-REINFORCED BEAMS

The following examples illustrate the calculation procedures for determining the nominal strengths of FRP-reinforced beams. Examples are provided in both U.S. standard and SI Units. For clarity in understanding the applications of several equations developed in the preceding section, Example 6.1 and Example 6.2 present calculations in a step-by-step format.

Note that a complete design of a concrete beam also includes a check for serviceability criteria. To preserve the completeness of the calculations, these same examples have been used in Chapter 8 to illustrate calculations related to their serviceability checks.

#### 6.4.5.1 Examples in U.S. Standard Units

*Example 6.1: To check the adequacy of a given beam for flexural strength and crack width.* A rectangular 12 in × 21 in beam is used to support a 6-in-thick concrete floor having a live load of 50 lb/ft². The beams are spaced at 12 ft on centers and

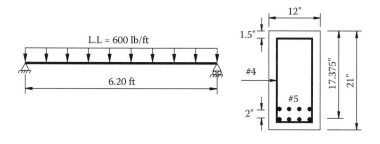

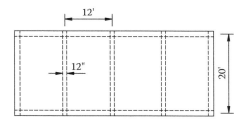

**FIGURE E6.1** Floor plan and beam cross section.

span 20 ft as shown in Figure E6.1. The tensile reinforcement consists of eight No. 5 GFRP bars arranged in two rows. Check the adequacy of a typical interior beam for moment strength. Assume $f_c' = 4000$ psi, $f_{fu}* = 90$ ksi, and $E_f = 6500$ ksi. Show all calculations step-by-step.

*Solution: Step 1.* Calculate service loads.

$$\text{Tributary width of beam} = 12 \text{ ft}$$

$$D_{slab} = (6/12)(1)(12)(150) = 900 \text{ lb/ft}$$

$$D_{beam} = (12)(21)(150)/144 = 262.5 \text{ lb/ft}$$

$$D = 900 + 262.5 = 1162.5 \text{ lb/ft}$$

$$L = (50)(12) = 600 \text{ lb/ft}$$

$$w_{ser} = D + L = 1162.5 + 600 = 1762.5 \text{ lb/ft}$$

*Step 2.* Calculate factored loads.

$$\text{Load combination 1: } 1.4D = 1.4(1162.5) = 1628 \text{ lb/ft}$$

$$\text{Load combination 2: } 1.2D + 1.6L = 1.2(1162.5) + 1.6(600) =$$
$$2355 \text{ lb/ft} \approx 2.36 \text{ k/ft (governs)}$$

$$M_u = w_u l^2/8 = (2.36)(20)^2/8 = 118 \text{ k-ft}$$

*Step 3.* Determine the tensile design strength of GFRP reinforcement, $f_{fu}$.

$$f_{fu} = C_E f_{fu}^*$$ (ACI 440 Equation 7-1)

For an interior conditioned space, the environmental modification factor, $C_E$, is 0.8 (ACI 440 Table 7.1).

$$f_{fu} = (0.8)(90) = 72 \text{ ksi}$$

*Step 4.* Calculate the reinforcement ratio, $\rho_f$. First, determine $d$. For a beam with two layers of tension reinforcement,

$d = h$ – clear cover – diameter of stirrup – diameter of tensile reinforcing bar –
1/2 (2 in. clear distance between the two layers of bars)
$= 21 - 1.5 - 0.5 - 0.625 - 1/2(2) = 17.375$ in. $\approx 17.38$ in.

$$A_f = 8(0.31) = 2.48 \text{ in}^2$$

$$\rho_f = \frac{A_f}{bd} = \frac{2.48}{(12)(17.38)} = 0.012$$

*Step 5.* Calculate the stress in GFRP reinforcement from ACI 440 Equation 8-4d.

$$f_f = \sqrt{\frac{(E_f\varepsilon_{cu})^2}{4} + \frac{0.85\beta_1 f_c'}{\rho_f} E_f\varepsilon_{cu}} - 0.5E_f\varepsilon_{cu} < f_{fu}$$

$$= \sqrt{\frac{[(6500)(0.003)]^2}{4} + \frac{0.85(0.85)(4)}{0.012}(6500)(0.003)} - 0.5(6500)(0.003)$$

$$= 59.5 \text{ ksi} < 72 \text{ ksi, OK}$$

*Step 6.* Calculate the strength reduction factor $\phi$ from ACI 440 Equation 8-7. First, calculate $\rho_{fb}$ from ACI 440 Equation 8.3.

$$\rho_{fb} = 0.85\beta_1 \frac{f_c'}{f_{fu}} \frac{E_f\varepsilon_{cu}}{E_f\varepsilon_{cu} + f_{fu}}$$ (ACI 440 Equation 8-3)

For $f_c' = 4000$ psi, $\beta_1 = 0.85$, and $f_{fu} = 72$ ksi (calculated earlier). Therefore,

$$\rho_{fb} = (0.85)(0.85)\frac{4}{72}\left[\frac{(6500)(0.003)}{(6500)(0.003)+72}\right] = 0.0086$$

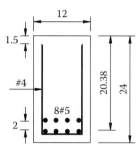

All dimensions are in inches.

**FIGURE E6.2** Beam cross section.

$$1.4\rho_{fb} = 1.4(0.0086) = 0.012$$

Because $(\rho_f = 0.012) \geq (1.4\rho_{fb} = 0.012)$, $\phi = 0.7$ (ACI 440 Equation 8-7).

*Step 7.* Calculate the nominal strength of beam from ACI 440 Equation 8-5.

$$M_n = \rho_f f_f\left(1 - 0.59\frac{\rho_f f_f}{f_c}\right)bd^2$$

$$= \frac{(0.012)(59.5)}{12}\left[1 - 0.59\frac{(0.012)(59.5)}{4}\right](12)(17.38)^2 \quad \text{(ACI 440 Equation 8-5)}$$

$$= 193 \text{ k-ft}$$

*Step 8.* Calculate $\phi M_n$.

$$\phi M_n = 0.7(193) = 135 \text{ k-ft} > M_u = 118 \text{ k-ft}.$$

Therefore, the beam satisfies the strength requirements: $\phi M_n = 135$ k-ft $> M_u = 118$ k-ft.

*Example 6.2: Determination of the nominal strength of a beam and crack width when the strength reduction factor $\phi$ is less than 0.7.* A 12 in. × 24 in. rectangular concrete beam is reinforced with eight No. 5 GFRP bars arranged in two layers as shown in Figure E6.2. The beam carries a superimposed dead load of 900 lb/ft in addition to its own weight and a floor live load of 600 lb/ft. Check the adequacy of this beam for a simple span of 20 ft for strength. Assume $f_c' = 4000$ psi, $f_{fu}^* = 90$ ksi, and $E_f = 6500$ ksi. Show all calculations step-by-step.
*Solution: Step 1.* Calculate service loads.

$$D_{beam} = \frac{(12)(24)}{14}(150) = 300 \text{ lb/ft}$$

$$D_{super} = 900 \text{ lb/ft}$$

$$D = 900 + 300 = 1200 \text{ lb/ft}$$

$$L = 600 \text{ lb/ft}$$

$$w_{ser} = D + L = 1200 + 600 = 1800 \text{ lb/ft} = 1.8 \text{ k/ft}$$

*Step 2.* Calculate factored loads.

Load combination 1: $1.4D = 1.4(1200) = 1680$ lb/ft

Load combination 2: $1.2D + 1.6L = 1.2(1200) + 1.6(900)$
$= 2400$ lb/ft $= 2.4$ k/ft (governs)

$$M_u = w_u l^2/8 = (2.4)(20)^2/8 = 120 \text{ k-ft}$$

*Step 3.* Determine the tensile design strength of GFRP reinforcement, $f_{pu}$.

$$f_{fu} = C_E f_{fu}{}^* \qquad \text{(ACI 440 Equation 7-1)}$$

For an interior conditioned space, the environmental modification (reduction) factor, $C_E$, is 0.8 (ACI 440 Table 7.1).

$$f_{fu} = (0.8)(90) = 72 \text{ ksi}$$

*Step 4.* Calculate the reinforcement ratio, $\rho_f$. First, determine $d$. For a beam with two layers of tension reinforcement,

$d = h -$ clear cover $-$ diameter of stirrup diameter of tensile reinforcing bar $-$
1/2 (2 in. clear distance between the two layers of bars)
$= 24 - 1.5 - 0.5 - 0.625 - 1/2(1) = 20.375$ in. $\approx 20.38$ in.

$$A_f = 8(0.31) = 2.48 \text{ in.}^2$$

$$\rho_f = \frac{A_f}{bd} = \frac{2.48}{(12)(20.38)} = 0.01$$

*Step 5.* Calculate the stress in GFRP reinforcement from ACI 440 Equation 8-4d.

$$f_f = \sqrt{\frac{(E_f \varepsilon_{cu})^2}{4} + \frac{0.85\beta_1 f_c'}{\rho_f} E_f \varepsilon_{cu}} - 0.5 E_f \varepsilon_{cu} < f_{fu}$$

$$= \sqrt{\frac{[(6500)(0.003)]^2}{4} + \frac{0.85(0.85)(4)}{0.010}(6500)(0.003)} - 0.5(6500)(0.003)$$

$= 66$ ksi $< 72$ ksi, OK

*Step 6.* Calculate the strength reduction factor $\phi$ from ACI 440 Equation 8-7. First, calculate $\rho_{fb}$ from ACI 440 Equation 8.3.

$$\rho_{fb} = 0.85\beta_1 \frac{f_c'}{f_{fu}} \frac{E_f \varepsilon_{cu}}{E_f \varepsilon_{cu} + f_{fu}} \qquad \text{(ACI 440 Equation 8-3)}$$

For $f_c' = 4000$ psi, $\beta_1 = 0.85$, and $f_{fu} = 72$ ksi (calculated earlier). Therefore,

$$\rho_{fb} = (0.85)(0.85)\frac{4}{72}\left[\frac{(6500)(0.003)}{(6500)(0.003)+72}\right] = 0.0086$$

$$1.4\rho_{fb} = 1.4(0.0086) = 0.012 > \rho_f = 0.010.$$

Because $(\rho_{fb} = 0.0086) < (\rho_f = 0.010) < (1.4\rho_{fb} = 0.012)$,

$$\phi = \frac{\rho_f}{2\rho_{fb}} = \frac{0.01}{2(0.0086)} = 0.58 \qquad \text{(ACI 440 Equation 8-7)}$$

*Step 7.* Calculate the nominal strength of beam from ACI 440 Equation 8-5.

$$M_n = \rho_f f_f\left(1 - 0.59\frac{\rho_f f_f}{f_c'}\right)bd^2$$

$$= \frac{(0.010)(66)}{12}\left[1 - 0.59\frac{(0.010)(66)}{4}\right](12)(20.38)^2 \qquad \text{(ACI 440 Equation 8-5)}$$

$$= 247.4 \text{ k-ft}$$

*Step 8.* Calculate $\phi M_n$.

$$\phi M_n = 0.58(247.4) = 143.5 \text{ k-ft} > M_u = 120 \text{ k-ft.}$$

The beam satisfies the strength requirements: $\phi M_n = 143.5$ k-ft $> M_u = 120$ k-ft.

*Example 6.3: Determination of nominal strength of a beam reinforced with CFRP bars.* An 8 in. × 15 in. rectangular concrete beam is reinforced with three No. 6 CFRP bars and No. 3 stirrups as shown in Figure E6.3 The beam carries a superimposed dead load of 875 lb/ft in addition to its own weight and a floor live load of 2500 lb/ft over a span of 10 ft. Check the adequacy of this beam for flexural strength. Assume $f_c' = 4000$ psi, $f_{fu}^* = 240$ ksi, and $E_f = 20,000$ ksi.
*Solution*: Service loads:

$$D_{beam} = \frac{(8)(15)}{144}(150) = 125 \text{ lb/ft}$$

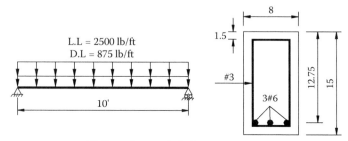

All dimensions are in inches.

**FIGURE E6.3** Beam cross section.

$$D_{super} = 875 \text{ lb/ft}$$

$$D = 125 + 875 = 1000 \text{ lb/ft}$$

$$L = 2500 \text{ lb/ft}$$

$$w_{ser} = 1000 + 2500 = 3500 \text{ lb/ft} = 3.5 \text{ k/ft}$$

Factored loads:

Load combination 1: $1.4D = 1.4(1000) = 1400 \text{ lb/ft}$

Load combination 2: $1.2D + 1.6L = 1.2(1000) + 1.6(2500)$
$= 5200 \text{ lb/ft} = 5.2 \text{ k/ft (governs)}$

$$M_u = w_u l^2/8 = (5.2)(10)^2/8 = 65 \text{ k-ft}$$

$$f_{fu} = C_E f_{fu}^* \qquad \text{(ACI 440 Equation 7-1)}$$

For an interior conditioned space, the environmental modification (reduction) factor, $C_E$, is 1.0 (ACI 440 Table 7.1).

$$f_{fu} = (1.0)(240) = 240.0 \text{ ksi}$$

Calculate the reinforcement ratio, $\rho_f$. First, determine $d$.

$d = h$ – clear cover – diameter of stirrup – 1/2 (diameter of tensile reinforcing bar)
$= 15 - 1.5 - 0.375 - 0.75/2 = 12.75 \text{ in.}$

$$A_f = 3(0.44) = 1.32 \text{ in.}^2$$

$$\rho_f = \frac{A_f}{bd} = \frac{1.32}{(8)(12.75)} = 0.0129$$

Calculate the stress in CFRP reinforcement from ACI 440 Equation 8-4d.

$$f_f = \sqrt{\frac{(E_f \varepsilon_{cu})^2}{4} + \frac{0.85 \beta_1 f_c'}{\rho_f} E_f \varepsilon_{cu}} - 0.5 E_f \varepsilon_{cu} < f_{fu}$$

$$= \sqrt{\frac{[(20,000)(0.003)]^2}{4} + \frac{0.85(0.85)(4)}{0.0129}(20,000)(0.003)} - 0.5(20,000)(0.003)$$

$$= 89.8 \text{ ksi} < 240 \text{ ksi, OK}$$

Calculate the strength reduction factor $\phi$ from ACI 440 Equation 8-7. First, calculate $\rho_{fb}$ from ACI 440 Equation 8.3.

$$\rho_{fb} = 0.85 \beta_1 \frac{f_c'}{f_{fu}} \frac{E_f \varepsilon_{cu}}{E_f \varepsilon_{cu} + f_{fu}} \qquad \text{(ACI 440 Equation 8-3)}$$

For $f_c' = 4000$ psi, $\beta_1 = 0.85$, and $f_{fu} = 240$ ksi. Therefore,

$$\rho_{fb} = (0.85)(0.85) \frac{4}{240} \left[ \frac{(20,000)(0.003)}{(20,000)(0.003) + 240} \right] = 0.0024$$

$$1.4 \rho_{fb} = 1.4(0.0024) = 0.0034$$

Because $(\rho_f = 0.0129) > (1.4 \rho_{fb} = 0.0034)$, $\phi = 0.7$ (ACI 440 Equation 8-7).

Calculate the nominal strength of beam from ACI 440 Equation 8-5.

$$M_n = \rho_f f_f \left( 1 - 0.59 \frac{\rho_f f_f}{f_c'} \right) bd^2$$

$$= \frac{(0.012)(89.8)}{12} \left[ 1 - 0.59 \frac{(0.0129)(89.8)}{4} \right] (8)(12.75)^2 \quad \text{(ACI 440 Equation 8-5)}$$

$$= 104.1 \text{ k-ft}$$

Calculate $\phi M_n$.

$$\phi M_n = 0.7(104.1) = 72.9 \text{ k-ft} > M_u = 65 \text{ k-ft.}$$

The beam has adequate flexural strength.

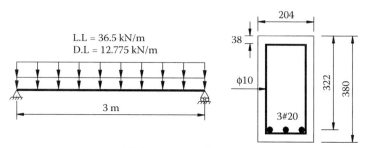

All dimensions are in mm.

**FIGURE E6.4** Beam cross section

### 6.4.5.2  Examples in SI Units

*Example 6.4: Determination of nominal strength of a beam reinforced with CFRP bars.* A 204 mm × 380 mm rectangular concrete beam is reinforced with three $\phi$ 20 mm CFRP bars and No. 3 stirrups as shown in Figure E6.4. The beam carries a superimposed dead load of 12.775 kN/m in addition to its own weight and a floor live load of 36.5 kN/m over a span of 3 m. Check the adequacy of this beam for flexural strength. Assume $f_c' = 27.6$ N/mm$^2$, $f_{fu}* = 1654.8$ N/mm$^2$, and $E_f = 137.9$ kN/mm$^2$.

*Solution*: Service loads:

$$D_{beam} = \frac{(204)(380)(23.56)}{(1000)(1000)} = 1.82 \text{ kN/m}$$

$$D_{super} = 12.775 \text{ kN/m}$$

$$D = 1.82 + 12.775 = 14.6 \text{ kN/m}$$

$$L = 36.5 \text{ kN/m}$$

$$w_{ser} = 14.6 + 36.5 = 51.1 \text{ kN/m}$$

Factored loads:

Load combination 1: $1.4D = 1.4(14.6) = 20.44$ kN/m

Load combination 2: $1.2D + 1.6L = 1.2(14.6) + 1.6(36.5) = 75.92$ kN/m (governs)

$$M_{ser} = \frac{w_{ser}l^2}{8} = \frac{(75.92)(3.0)^2}{8} = 85.41 \text{ kN/m}$$

$$f_{fu} = C_E f_{fu}* \qquad\qquad \text{(ACI 440 Equation 7-1)}$$

For an interior conditioned space, the environmental modification (reduction) factor, $C_E$, is 1.0 (ACI 440 Table 7.1).

$$f_{fu} = (1.0)(1654.8) = 1654.8 \text{ N/mm}^2$$

Calculate the reinforcement ratio, $\rho_f$. First, determine $d$.

$$d = h - \text{clear cover} - \text{diameter of stirrup} - 1/2 \text{ (diameter of tensile reinforcing bar)}$$
$$= 380 - 38 - 10 - 20/2 = 322 \text{ mm}$$

$$A_f = 3(314) = 942 \text{ mm}^2$$

$$\rho_f = \frac{A_f}{bd} = \frac{942}{(204)(322)} = 0.0143$$

Calculate the stress in CFRP reinforcement from ACI 440 Equation 8-4d.

$$f_f = \sqrt{\frac{(E_f\varepsilon_{cu})^2}{4} + \frac{0.85\beta_1 f_c'}{\rho_f} E_f\varepsilon_{cu}} - 0.5E_f\varepsilon_{cu} < f_{fu}$$

$$= \sqrt{\frac{(137.9)(0.003)^2}{4} + \frac{0.85(0.8692)(0.0276)}{0.0143}(137.9)(0.003)} - 0.5(137.9)(0.003)$$

$$= 0.5886 \text{ kN/mm}^2 \leq 1.654 \text{ kN/mm}^2, \text{ OK}$$

Calculate the strength reduction factor $\phi$ from ACI 440 Equation 8-7. First, calculate $\rho_{fb}$ from ACI 440 Equation 8.3.

$$\rho_{fb} = 0.85\beta_1 \frac{f_c'}{f_{fu}} \frac{E_f\varepsilon_{cu}}{E_f\varepsilon_{cu} + f_{fu}} \qquad \text{(ACI 440 Equation 8-3)}$$

For $f_c' = 27.6$ N/mm2, $\beta_1 = 0.8692$, and $f_{fu} = 1654.8$ N/mm². Therefore,

$$\rho_{fb} = (0.85)(0.8692)\frac{27.6}{1654.8}\left[\frac{(137,900)(0.003)}{(137,900)(0.003)+1654.8}\right] = 0.0025$$

$$1.4\rho_{fb} = 1.4(0.0025) = 0.0034 > 1.4\rho_f = 0.002$$

Because $(\rho_f = 0.0129) > (1.4\rho_{fb} = 0.0034)$, $\phi = 0.7$ (ACI 440 Equation 8-7).

Calculate the nominal strength of beam from ACI 440 Equation 8-5.

$$M_n = \rho_f f_f \left( 1 - 0.59 \frac{\rho_f f_f}{f_c'} \right) bd^2$$

$$= (0.0143)(588.6) \left[ 1 - 0.59 \frac{(0.0143)(588.6)}{27.6} \right] (204)(322)^2$$

$$= 146.0 \text{ kN-m}$$

Calculate $\phi M_n$.

$$\phi M_n = 0.7(146.0) = 102.2 \text{ kN-m} > M_u = 85.41 \text{ kN-m}$$

The beam has adequate flexural strength.

## 6.5 MINIMUM AND MAXIMUM FRP REINFORCEMENT RATIOS

### 6.5.1 MINIMUM FRP REINFORCEMENT

Fundamentally speaking, all reinforced concrete members must be minimally reinforced to preclude the possibility of sudden collapse, i.e., all concrete beams must have sufficient reinforcement to ensure that $\phi M_n \geq M_{cr}$, where $M_{cr}$ is the cracking moment (which corresponds to a concrete stress equal to the modulus of rupture of concrete, $f_r$).

When a small amount of tensile reinforcement is used in a concrete beam, the computed nominal strength of the cracked reinforced section may be less than its nominal strength corresponding to the modulus of rupture. For FRP-reinforced members designed for tension failure (for FRP rupture, $\rho_f < \rho_{fb}$), a minimum amount of reinforcement is necessary to preclude the possibility of the sudden collapse of the beam following the cracking of concrete. The provisions in ACI 318 for minimum steel reinforcement are based on this concept, which are modified for FRP-reinforced beams.

The minimum tensile reinforcement area for FRP-reinforced beams is obtained by multiplying the minimum reinforcement equations of ACI 318 (see Chapter 10 of ACI 318) for steel-reinforced concrete by 1.8, which is the ratio of the strength reduction factor of 0.9 for steel-reinforced concrete beams and 0.5 as the strength reduction factor for tension-controlled FRP-reinforced concrete beams (0.90/0.50 = 1.8). The modified relationships for minimum FRP reinforcement are given by Equation 6.35, which must not be less that that given by Equation 6.36:

$$A_{f,\min} = \left( \frac{0.9}{0.5} \right) \frac{3\sqrt{f_c'}}{f_{fu}} b_w d$$

$$= \frac{5.4\sqrt{f_c'}}{f_{fu}} b_w d$$

(6.35)

$$A_{f,\min} \ge \left(\frac{0.9}{0.5}\right)\frac{200}{f_{fu}}b_w d$$

$$= \frac{360}{f_{fu}}b_w d \qquad (6.36)$$

Note that for $f_c' = 4440$ psi, Equation 6.35 and Equation 6.36 give the same values of $A_{f,\min}$. According ACI 318 Commentary, when $f_c'$ higher than about 5000 psi is used, $A_{f,\min}$ given by Equation 6.36 may not be sufficient and Equation 6.35 should be used.

For a FRP-reinforced T-section having its flange in tension, the modified minimum area of reinforcement is:

$$A_{f,\min} = \left(\frac{0.9}{0.5}\right)\frac{6\sqrt{f_c'}}{f_{fu}}b_w d$$

$$= \frac{10.8\sqrt{f_c'}}{f_{fu}}b_w d \qquad (6.37)$$

For tension-controlled FRP-reinforced concrete beams, ACI 440 recommends using $A_{f,\min}$ equal to the greater of the values given either by Equation 6.36 or Equation 6.37, with $b_w$ set equal to the width of the rectangular beams, or the width of the web of a T-beam.

For compression failures ($\rho_f > \rho_{fb}$), the minimum amount of reinforcement requirement to avoid sudden failure due to cracking is automatically satisfied (because when $\rho_f > \rho_{fb}$, the $A_f$ is too large for rupture to occur and hence, no cracking) so that a check for $A_f$ is not necessary. This also means that a check by Equation 6.35 and Equation 6.36 is required only if $\rho_f < \rho_{fb}$.

### 6.5.2 MAXIMUM FRP REINFORCEMENT

Because FRP-reinforced concrete beams are designed to fail primarily in compression, no upper limit exists on the amount of FRP reinforcement that can be provided in such members (a high amount of FRP reinforcement ensures that it does not fail suddenly or rupture). Note that this absence of an upper limit on FRP reinforcement is in direct contrast to those of ACI 318 for steel-reinforced concrete members because in the latter case, the reinforcement must yield (to ensure tension or ductile failure).

## 6.6 TEMPERATURE AND SHRINKAGE REINFORCEMENT

The stiffness and strength of FRP reinforcing bars influence the crack control behavior of concrete members as a result of shrinkage and temperature effects. Flexural reinforcement controls shrinkage and temperature cracks perpendicular to the member span. Hence, shrinkage and temperature reinforcement are most useful in the direction perpendicular to the span.

ACI 318 provisions recommend a minimum steel reinforcement ratio of 0.0020 when using Grade 40 or Grade 50 deformed steel bars and 0.0018 when using Grade 60 deformed bars or welded fabric. ACI 318 limits the spacing of shrinkage and temperature reinforcement to smaller of 18 in. or five times the member thickness.

ACI 318 states that for slabs with steel reinforcement having a yield stress exceeding 60 ksi (414 MPa) measured at a yield strain of 0.0035, the ratio of reinforcement to gross area of concrete should be at least $0.0018 \times 60/f_y$ (where $f_y$ is in ksi) but not less than 0.0014. Due to the lack of test data, ACI 440 recommends the temperature and shrinkage steel ratio ($\rho_{f,ts}$) as a function of strength and stiffness of FRP reinforcement and the stiffness of steel reinforcement (Equation 6.38), with the stipulation that the ratio of temperature and shrinkage reinforcement should not be taken less than 0.0014 (the value specified by ACI 318 for steel shrinkage and temperature reinforcement):

$$\rho_{f,ts} = 0.0018 \times \frac{60,000}{f_{fu}} \frac{E_s}{E_f} \qquad (6.38)$$

The ACI 440 recommended spacing of shrinkage and temperature FRP reinforcement is limited to three times the slab thickness or 12 in. (300 mm), whichever is less.

## 6.7 ENERGY ABSORPTION IN FRP-REINFORCED MEMBERS: DUCTILITY AND DEFORMABILITY FACTORS

Ductility is an important consideration in design of all structures, particularly when they are subjected to overloads. In view of its significance, a discussion on the basic concepts of ductility and an analogous concept, deformability, is presented.

### 6.7.1 DUCTILITY FACTOR

All materials deform under loads. These deformations can be elastic or plastic, small or large. A material that is capable of undergoing a large amount of plastic deformation is said to be *ductile*. Examples of ductile materials include mild steel and aluminum and some of their alloys, copper, magnesium, lead, molybdenum, nickel, brass, bronze, nylon, Teflon, and many others. A material that is capable of undergoing very little plastic deformation before rupture is said to be *brittle*. Ordinary glass is a nearly ideal brittle material because it exhibits almost no ductility whatsoever (i.e., no plastic deformation before its failure load).

After attaining certain levels of deformations, all materials fail. A structural material is considered to have failed when it becomes incapable of performing its design function, either through fracture as in brittle materials or by excessive deformation as in ductile materials. Ductile materials are capable of absorbing large amounts of energy prior to failure; they are capable of undergoing large deformations before rupture. Because elastic deformations of materials are generally small, the usual measure of ductility (or brittleness) is the total percent elongation up to rupture of a tensile test specimen (typically a 2-in. gage length having a diameter of 0.505 in.). Sometimes the percent reduction of cross-sectional area of a tensile test specimen after rupture is also used as a measure of ductility. A very ductile material such as structural steel can have a longitudinal strain (or elongation) of 30% at rupture, whereas a brittle material such as gray cast iron or glass will have relatively little elongation at rupture (e.g., the percent elongation for gray cast iron is less than 1% and for polystyrene it is 2%). For some materials, percent elongation can be quite large — for example, 350% for polyethylene and 800% for rubber (natural or molded) [Byars and Snyder 1975].

Ductility is an important attribute of materials in that visible deformations can occur if the loads become too large, thus providing an opportunity to take remedial action before fracture occurs. In the context of such structural members as slabs, beams, and columns, the concept of ductility generally applies — because of inherent ductility, large deformations can become visible, giving ample warning of an impending structural failure. This is particularly important for the performance of structures in high seismic regions where structures must undergo large cyclic displacements (often inelastic) without structural collapse. Satisfactory structural response under such loading conditions relies on the capacity of a structure (or structural members) to displace inelastically through several cycles of response without a significant degradation of strength or stiffness is a quality termed *ductility* [Pristley et al. 1996].

Because of its significance, ductility must be quantified. The usual practice is to express ductility as a *ductility factor* or a *ductility index*. Referenced to displacements, the ductility factor is often mathematically defined as the ratio of maximum displacement to displacement at yield [Pristley et al. 1996]:

$$\mu_\Delta = \frac{\Delta_m}{\Delta_y} \qquad (6.39)$$

where

$\mu_\Delta$ = the ductility factor (the subscript refers to displacement)
$\Delta_m$ = the maximum displacement (inelastic response)
$\Delta_y$ = the displacement at yield

The maximum displacement quantity, $\Delta_m$ (inelastic response), in Equation 6.39 can be any prescribed value greater than yield displacement ($\Delta_y$) so that the ductility factor is always greater than unity. Note that the key word here is inelastic response, because $\Delta_m > \Delta_y$. For example, for seismic design considerations $\Delta_m$ can be defined as the maximum displacement (e.g., post-yielding, post-buckling) expected under the design-level earthquake. For a different design consideration, such as in concrete structures, the maximum displacement can be the displacement that may be attained at the ultimate force level, $\Delta_u$, and the corresponding ductility factor can be defined as

$$\mu_\Delta = \frac{\Delta_u}{\Delta_y} \qquad\qquad (6.40)$$

Although Equation 6.39 and Equation 6.40 define ductility factors in terms of displacements in the context of overall structural response (structure or structural system displacement ductility factor), they can also be expressed to define a member response to loads (member displacement ductility factor). In general, ductility can be defined in a number of different ways depending on design considerations. Measures of ductility can be expressed as displacement ductility, rotational (curvature being the rotation over the depth of a section) ductility, and strain ductility to quantify the structural response under maximum loads (e.g., ultimate loads, or design-level earthquake loads), which are useful indicators of structural response. The relationship between the curvature and displacement ductility factors depends on the structural geometry and is important for determining safe levels of inelastic displacements for the structure as a whole. A large number of concrete structures (buildings and bridges) in California either collapsed completely or were significantly damaged during earthquakes in 1933, 1971, 1987, 1989, and 1994; the main reason was a lack of ductility. Providing sufficient ductility or improving ductility in structures that either lack or have a little of it is of utmost significance to structural engineers. A comprehensive discussion ductility factors has been presented by Pristley, Seible, and Calvi [1996].

The ductility factor (ductility index) of concrete beams reinforced with steel bars (member ductility factor) can be defined as the ratio of deformation at ultimate to that at yield. From a design perspective, the ductility index of a concrete beam reinforced with steel bars provides a measure of the energy absorption capability [Jaeger et al. 1996; Naaman and Jeong, 1995].

Because deflections, curvatures, and rotations in a beam are all proportional to the moment, the ductility factor for materials such as metals that exhibit a post-yielding plateau, can be expressed in terms of the ratio of any of these quantities. Thus,

$$\text{Ductility factor} = \frac{\text{Deflection (or curvature, or rotation) at ultimate}}{\text{Deflection (or curvature, or rotation) at steel yield}} \qquad (6.41)$$

## 6.7.2 DEFORMABILITY FACTOR

The concept of ductility was developed as a measure to determine the post-yielding deformational response energy absorption capability of ductile materials. This is the conventional concept of ductility. Essentially, it grew from the theory of plasticity of steel. Ductile materials characteristically exhibit an elastic response up to yield followed by inelastic deformation. Because nonductile or brittle materials such as FRPs typically do not exhibit such a response, the conventional concept of ductility cannot be used for these materials. With the exception of a few types of FRPs that have low ductility (they exhibit linear deformations followed by small nonlinear deformations), these materials neither yield nor exhibit the inelastic behavior characteristic of steel. The stress-strain relationship for FRPs referred to in this book is essentially linear; they rupture without any warning at ultimate loads. For beams reinforced with FRP materials, the concept of *deformability* rather than ductility is introduced (see next section) to measure the energy absorbing capacity. The two concepts are analogous in the sense that both relate to the energy absorbing capacities of beams at ultimate loads, with the exception that the behavior of steel reinforcement is different from that of FRP reinforcement.

The deformability of a FRP-reinforced beam depends on:

1. Elongation of FRP bars at different locations
2. Confinement effect
3. Extent of cracking
4. Bond between bar and concrete
5. Frictional force development along diagonal and wedge cracks.

The *deformability factor* (DF) of FRP-reinforced concrete members can be defined by Equation 6.42:

$$\text{Deformability factor} = \frac{\text{Energy absorption (area under moment} - \text{curvature curve) at ultimate value}}{\text{Energy absorption (area under moment} - \text{curvature curve) at a limiting curvature}} \quad (6.42)$$

The concept of the limiting curvature introduced in Equation 6.42 and eventual quantification of curvature limit is based on the serviceability criteria for both deflection and crack width as specified in ACI 318:

1. The serviceability deflection limit of span/180
2. The crack-width limit of 0.016 in

Curvature at the load corresponding to limiting deflection or crack-width can be determined by calculating tensile strain, $\varepsilon_f$, in FRP bars based on the stress induced by the applied loads, $f_f$, and the depth of neutral axis, $c$ (so that $\varepsilon_f = f_f/E_f$). Curvature ($\varphi$) can be calculated using one of the expressions in Equation 6.43:

$$\varphi = \begin{cases} (\varepsilon_f/d - c) \\ (\varepsilon_f + \varepsilon_c)/d \\ \varepsilon_c/c \end{cases} \quad (6.43)$$

where

$\varepsilon_f$ = the tensile strain in FRP reinforcement
$\varepsilon_c$ = the strain in extreme concrete fiber in compression
$c$ = the depth of neutral axis from the extreme compression fibers
$d$ = the effective depth of the beam

Experimental research has been conducted for evaluating the energy absorption characteristics of GFRP-reinforced concrete beams, failing in both tension or compression failure modes [GangaRao and Faza 1993; Vijay and GangaRao 1999]. Based on moment-curvature diagrams of GFRP-reinforced beams, the maximum unified curvature limit at service load corresponding to the two serviceability limits of deflection and crack-width was determined to be limited to $\varphi = (0.005/d)$ rad/unit depth [Vijay and GangaRao 1996; Vijay 1999]. A curvature limit exceeding $(0.005/d)$ generally failed to satisfy either the deflection or the crack-width criterion for beams with span/depth ratios of 8 to 13. Comparing the numerator of the curvature $(\varphi)$ limit $(0.005/d)$ with the curvature equation $\varphi = (\varepsilon_f + \varepsilon_c)/d$, observe that the sum of strain values in the outermost concrete fiber $(\varepsilon_c)$ and FRP reinforcement $(\varepsilon_f)$ exceeding 0.005 may not typically satisfy deflection or crack-width criteria as specified above.

Energy absorption, given by the area under the moment-curvature curve (or the area under the moment-deflection curve) of a GFRP-reinforced concrete beam in compression failure is shown in Figure 6.3 [Vijay and GangaRao 1999]. Each beam was loaded and unloaded in five to six cycles with each cycle having a higher load than the previous. A high amount of energy absorption was observed during the final cycle of loading. Gradual member failure was observed in beams with compression failure. A high deformability factor in compression failure of GFRP-reinforced beams during the final load cycle can be attributed to several factors, such as:

1. Plastic hinge formation
2. Confinement
3. Significant concrete cracking in compression zone
4. Stress redistribution

Deformability factors in compression failures based on the unified serviceability approach were observed to be in the range of 7 to 14.

In tension failure cases, deformability factors were observed to be in the range of 6 to 7. These values can differ with extremely low or high reinforcement ratios. GFRP rebar rupture under tension failures cause sudden member collapse. Though the tension failure in a GFRP-reinforced T-beam is less catastrophic as compared to rectangular beams (due to load distribution between the slab and beam), to expect that failure under a given set of loading is imminent even in a T-beam would be prudent [GangaRao et al. 2000]. Based on experimental results, a $c/d$ ratio in the range of 0.15 to 0.30 appears to be a reasonable design choice for concrete beams and slabs to achieve a deformability factor of 6 or higher.

Tests indicate that the total energy absorbed in GFRP-reinforced beams is higher than in comparable steel-reinforced beams in many instances [Vijay and GangaRao 2001; Vijay 1999]. Higher energy recovery (i.e., most of the deflections are recovered) even at very high rotational values (near failure) has been observed to result in less structural distress in GFRP-reinforced beams than in steel-reinforced concrete beams.

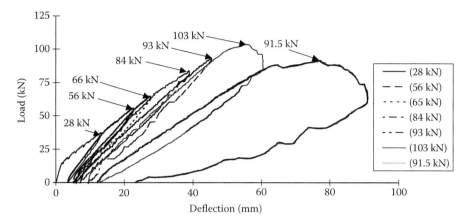

**FIGURE 6.3** Load vs. deformation of a GFRP-reinforced beam (compression failure after several cycles of loading).

### 6.7.3 COMPARISON OF DEFORMABILITY FACTORS

A comparison of deformability factors in compression failures of GFRP-reinforced concrete beams and steel-reinforced concrete beams indicates that the values are similar. Park and Paulay [1975] have reported a deformability factor of 10.2 for steel-reinforced beams with compression failure. From Park and Paulay's results, the serviceability load was calculated based on factored loads (using a load factor of 1.4 for dead load and 1.7 for live load) and a material strength reduction factor of 0.9 for flexure. In addition, the assumption of dead load fraction between 20% and 40% of total load leads to a serviceable load range of 55% and 57% of the maximum load as the limiting load value for establishing the deformability factor. An analysis by Vijay and GangaRao [2001] of the results of Ozbolt et al. [2000] on FRP-reinforced beams with depths of 203 mm (8 inches) and 406 mm (16 inches) and failing in compression indicates a deformability factor of about 11. These values are in the range of deformability factors 7 to 14 found in GFRP-reinforced concrete beams failing in compression. Example 6.5 presents the application of the concept of deformability factor in design.

*Example 6.5*: Figure E6.5 shows the moment-curvature relationship (an idealized bilinear curve) of an FRP-reinforced concrete beam having an effective depth of 16 inches. Calculate the deformability factor (DF) for this beam based on the limiting curvature value of $0.005/d$.

*Solution*: Using the suggested limiting curvature $0.005/d$ for an FRP reinforced concrete beam having a depth of 16 in., the limiting curvature values is

Limiting curvature value = $0.005/d = 0.005/16 = 0.000313$ rad/in.

The *deformability factor* of a concrete member reinforced with FRP bars is defined as:

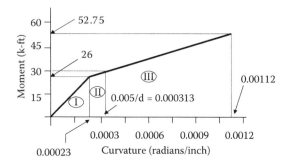

**FIGURE E6.5** Moment-curvature diagram of a FRP bar-reinforced concrete beam.

$$DF = \frac{[\text{Energy absorption (area under moment} - \text{curvature curve) at ultimate value}]}{[\text{Energy absorption (area under moment} - \text{curvature curve) at a limiting curvature}]}$$

$$DF = \frac{\text{Sum of areas (I + II + III)}}{\text{Sum of areas (I + II)}}$$

Different areas under moment curvature diagram (Figure E6.5) are:

Area I of moment curvature diagram = 0.5(0.00023)(26) = 0.00299

Area II of moment curvature diagram =
0.5(26 + 30)(0.000313 – 0.00023) = 0.002324

Area III of moment curvature diagram =
0.5(30 + 52.75)(0.00112 – 0.000313) = 0.03339

$$DF = \frac{0.00299 + 0.002324 + 0.03339}{0.00299 + 0.002324} = 7.28$$

*Commentary*: In a realistic beam reinforced with FRP bars, the bifurcation point shown in the bilinear moment-curvature would not be as clearly delineated as indicated in Area I of Figure E6.5. Near the ultimate moment, a larger energy absorption has been observed in tests of FRP-reinforced beams. The magnitude of energy absorption depends on the failure mode (tension- or compression-controlled) and confining effects provided by stirrups. The DF of steel reinforced-concrete beams with a compression-controlled failure mode was found to be about 10.2, whereas the DFs of FRP reinforced-concrete beams having compression-controlled failure mode was found to be in the range of 6.70 to 13.9 [Vijay and GangaRao 2001].

## 6.8 SHEAR STRENGTH OF FRP-REINFORCED BEAMS

Shear failure modes of members with FRP as shear reinforcement are classified as either shear-tension failure mode or shear-compression failure mode. The shear-tension failure mode is characterized by FRP shear reinforcement rupture, and the failure modes are described as brittle. The shear-compression failure mode is observed to be controlled by the crushing of the concrete web and the failure is relatively less brittle accompanied by larger deflections. These failure modes depend on the shear reinforcement index $\rho_{fv}E_f$, where $\rho_{fv}$ is the ratio of FRP shear reinforcement, $A_{fv}/b_w s$, where $s$ is the spacing of shear reinforcement. With increasing value of $\rho_{fv}E_f$, the shear capacity in shear-tension increases and the failure mode changes from shear-tension to shear-compression.

ACI 440 addresses use of FRP stirrups and continuous rectangular spirals as a shear reinforcement for beams. Punching shear issues related to FRP-reinforced members have not been extensively researched.

### 6.8.1 Shear Design Philosophy

The design of FRP shear reinforcement can be carried out based on the strength design method of ACI 318. The strength-reduction factor given by ACI 318 for calculating the nominal shear capacity of steel-reinforced concrete members can be also used for FRP-reinforced members.

### 6.8.2 Considerations for Using FRP Reinforcement in Shear

When using FRP as shear reinforcement, the following factors must be considered.

1. FRPs have a higher tensile strength than that of steel and many FRP types do not exhibit yield behavior.
2. Some FRPs have a relatively low modulus of elasticity.
3. FRPs have low dowel resistance.
4. The tensile strength of the bent portion of a FRP bar is less than about 40% of the straight portion (refer to Equation 6.4 on shear strength of bent portions).

### 6.8.3 Limits on Tensile Strain of Shear Reinforcement

The total shear strength of a FRP-reinforced member, obtained as the sum of shear strengths of concrete and reinforcement capacities, is valid when shear cracks are adequately controlled. For this assumption to be valid, the tensile strain in FRP shear reinforcement is limited to ensure that the ACI design approach is applicable. The design strength of FRP shear reinforcement is chosen as the smaller of the stress corresponding to $0.004E_f$ (i.e., corresponding to a strain of 0.004 in FRP shear reinforcement) or the strength of the bent portion of the stirrups, $f_{fb}$.

### 6.8.4 SHEAR STRENGTH OF FRP-REINFORCED MEMBERS

The estimation of shear strength of FRP-reinforced concrete is based on the additive model suggested by ACI 318. The nominal shear strength of a steel-reinforced concrete cross section, $V_n$, is taken as the sum of the shear resistance provided by concrete ($V_c$) and shear reinforcement ($V_s$):

$$V_n = V_c + V_s \qquad (6.44)$$

where $V_c$ is the shear contribution of concrete in a steel-reinforced concrete beam; its value, as specified in ACI 318, is given by Equation 6.45:

$$V_c = 2\sqrt{f_c'}b_w d \qquad (6.45)$$

For FRP-reinforced concrete members, Equation 6.44 can be expressed as

$$V_n = V_{cf} + V_f \qquad (6.46)$$

so that

$$V_f = V_n - V_{cf} \qquad (6.47)$$

where

$V_{cf}$ = the shear strength provided by concrete in the FRP-reinforced beam
$V_f$ = the shear strength provided by the FRP shear reinforcement

After cracking, FRP-reinforced members have smaller depths to a neutral axis (measured from the extreme compression fibers) as compared with those in steel-reinforced concrete beams. This is because of the relatively lower stiffness ($E_f$) of FRP reinforcement (e.g., compare approximately 6000 ksi for GFRP to 29,000 ksi for steel), which results in higher strains in the tension zone (i.e., $f_f/E_f$ is large), leading to wide and long flexural cracks that extend along the depth of the member. The smaller depth of the neutral axis results in a lower axial stiffness (the product of the reinforcement area and the modulus of elasticity, $A_f E_f$). Hence, the shear resistance offered by both the aggregate interlock and the compressed concrete, $V_{cf}$, is reduced.

The shear strength of flexural concrete members with FRP longitudinal reinforcement and no shear reinforcement has indicated a lower shear strength than a similarly steel-reinforced member without any shear reinforcement. Due to the lower strength and stiffness of FRP bars in the transverse direction, their role toward dowel action is also expected to be less than that of an equivalent steel area.

### 6.8.4.1 Contribution of Concrete, $V_{cf}$

According to ACI 440, the shear strength contribution of concrete in an FRP-reinforced beam, $V_{cf}$, is determined from Equation 6.48. The parenthetical quantity on the right side of Equation 6.48 represents the ratio of the axial stiffness of FRP reinforcement to that of steel reinforcement, which accounts for the reduced axial stiffness of the FRP reinforcement $(A_f E_f)$ as compared to that of steel reinforcement $(A_s E_s)$. Note that because $E_f \ll E_s$, this ratio is always less than unity and $V_{c,f} < V_c$.

$$V_{c,f} = \left( \frac{A_f E_f}{A_s E_s} \right) V_c = \left( \frac{\rho_f E_f}{\rho_s E_s} \right) V_c \leq V_c \tag{6.48}$$

For a practical design, the value of $\rho_s$ is taken as $0.5\rho_{s,max}$ or $0.375\rho_b$. Considering a typical steel yield strength of 60 ksi for flexural reinforcement, ACI 440 suggests Equation 6.49 for determining $V_{c,f}$.

$$V_{c,f} = \left( \frac{\rho_f E_f}{90 \beta_1 f_c'} \right) V_c \leq V_c \tag{6.49}$$

Note again that the value of the parenthetical quantity on the right hand side of Equation 6.49 is always less than unity.

### 6.8.4.2 Shear Contribution of FRP Stirrups, $V_f$

The contribution of FRP stirrups to shear is determined by an approach similar to that suggested by ACI 318 for determining the shear contribution from steel stirrups, i.e., by substituting properties of shear steel reinforcement with those of FRP shear reinforcement. Thus, the shear resistance, $V_f$, provided by FRP stirrups placed perpendicular to the axis of the member can be determined from Equation 6.50:

$$V_f = \frac{A_{fv} f_{fv} d}{s} \tag{6.50}$$

Equation 6.50 can also be expressed as

$$\frac{A_{fv}}{s} = \frac{V_f}{f_{fv} d} \tag{6.51}$$

In Equation 6.50 and Equation 6.51, the stress in the FRP shear reinforcement, $f_{fv}$, is limited to a value corresponding to a strain of 0.004 to restrict shear crack widths, to maintain the shear integrity of the concrete, and to avoid failure at the bent portion of the FRP stirrup. The strain value of 0.004 is a modification over the strain value of 0.002 recommended in earlier versions of ACI 440 and represents a strain that prevents the degradation of aggregate interlock and corresponding concrete shear. Note that a strain value of 0.004 represents a strain value for which composite-material jackets for confining concrete in columns have generally been designed [Priestly et al. 1996]. Based on this premise, the design stress in the FRP shear reinforcement at the ultimate load condition is limited to a value given by Equation 6.52:

$$f_{fv} = 0.004E_f \leq f_{fb} \qquad (6.52)$$

However, recognizing that the strain to failure in FRP bars under tension can be as high as 0.03 is important, in which case a designer would be required to modify Equation 6.52 accordingly.

For a beam subjected to a shear demand (i.e., design shear) of $V_u$,

$$V_u \leq \phi(V_{c,f} + V_f) \qquad (6.53)$$

so that the shear strength to be provided by the FRP shear reinforcement can be expressed as

$$V_f = \frac{V_u - \phi V_{c,f}}{\phi} \qquad (6.54)$$

where $\phi = 0.85$.

For designing shear reinforcement, one typically needs to determine the spacing and the size of the FRP stirrup. These values can be obtained by equating Equation 6.51 and Equation 6.54:

$$\frac{A_{fv}}{s} = \frac{(V_u - \phi V_{c,f})}{\phi f_{fv} d} \qquad (6.55)$$

Knowing the size of the FRP bar selected for stirrups, their spacing can be determined from Equation 6.55.

Alternative forms of shear reinforcement may be used at designer's discretion. Again, the strength contribution of FRP shear reinforcement is determined based on the ACI 318 approach. If inclined stirrups (instead of stirrups oriented perpendicular to the beam axis) are used, their shear strength contribution can be determined from Equation 6.56:

$$V_f = \frac{A_{fv} f_{fv} d}{s} (\sin\alpha + \cos\alpha) \qquad (6.56)$$

Similarly, if continuous FRP rectangular spirals are used as shear reinforcement with a pitch $s$ and angle of inclination $\alpha$, then contribution of the FRP spirals can be determined from Equation 6.57:

$$V_f = \frac{A_{fv} f_{fv} d}{s} (\sin\alpha) \qquad (6.57)$$

### 6.8.5 MAXIMUM AMOUNT OF SHEAR REINFORCEMENT

Shear failure in concrete beams can occur due to the crushing of the web. However, the correlation between rupture and crushing failure is not fully understood. As a conservative approach, ACI 440 recommends limiting the shear strength contribution of FRP stirrups in all cases as given by Equation 6.58:

$$V_{f,\max} \leq \left( \frac{\rho_f E_f}{90 \beta_1 f_c'} \right) 8 \sqrt{f_c'} b_w d \qquad (6.58)$$

If the shear that must be resisted by reinforcement exceeds that given by Equation 6.58, the beam configuration must be modified (i.e., modify $b_w$ or $d$, or both).

### 6.8.6 MINIMUM AMOUNT OF SHEAR REINFORCEMENT

ACI 440 provisions for avoiding sudden formation of cracks that can lead to brittle shear failure in concrete members are similar to those specified in ACI 318. For steel-reinforced concrete beams, the ACI 318 minimum shear reinforcement is given by Equation 6.59:

$$A_{v,\min} = \frac{50 b_w s}{f_y} \qquad (6.59)$$

For FRP-reinforced beams, the minimum amount of shear reinforcement can be determined from Equation 6.59 by substituting the properties of steel reinforcement with those of FRP.

$$A_{fv,\min} = \frac{50 b_w s}{f_{fv}} \qquad (6.60)$$

Note that the amount of minimum shear reinforcement given by Equation 6.60 is independent of the strength of concrete. Originally derived for steel-reinforced members, Equation 6.59 is more conservative for FRP-reinforced members. For example, for a concrete section with GFRP longitudinal reinforcement, the minimum shear reinforcement given by Equation 6.60 could provide a shear strength of $3V_c$ or greater. If steel stirrups are used, the minimum reinforcement provides a shear strength that varies from $1.50V_c$ when $f_c'$ is 2500 psi to $1.25V_c$ when $f_c'$ is 10,000 psi.

### 6.8.7 DETAILING OF SHEAR STIRRUPS

For vertical FRP shear reinforcement, ACI 440 recommends the spacing to be same as those for vertical steel stirrups specified in ACI 318: the smaller of $d/2$ or 24 in. This provision ensures that each shear crack is intercepted by at least one stirrup.

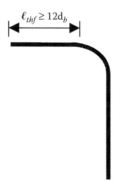

**FIGURE 6.4** Minimum tail length for FRP stirrups.

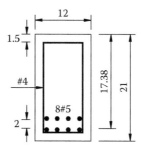

**FIGURE E6.6a** Beam cross section.

In addition to the closing of FRP stirrups with 90° hooks, ACI 440 recommends a minimum $r_b/d_b$ ratio of three for bends to avoid failure at low load levels.

For hooked FRP bars, the tensile force in a vertical stirrup leg is transferred to the concrete through the tail beyond the hook as shown in Figure 6.4. A minimum tail length of $12d_b$ is recommended by ACI 440.

*Example 6.6: Shear strength of FRP-reinforced beams.* A rectangular 12 in. × 21 in. beam is used to support a 6-in. thick concrete floor having a live load of 50 lb/ft². The beams are spaced at 12 ft on centers and span 20 ft as shown in Figure E6.1a. The tensile reinforcement consists of eight No. 5 GFRP bars arranged in two rows as shown in Figure E6.6a. Check the adequacy of a typical beam for shear and provide FRP shear reinforcement as required. Assume $f_c' = 4000$ psi, $f_{fu}* = 90$ ksi, and $E_f = 6500$ ksi. Show all calculations step-by-step.

*Solution:* The beam described in this example is the same as in Example 6.1; the following information is obtained from that example.

$$b = 12 \text{ in.}, \ h = 21 \text{ in.}, \ d = 18.38 \text{ in.}, \ A_f = 2.48 \text{ in}^2, \ \rho_f = 0.012, \ f_c' = 4 \text{ ksi},$$
$$f_{fu} = 72 \text{ ksi}, \ E_f = 6500 \text{ ksi}, \ w_u = 2.36 \text{ k/ft}, \ l = 20 \text{ ft}$$

Calculate shear due to factored at a distance $d$ (= 17.38 in.) from the support.

$$V_u = \frac{w_u l}{2} - w_u x = \frac{(2.36)(20)}{2} - (2.36)\left(\frac{17.38}{12}\right) = 20.18 \text{ kips}$$

Determine the shear strength provided by concrete. Calculate first the shear strength of a steel-reinforced concrete beam, $V_c$, from Equation 6.45:

$$V_c = 2\sqrt{f'_c} b_w d = \frac{2(\sqrt{4000})(12)(17.38)}{1000} = 26.4 \text{ kips}$$

Check the maximum shear permitted to be carried by the beam: Calculate the shear strength reduction factor, $\phi_{c,f}$, the parenthetical quantity in Equation 6.49:

$$\phi_{c,f} = \frac{\rho_f E_f}{90\beta_1 f'_c} = \frac{(0.012)(6500)}{(90)(0.85)(4)} = 0.255$$

The shear contribution of concrete in a FRP reinforced beam can be determined from Equation 6.49:

$$V_{cf} = \phi_{c,f}(V_c) = (0.255)(26.4) = 6.73 \text{ kips} < V_u = 20.18 \text{ kips}$$

Because the shear strength provided by concrete (6.73 kips) is less than the maximum shear (20.18 kips) in the beam, transverse (i.e., shear) reinforcement must be provided. Calculate the shear that must be carried by the transverse reinforcement (Equation 6.54):

$$V_f = \frac{V_u - \phi V_{cf}}{\phi} = \frac{20.18 - 0.85(6.73)}{0.85} = 17.01 \text{ kips}$$

Check the maximum shear permitted to be carried by the shear reinforcement (Equation 6.58):

$$\frac{\rho_f E_f}{90\beta_1 f'_c} 8\sqrt{f'_c} b_w d = \frac{(0.012)(6500)}{(90)(0.85)(4)}\left[\frac{8\sqrt{4000}(12)(17.38)}{1000}\right]$$

$$= 26.9 \text{ kips} > V_f = 17.01 \text{ kips, OK}$$

Therefore, the beam cross section is adequate for shear, provided it is reinforced adequately for shear. The amount of FRP shear reinforcement will be determined from Equation 6.55. A No. 4 GFRP bar will be used, so that for a vertical, two-legged, closed stirrup, $A_f = 0.4$ in². To use Equation 6.55, $f_{fb}$ should be determined first from Equation 6.4 and then $f_{fv}$ from Equation 6.52.

For a No. 4 stirrup, $d_b = 0.5$ in., $r_b = 3d_b$ $3(0.5) = 1.5$ in.

$$f_{fb} = \left(0.05\frac{r_b}{d_b} + 0.3\right)f_{fu} = \left[0.05\left(\frac{1.5}{0.5}\right) + 0.3\right](72) = 32.4 \text{ ksi} \le f_{fu} = 72 \text{ ksi, OK}$$

The design stress in FRP reinforcement, $f_{fv}$, should not exceed

$$f_{fv} = 0.004E_f = 0.004(6500) = 26 \text{ ksi} < f_{fb} = 32.4 \text{ ksi}$$

Therefore, use $f_{fv} = 26$ ksi. The spacing, $s$, is calculated from Equation 6.55:

$$s = \frac{\phi A_{fv} f_{fv} d}{(V_u - \phi V_{c,f})} = \frac{(0.85)(0.4)(26)(17.38)}{[20.18 - (0.85)(6.73)]} = 10.63 \text{ in.}$$

Check the maximum spacing that can be permitted based on the value of $V_f$:

$$\frac{\rho_f E_f}{90\beta_1 f_c'}4\sqrt{f_c'}b_w d = \frac{(0.012)(6500)}{(90)(0.85)(4)}\left[\frac{4\sqrt{4000}(12)(17.38)}{1000}\right]$$

$$= 13.45 \text{ kips} < V_f = 16.63 \text{ kips}$$

Therefore, the maximum spacing of stirrups must be limited to $d/4 = 17.38/4 = 4.3$ in.
Determine the distance from the support at which $V_u = \phi V_c$ and $1/2\phi V_c$; this is the segment of the beam where minimum shear reinforcement must be provided (ACI 318-02 Sect. 11.5.5).

$$\phi V_c = 0.85(6.73) = 5.72 \text{ kips}$$

$$1/2\phi V_c = 1/2(5.72) = 2.86 \text{ kips}$$

With a factored uniform load $w_u = 2.36$ k/ft, a factored shear of $\phi V_c = 5.72$ kips occurs at $x = 5.72/2.36 = 2.42$ ft from midspan, i.e., at $10 - 2.42 = 7.58$ ft from the centerline of support. Likewise, a factored shear of $1/2\phi V_c = 3.02$ kips occurs at $x = 2.86/2.36 = 1.21$ ft from midspan, i.e., at $10 - 1.21 = 8.79$ ft from the centerline of support. Theoretically, no shear reinforcement is required in a length of 1.21 ft on each side of the midspan; however, minimum shear reinforcement will be provided in this segment of the beam. Thus, minimum shear reinforcement must be provided in the beam segment defined by $x = 7.58$ ft to $x = 10$ ft (midspan) from the centerline of each support.
Determine the maximum spacing for No. 4 closed stirrup from ACI 440 Equation 9-7:

$$s = \frac{A_v f_{fv}}{50 b_w} = \frac{(0.4)(26,000)}{50(12)} = 17.33 \text{ in.}$$

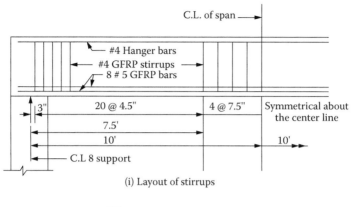

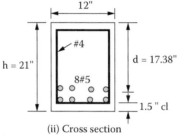

**FIGURE E6.6b** Details of shear reinforcement.

Check the maximum permitted spacing as per ACI 318-02:

$$s = \frac{A_v f_{fv}}{0.75 b_w \sqrt{f_c'}} = \frac{(0.4)(26,000)}{0.75(12)\sqrt{4000}} = 18.27 \text{ in.} > 17.33 \text{ in.}$$

Provide the following spacing of stirrups:

| Distance from Support | Calculated Spacing | Actual Spacing |
|---|---|---|
| 0–7.58 ft | 4.3 in. (= $d/4$) | 4.5 in.* |
| 7.58 ft–10 ft (midspan) | 17.33 in. | 8.5 in. |

\* Conservatively 4 inch can be used as the spacing.

Details of shear reinforcement are shown in Figure E6.6b.

## COMMENTARY

In the above example, it is assumed that the entire beam (i.e., end to end) is uniformly loaded with both dead and live load, in which case the shear is maximum at the ends and zero at the mid-span; the beam has been designed for shear for this loading condition. However, this loading condition does not produce maximum shear at the interior points. For such points, the maximum shear would be obtained when the uniform live load is placed from the point in question to the most distant end support. ACI 318 does not specify variable positioning of live load to produce maximum shear *in span* for *simple beams* although it does require consideration of variable positioning of loading for continuous spans. In any case, it would be prudent for engineers to exercise sound judgment and position live load on the beam to produce maximum shear at critical points and design the beam accordingly; *that* certainly is the Code's intent.

## REFERENCES

Al-Salloum, Y.A., Sayed, S.H., and Almusallam, T.H. *Some Design Considerations for Concrete Beams Reinforced with GFRP Bars*, First International Conference on Composite Infrastructure, *ICCI-96*, 1996, pp. 318–331.

American Concrete Institute (ACI), *American Concrete Institute Building Code Requirements for Reinforced Concrete*, ACI 318-95, Farmington Hills, MI: American Concrete Institute, 1995.

American Concrete Institute (ACI), *American Concrete Institute Building Code Requirements for Reinforced Concrete*, ACI 318-99, Farmington Hills, MI: American Concrete Institute, 1999.

American Concrete Institute (ACI), *American Concrete Institute Building Code Requirements for Reinforced Concrete*, ACI 318-02, Farmington Hills, MI: American Concrete Institute, 2002.

American Concrete Institute (ACI), *Guide for the Design and Construction of Concrete Reinforced with FRP Bars*, ACI Committee 440, ACI 440.1R-01, Farmington Hills, MI: American Concrete Institute, 2001.

American Concrete Institute (ACI), *Guide for the Design and Construction of Concrete Reinforced with FRP Bars*, ACI Committee 440, ACI 440.1R-03, Farmington Hills, MI: American Concrete Institute, 2003.

Bakis, C.E., Nanni, A., and Terosky, J.A., Smart, pseudo-ductile reinforcing rods for concrete; manufacture and test, *Proc. ICCI '96*, Saadatmanesh, H. and Ehsani, M.R. (Eds.), Tucson, AZ., January 15–17, 1996.

Belarbi, A., Chandrashekhara, K., and Warkins, S.E., Smart composite rebars with enhanced ductility, *Proc. 11th Conf. Eng. Mech.*, Ft. Lauderdale, FL, May 19–22, 1996.

Benmokrane, B., Challal, O., and Masmoudi, R. Flexural response of concrete beams reinforced with GFRP bars, *ACI Structural Journal*, 93(1): 46–55, January-February 1996.

Brown, V., Sustained load deflections in GFRP-reinforced concrete beams, *Proc. 3rd Intl. Symp. Non-Metallic (FRP) Reinforcement for Concrete Structures (FRPRCS-3)*, Vol. 2, Japan Concrete Institute, Sapporo, Japan, 1997.

Brown, L.V. and Bartholomew, C. FRP reinforcing bars in reinforced concrete members, *ACI Materials Journal*, 90(1): 34–39, January-February 1993.

Byars, E.F. and Snyder, R.D. *Engineering Mechanics of Deformable Bodies*, New York: Intext Educational Publishers, 1975.

Canadian Standards Association (CSA), Fiber reinforced structures and commentary, *Canadian Highway Bridge Design Code (CHBDC)*, draft Chapter 16, 1996 (final report completed in 1998).

Cozensa, E., Manfredi, G., and Realfonzo, R. Behavior and modeling of bond of FRP rebars to concrete, *Journal of Composites for Construction*, ASCE, 1(2): 40–51, 1997.

Dowden, D.M. and Dolan, C.W., Comparison of experimental shear data with code predictions for FRP prestressed beams, *Proc. 3rd Intl. Symp. Non-Metallic (FRP) Reinforcement for Concrete Structures (FRPRCS-3)*, Vol. 2, Japan Concrete Institute, Sapporo, Japan, 1997.

Faza, S.S. and GangaRao, H.V.S., Theoretical and experimental correlation of behavior of concrete beams reinforced with fiber reinforced plastic rebars, *Fiber-Reinforced-Plastic Reinforcement for Concrete Structures*, SP-138, Farmington Hills, MI: American Concrete Institute, 1993.

FRPRCS-IV, *Proc. 4th Intl. Symp. Fiber Reinforced Polymer Reinforcement for Reinforced Concrete Structures*, ACI Sp-188, 1999.

GangaRao, H.V.S. and Faza, S.S. Bending and Bond Behavior and Design of Concrete Beams Reinforced with Fiber Reinforced Plastic Rebars, Research Report CFC-92-142, Submitted to West Virginia Department of Highways, Charleston, WV, 1991.

GangaRao, H.V.S., Vijay, P.V, Kalluri, R., and Taly, N.B., Behavior of concrete T-beams reinforced with GFRP bars under bending, *Proc. ACMBS-III*, Ottawa, Canada, August 2000.

Harris, H.G., Somboonsong, F.K. Ko, and Huesgen, R., A second generation ductile hybrid fiber reinforced polymer (FRP) for concrete structures, *Proc. ICCI '98*, Saadatmanesh, H. and Ehsani, M.R. (Eds.), Tucson, AZ, 1998.

Huesgen, R., Flexural behavior of ductile hybrid FRP bars in singly reinforced concrete beams, M.Sc. thesis, Philadelphia: Department of Civil and Architectural Engineering, Drexel University, 1997.

*ICCI '96, Proc. ICCI '96*, Tucson, AZ, January 15–17, 1996.

Jaeger, L.G., Mufti, A.A. and Tadros, G., The concept of the overall performance factor in rectangular section reinforced concrete members, *Proc. 3rd Intl. Symp. Non-Metallic (FRP) Reinforcement for Concrete Structures (FRPRCS-3)*, Vol. 2, Japan Concrete Institute, Sapporo, Japan, 1997.

Japanese Society of Civil Engineers (JSCE), Serviceability limit states, recommendations for design and construction of concrete structures using continuous fiber reinforcing materials, Machida, A. (Ed.), *JSCE Research Committee on Continuous Fiber Reinforcing Materials, Concrete Engineering Series 23*, JSCE, October 1997.

Kage, T., Masuda, Y., Tanano, Y., and Sato, K., Long-term deflection of continuous fiber reinforced concrete beams, *Proc. 2nd Intl. RILEM Symp. Non-Metallic (FRP) Reinforcement for Concrete Structures (FRPRCS-2)*, Gent, Belgium, 1995.

Masmoudi, R., Benmokrane, B., and Challal, O., Cracking behavior of concrete beams reinforced with FRP rebars, *Proc. ICCI '96*, Tucson, AZ, January 15–17, 1996.

Naaman, A.E. and Jeong, S.M., Structural ductility of concrete beams prestressed with FRP tendons, Non-metallic (FRP) Reinforcement for Concrete Structures, *Proc. 2nd Intl. RILEM Symp. (FRPRCS-2)*, Taewere, L. (Ed.), London: E & FN Spon, Publishers, 1995.

Nanni, A., Henneke, M.J., and Okamoto, T., Tensile properties of hybrid rods for concrete reinforcement, *Construction and Building Matls.*, 8, 1, 27–34, 1994a.

Nanni, A., Henneke, M.J., and Okamoto, T., Behavior of hybrid rods for concrete reinforcement, *Construction and Building Matls.*, 8, 2, 89–95, 1994b.

Nawy, E.W. and Neuwerth, A.M. Fiberglass reinforced concrete beams and slabs, *Journal of Structural Division*, ASCE, February, pp. 421–440, 1977.

Nawy, E.W. and Neuwerth, A.M. Behavior of fiberglass reinforced concrete beams, *Journal of Structural Division*, American Society Civil Engineers, September, pp. 2203–2215, 1971.

Ozbolt, J., Mestrovic, D., Li, Y.J., and Eligehausen, R., Compression failure of beams made of different concrete types and sizes, *J. Struct. Eng.,* 200–209, February 2000.

Park, R. and Paulay, T., *Reinforced Concrete Structures*, New York: John Wiley & Sons, 1975.

Priestley, M.J.N., Seible, F., and Calvi, G.M. *Seismic Design and Retrofit of Bridges*, New York: John Wiley & Sons, 1996.

*Proc. ACMBS-II*, Montreal, Canada, August 1996.

*Proc. ACMBS-III*, Ottawa, Canada, August 2000.

Razaqpur, A.G., and Ali, M.M., A new concept for achieving ductility in FRP reinforced concrete, *Proc. ICCI '96*, Saadatmanesh, H. and Ehsani, M.R. (Eds.), Tucson, AZ, January 15–17, 1996.

Somboonsong, W., Development of a class of ductile hybrid FRP reinforcement bar for new and repaired concrete structures, Ph.D. thesis, Philadelphia: Department of Civil and Architectural Engineering, Drexel University, August 1997.

Sonobe, Y., Fukuyama, H., Okamoto, T., Kani, N., Kimura, K., Kobayashi, K., Masuda, Y., Matsuzaki, Y., Nochizuki, S., Nagasaka, T,. Shimizu, A., Tanano, H., Tanigaki, M., and Teshigawara, M., Design guidelines of FRP reinforced concrete building structures, *J. Composites for Construction*, ASCE, 90–115, August 1997.

Tamuzs, V., Tepfers, R., Apints, R., Vilks, U., and Modniks, J., Ductility of hybrid fiber composite reinforcement bars for concrete, *Proc. ICCI '96*, Saadatmanesh, H. and Ehsani, M.R. (Eds.), Tucson, AZ, January 15–17, 1996.

Theriault, M. and Benmokrane, B. Effects of FRP reinforcement ratios and concrete strength on flexural behavior of concrete beams, *Journal of Composites for Construction*, ASCE, February, pp. 7–16, 1998.

Vijay, P.V., Aging Behavior of Concrete Beams Reinforced with GFRP Bars, Ph.D. dissertation, West Virginia University, Morgantown, May 1999.

Vijay, P.V. and GangaRao, H.V.S., Bending behavior and deformability of glass fiber-reinforced polymer reinforced concrete members, *ACI Struct. J.*, 98(6), 834–842, November-December 2001.

Vijay, P.V. and GangaRao, H.V.S., A unified limit state approach using deformability factors in concrete beams reinforced with GFRP bars, *Materials for the New Millennium*, 4th Matl. Conf., ASCE, Vol. 1, Washington DC, 1996.

Vijay, P.V. and GangaRao, H.V.S., Creep behavior of concrete beams reinforced with GFRP bars, *Proc. 1st Intl. Conf. Durability of Fiber Reinforced Polymer (FRP) Composites for Construction (CDCC-1998)*, 1998.

Vijay, P.V. and GangaRao, H.V.S., Development of fiber reinforced plastics for highway applications (Task A-2)-Aging Behavior of Concrete Beams Reinforced with GFRP Bars, *WVDOH RP #T-699-FRP-1*, Charleston, WV: West Virginia Department of Transportation, May 1999.

Zhao, W., Pilakoutas, K., and Waldron, P. FRP Reinforced Concrete: Cracking Behavior and Determination, *Non-Metallic (FRP) Reinforcement for Concrete Structures*, Proceedings of the Third International Symposium, Vol. 2, October 1997, pp. 439–446.

# 7 Bond and Development Length

## 7.1 INTRODUCTION

The bending strength of a concrete beam reinforced with a bar — FRP or steel — depends upon the compressive strength in concrete, and development of ultimate tensile strength in a bar, which in turn depends on the strain compatibility of the reinforcing bar with the surrounding concrete. The two materials (bar and concrete) must act in unison to develop full structural compositeness and efficiently resist external loads. Thus, the tension force resisted by the reinforcing bar, which induces bond stress, must be in equilibrium with the compression force resisted by the concrete.

*Bond stress* is defined as the local horizontal shear force (contiguous to the outer surface of the bar) per unit area of the bar perimeter. The bond stress is transferred from the concrete to the bar at its interface. Bond stress varies along the bar length and is a function of bending moment variation or other force variations such as the net pulling force. To attain full bending capacity of a concrete section, the reinforcing bar must deform (elongate or shorten) along with the surrounding concrete without any interfacial slippage or separation due to external loads. Bond stresses are induced at the concrete interface due to the forces developed in tension or compression reinforcement under external loads. In addition, adequate confinement of concrete with the help of reinforcing bars and stirrups must be present to resist the design loads [Malvar 1992].

The Young's modulus and ultimate strength of the FRP bar are much greater than those of the typical structural concrete. These properties help to increase the bending capacity of a reinforced concrete section by several times over an unreinforced concrete section. An increase in the bending capacity is achieved if premature slippage between a reinforcing bar and the surrounding concrete is prevented. The strain-to-failure ratio of the bar (around 0.015 to 0.025) is several times greater than the concrete failure strain (around 0.003 to 0.0045); however, proper confinement of the concrete prevents catastrophic failure of concrete members and provides more than adequate warning before failure [Eshani et al. 1993; Malvar 1992; Vijay and GangaRao 1996]. The influence of hoop reinforcement on the stress-strain curve for concrete, especially on strain-to-failure of concrete, could be as high as five times that of strain-to-failure in plain concrete. Of course, the strain-to-failure variation depends on the ratio of confining steel spacing (pitch) to the portion of confined concrete core. The behavior of confined concrete under axial and flexural stress has been discussed by Park and Paulay [1975].

Bond behavior is one of the many aspects of performance of FRP bars in a concrete environment that needs a thorough evaluation and comprehensive understanding under various thermomechanical loads, and physical and chemical aging of the bond under varying environmental conditions. Bond strength is controlled by many factors, chiefly:

1. The ability to develop interlocking stresses under tension, including mechanical frictional resistance and chemical adhesion at the interface to resist slippage;
2. The effect of concrete strength in controlling failure modes, including horizontal shear-frictional resistance offered by the bar and concrete at the interface against slip;
3. The effect of concrete shrinkage around the bar; and
4. The diameter, surface texture, spacing, cover, stiffness, and bundling of FRP bars.

As with any composite material, understanding the interaction between the reinforcements (fibers) and the matrix is important to understand the failure of the FRP composite. The design of concrete members reinforced with FRP or steel bars is based on the assumption of perfect bond (no slip) between the bars and concrete, i.e., they act in unison in terms of tensile, compressive, or bending deformations. Generally, concrete is assumed to exert horizontal shear-type stresses on the surface of the bars. This interaction provides the mechanism by which forces are transferred internally from concrete to reinforcing bars. In the absence of these resisting horizontal shear-type stresses (i.e., bond stresses), the reinforcing bars will remain unstressed and would contribute minimally to the flexural strength of the concrete beam. Upon loading, such a beam will fail in a brittle manner, with the concrete tensile stress reaching the modulus of rupture. Thus, in a reinforced concrete beam the bond between the reinforcing bars and concrete must be present to resist the applied forces by these bars to avoid brittle failure.

The bond stresses induced in a reinforced concrete beam can be classified into two types: (1) flexural bond stress (a change in stress along the bar length due to a bending moment variation between the adjacent cracks is equated to bond stress resistance), and (2) anchorage bond stress (an anchorage pull force on the bar is equated to bond resistance encountered by the bar in concrete). These two topics are discussed in Section 7.2 and Section 7.4.

The length of bar that offers the bond resistance between two adjacent cracks in concrete is referred to as the *development length*, $l_{bf}$. The development length, which can be expressed as a function of the bar size and its yield or ultimate strength, determines the bar's resistance to slippage. It is also a function of many other parameters, such as the compressive strength of concrete and the bond strength. Consequently, the development length dictates, to some extent (others being bending, shear, and so on), the bending resistance or failure capacity of a concrete beam reinforced with a FRP or steel bar. In addition, the development length varies with bar surface texture, spacing, stiffness, and bundling configurations. These values have been determined primarily from experimental data [ACI 440.1R-03; Vijay and Ganga-Rao 1999]. Research on the bond strength of FRP bars is discussed in Section 7.4.

## 7.2  BOND BEHAVIOR OF STEEL BARS

Provisions for the bond of steel-reinforced concrete beams in the ACI Code have changed over time [ACI 318-02]. Prior to 1971, bond stresses were assumed to be caused by only the change in moment from section to section along the span length. Expressions for the average bond stress, $u_{avg}$, at a section in a beam were related to shear at that section, as given by Equation 7.1:

$$u_{avg} = \frac{V}{jd\Sigma o} \qquad (7.1)$$

where

$V$ = the shear at the section under consideration

$\Sigma o$ = the sum of perimeters of all reinforcing bars resisting tension due to moment at the section

$jd$ = the lever arm of the moment-resisting couple

Equation 7.1 is based on the assumption that the moment gradient in a beam causes a proportional change in tensile stresses in the reinforcement, and this change is equilibrated by flexural stresses. This equation is no longer used for the design of reinforced concrete members except for reinforced masonry, for which allowable stress design principles are still in use [Taly 2001]. In reinforced concrete members, the presence of flexural cracks has been recognized to create large local variations in experimental bond stresses when compared with bond stresses computed at the crack location. This inconsistency resulted in the development of an alternate approach — the concept of development length — in which the actual length of anchorage (i.e., the distance between the point of maximum stress and the near end, i.e., location of minimum bond stress, of the bar) is compared with the minimum length to ensure the adequate anchorage required to prevent bar slippage before the ultimate design load is reached.

In recognition of the above stated inconsistencies, research is continuing for identifying the many factors that contribute to bond strength, which include [Cosenza et al. 1997]:

1. Strength of concrete
2. Chemical adhesion and interfacial elongation
3. Friction
4. Pressure of the hardened concrete against the bar due to the drying shrinkage of the concrete (i.e., the gripping effect)
5. Bearing of bar deformations (also called *lugs* or *ribs*) against the concrete

Presence of tensile force in the bars gives rise to Poisson's effect, which leads to a reduction in the bar diameters as they are tensioned and enables them to slip more easily. This means that if the reinforcing bars were to be straight, plain, and smooth, very little resistance to slippage will exist except mechanical friction and chemical bonding. The beams will then be only a little stronger than their unreinforced counterpart. Another parameter that influences the bond strength is the position of a reinforcing bar in a concrete beam. Tests have shown that the bond strength of top bars (i.e., bars positioned above a 12-in. or greater depth of concrete) is approximately 66% that of the bottom bars [Ehsani et al. 1993]. This happens as a consequence of the air and water rising from the freshly placed concrete beneath the bars and collecting on the bottom surface of the top bar. In addition, better compaction and higher hydrostatic pressure exerted on the bottom bar than on the top bar of the concrete beam leads to a better bond resistance provided by the bottom bar.

Bond resistance mechanisms in a concrete-steel bar environment can be classified in two distinct types: friction and bearing, the latter being the primary source of stress transfer. The development length, $l_d$, for deformed steel reinforcing bars with a nominal diameter $d_b$, is provided by ACI 318-02 (Equation 12-1), which is subject to a minimum development length of 12 in.

$$\frac{l_d}{d_b} = \frac{3}{40} \frac{f_y}{\sqrt{f_c'}} \frac{\alpha\beta\gamma\lambda}{\left(\dfrac{c + K_{tr}}{d_b}\right)}$$
(7.2)

and $l_d$ not less than 12 in.
where

$\alpha$ = reinforcement location factor
$\beta$ = coating factor
$\gamma$ = reinforcement size factor
$\lambda$ = lightweight aggregate concrete factor
$c$ = concrete cover (in.)

$K_{tr} = \dfrac{A_{tr} f_{yt}}{1500sn}$ transverse reinforcement index

$A_{tr}$ = total cross-sectional area of all transverse reinforcement that is within the spacing $s$ and that crosses the potential plane of splitting through the reinforcement being developed (in²)
$s$ = maximum center to center spacing of transverse steel within $l_d$ (in.)
$f_{yt}$ = specified yield strength of transverse reinforcement (psi)
$n$ = number of bars being spliced or developed along the plane of splitting

In Equation 7.2, the development length $l_d$ is defined as the minimum embedment length required to anchor a bar that is stressed to the yield point $f_y$ of steel. $K_{tr}$ is a factor that represents the contribution of confining reinforcement across potential splitting planes. The term $(c + K_{tr})/d_b$ cannot be greater than 2.5 or less than 1.5.

For values of $(c + K_{tr})/d_b$ less than 2.5, splitting failures are likely to occur, whereas for values greater than 2.5, pullout failure can be expected. Recognizing that bond failures are caused by the tensile cracking of concrete, bond strength is correlated with $\sqrt{f_c'}$, limited to a maximum of 100 psi per ACI 318-02, (Sect. 12.1.2). The coefficient $\alpha$ is the traditional *reinforcement location factor*, which is included in Equation 7.2 to account for the adverse effects of the top reinforcement casting position. The coefficient $\beta$ is a *coating factor* that accounts for the effects of epoxy coating application. The coefficient $\gamma$ is the *reinforcement size factor*, included in Equation 7.2 to reflect the more favorable performance of smaller-size reinforcing bars as compared with that of the larger-size bars. The coefficient $\lambda$ is the *lightweight concrete factor*, to reflect the generally lower tensile strength of lightweight concrete and the resulting reduced splitting strength that is important in the development of reinforcement. The *strength reduction factor*, $\phi$, does not appear in Equation 7.2 because an allowance for strength reduction is already included in developing this equation. Readers are referred to the ACI 318 Code (Sect. 9.3.3) and the ACI 318 Commentary (Sect. 12.2.2).

Equation 7.2 presents difficulties for practical use because of the difficulty associated with the determination of the term $(c + K_{tr})/d_b$ as both $c$ and $K_{tr}$ vary along the length of a member. However, many practical construction applications utilize bar spacings and cover values along with confining reinforcement (e.g., stirrups, ties, spirals, and so on) that result in the value of $(c + K_{tr})/d_b$ at least equal to 1.5. For example, a clear cover of the bar being developed (or spliced) of not less than $d_b$ yields $c = d_b$ (= spacing of $d_b/2 + d_b/2$). Also, the minimum transverse reinforcement (i.e., stirrups) corresponds to the value of $K_{tr} = 0.5d_b$. These values yield $(c + K_{tr})/d_b = 1.5$. In the case of a clear spacing of bars being developed (or spliced) at $2d_b$ or greater, and clear cover of miminum $d_b$ and no transverse reinforcement (i.e., no stirrups or ties), $c$ corresponds to $1.5d_b$ so that again $(c + K_{tr})/d_b = 1.5$. Furthermore, the ACI-specified minimum bar spacings and cover correspond to $c = 1.0d_b$ and $(c + K_{tr})/d_b = 1.0$. The bar size factor has been determined to be 0.8 for bars No. 6 and smaller and 1.0 for No. 7 and larger bars. With these considerations, ACI 318-02 Section 12.2.2 presents modified forms of Equation 7.2, i.e., without the $(c + K_{tr})/d_b$ term and the $\gamma$-factor, which are more user-friendly.

## 7.3  BOND BEHAVIOR OF FRP BARS

The forgoing brief discussion serves as background information on the bond behavior of steel bars and to see if the model represented by Equation 7.2 for steel reinforced concrete members can be applied for estimating bond stresses in members reinforced with FRP bars. The understanding of bond behavior of FRP bars is important because most bond behavior models for studying the bond of FRP bars have evolved from models that were previously developed for steel bars. Because these models for steel bars are well-established, they serve as convenient data to define the bond behavior of FRP bars. Whereas the use of both types of reinforcing bars is the same (i.e., to transfer tensile forces in concrete to FRP bars), marked differences exist in force transfer and failure mechanisms of steel and FRP bars

[Faza 1991; Faza and GangaRao 1990]. They arise from variations in the material properties as well as interaction mechanisms of concrete and reinforcing bars. The most fundamental difference is that steel is an isotropic, homogenous, and elasto-plastic material, whereas the FRP is anisotropic, nonhomogeneous, and brittle, having a Young's modulus one-fifth to one-sixth that of steel. The anisotropy of the FRP bar results from the fact that its shear and transverse properties are dependent on both resin and fiber type, even though the longitudinal properties are dominated by fibers [Cosenza et al. 1997]. Because material anisotropy leads to different physical and mechanical properties in both longitudinal and transverse directions, the anisotropic nature of the FRP materials needs to be accounted for in the development of design equations and in the understanding of failure mechanisms [Ganga-Rao et al. 2001].

The mechanical properties of the steel and FRP reinforcing bars are qualitatively and quantitatively different from each other [JSCE 1997]. Also, FRP bars produced by different manufacturers are different in that they involve different manufacturing processes for the outer surface and significant material differences in the longitudinal and transverse directions. Because of these and many other differences, the models used to predict the bond behavior of steel bars are not suitable for FRP bars.

The bond behavior of FRP bars in a concrete environment is controlled by many factors, including those that were discussed earlier in the context of steel bars. These factors mainly include the following:

1. Chemical adhesion
2. Friction due to the surface roughness of FRP bars leading to mechanical interlock of the FRP rods against the concrete
3. Hydrostatic pressure against the FRP rods from shrinkage of hardened concrete
4. Swelling of FRP rods due to temperature change and moisture absorption

In addition, the long-term bond behavior of FRP bars in a concrete environment can vary significantly depending upon environmental exposures such as:

1. Freeze-thaw effects
2. Wet-dry cycles
3. pH variation
4. Sustained stress (creep)
5. Combination of items 1 through 4

Experimental research has been carried out by many researchers in the last decade to investigate the bond behavior of FRP bars embedded in concrete [Vijay and GangaRao 1999]. The variables include types of bars characterized by surface configuration, quality and quantity of fibers, and various types of resins used as binders of fibers. Therefore, understanding the differences in the various types of commercially available FRP bars that have been investigated is instructive.

Various types of FRP bars used in reinforced concrete construction can be classified based on the type of fibers used in their manufacture: glass (glass fiber-

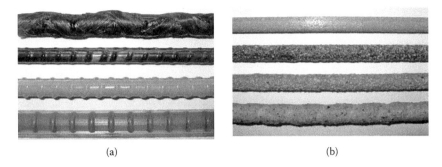

(a)                                                    (b)

**FIGURE 7.1** Various kinds of FRP bars investigated for bond: (a) ribbed bars (bottom to top): glass with nanocaly, glass, carbon, braided surface carbon bar; (b) sand coated surface bars (from bottom to top): glass with helical wrap, plain glass, plain carbon, and plain aramid.

reinforced polymer or GFRP); carbon (carbon fiber-reinforced polymer or CFRP); aramid (aramid fiber-reinforced polymer or AFRP); and polyvinyl alcohol (polyvinyl alcohol fiber reinforced FRP or VFRP). At a macro level, FRP bars can be classified with respect to their surface characteristics: smooth bars, i.e., bars having no surface indentations or undulation; and surface treated bars, e.g., with sand coating or helical wrapping. Several types of commercially available FRP bars are shown in Figure 7.1.

The use of smooth-surfaced bars can lead to partial compositeness, which is manifested by premature bond failure of reinforced concrete elements [Vijay and GangaRao 1999]. Producers of FRP bars have adopted various processes to improve bond characteristics, which can be classified into two main types: deforming the outer surface of bars (ribbed or braided) and roughening the surface of bars (sand-coated, Figure 7.1).

Yet another type of FRP bars is the hollow bar (Figure 7.2) that purports to have advantages of having a strength-to-weight ratio higher than that of the solid bars. The hollow bar can be used both as structural elements as well as conduits [Kachlakev 1997]. Two types of hollow GFRP bars have been investigated: bars

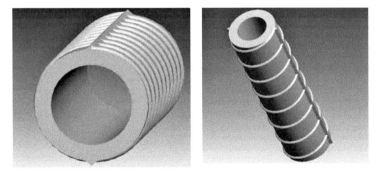

**FIGURE 7.2** Schematic representation of a hollow core FRP bar with ribbed surface. (From Ohnstad, T.S. and Kachlakev, D.I., "Hollow FRP rebar — Its unique manufacturing process and applications," *Proc. 46th Intl. SAMPE Symp., Vol. 46: A Materials and Processes Odyssey*, Paper code: 46-064, 2001. With permission.)

with unidirectional fibers and bars with fibers in an off-axis direction [Kachlakev and Lundy 1999].

Additionally, differences in various types of bars can exist in fiber-resin volume ratios and production methods, both of which significantly influence their mechanical properties. Recognizing that no standards exist for the manufacturing of FRP bars is important. For example, the deformed FRP bars include ribbed-type bars, indented rods, twisted strands, and spiral glued-type bars. Within each type, the bars are made of different materials (fibers and matrix) and have a different geometry (i.e., fiber diameter and orientation, spacing and size of ribs).

These differences in the physical and mechanical properties of various types of FRP bars introduces randomness in bond behavior of the FRP bars, leading to complexities in the development of a uniform quantitative bond-slip relationship [Cosenza, Manfredi, and Realfonzo 1996].

## 7.4   RESEARCH ON BOND STRENGTH OF FRP BARS

Much work has been conducted in the last two decades to investigate the mechanical behavior of FRP-reinforced concrete structures [Dolan, Rizkalla, and Nanni 1999; Nanni 1993; Yonizawa et al. 1993]. Considerable experimental research has been conducted to understand specifically the bond behavior of FRP bars in a concrete environment. This includes tests on beams and pullout specimens using many different types and sizes of bars [Bakis et al. 1983; Honma and Maruyama 1989; Ito et al. 1989; Soroushian et al. 1991; Cox and Herrmann 1992; Daniali 1992; Faoro 1992; Malvar 1992, 1994, 1995; Tepfers et al. 1992; Challal and Benmokrane 1993; Ehsani et al. 1993; Kanakubo et al. 1993; Laralde and Rodriguez 1993; Makitani et al. 1993; Mashima and Iwamoto 1993; Al-Zaharani 1995; Alunno et al. 1995; Boothby et al. 1995; Hattori et al. 1995; Noghabai 1995; Rossetti et al. 1995; Cosenza et al. 1995, 1996; Nanni et al. 1995; Al-Zaharani 1996; Al-Zaharani et al. 1996, 1999; A-Dulajian et al. 1996; Benmokrane et al. 1996; Tomosawa and Nakasuji 1997; Freimanis et al. 1998; Tighiourt et al. 1998; Guo and Cox 1999; Vijay and GangaRao 1999; Foccachi et al. 2000]. Reserach conducted for understanding the behavior of hollow FRP bars has been reported by Kachlakev [1997] and Kachlakev and Lundy [1999]. Most of this research involved pullout tests (Figure 7.3). The focus of this effort has been to understand the resisting mechanisms generated in the pullout tests, to formulate bond ($\tau$)-slip constitutive laws (Figure 7.4 and Figure 7.5), determining design criteria, and bond strength and stiffness of steel and FRP bars. Most tests were done under monotonic loading; only limited research has been conducted for cyclic loading [Bakis et al. 1998; Den Uijl 1995; Katz 2000]. Also, most of the tests were conducted studying the behavior of bars placed in concrete in a single layer; only very limited research has been conducted on bundled bars. Vijay and GangaRao [1999] have reported on the bond strength of bundled bars based on tests of two-bar, three-bar, and four-bar bundle configuration. A limited amount of analytical work has been carried out to model the bond behavior of FRP bars based on test results [Cox and Guo 1999; Cox and Yu 1999; Uppuluri et al. 1996; Yu and Cox 1999]. Bond behavior models for FRP bars have

**FIGURE 7.3** Pull-out tests for FRP bars. (From Vijay, P.V. and GangaRao, H.V.S., Development of fiber reinforced plastics for highway application: aging behavior of concrete beams with GFRP bars, CFC-WVU Report No. 99-265 (WVDOT Rep. No. 699-FRP1), Department of Civil and Environmental Engineering, Morgantown, WV: West Virginia University, 1999.)

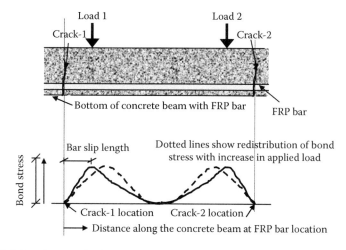

**FIGURE 7.4** Bond stress distribution in a FRP reinforced beam with cracks and bar slippage.

been developed by Malvar [1994], Cosenza, Manifredi, and Realfonzo [1995; 1996] and others [Tighiourt et al. 1998], which are based on the Eligehausen-Popov-Bertoro (BPE) model developed for steel bars [Ciampi et al. 1981; Eligehausen et al. 1983]. Figure 7.6 shows a comparison between the bond strength of steel bars and two types of commercially available FRP bars. A summary of the state-of-the-art on bond behavior of FRP reinforcing bars has been presented by Taly and GangaRao [2001].

In general, in a concrete-FRP bars environment, two types of bond resistance mechanisms can be identified: friction-type resistance and bearing-type resistance

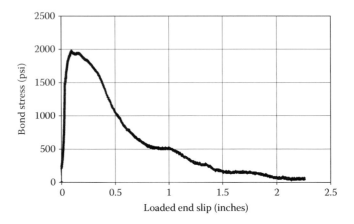

**FIGURE 7.5** Load-slip relation for a GFRP reinforced bar. (From Vijay, P.V. and GangaRao, H.V.S., Development of fiber reinforced plastics for highway application: aging behavior of concrete beams with GFRP bars, CFC-WVU Report No. 99-265 (WVDOT Rep. No. 699-FRP1), Department of Civil and Environmental Engineering, Morgantown, WV: West Virginia University, 1999.)

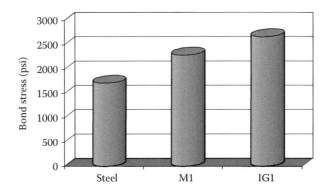

**FIGURE 7.6** Comparison of bond strength between steel, ribbed GFRP (M1) and sand coated (IG1) bars. (From Vijay, P.V. and GangaRao, H.V.S., Development of fiber reinforced plastics for highway application: aging behavior of concrete beams with GFRP bars, CFC-WVU Report No. 99-265 (WVDOT Rep. No. 699-FRP1), Department of Civil and Environmental Engineering, Morgantown, WV: West Virginia University, 1999.)

(when the mechanical interlock becomes dominant) [Makitani et al. 1993]. To simplify understanding of the research findings of the experimental investigation, the researchers have found it convenient to separate deformed bars (i.e., ribbed, twisted, braided) from other bars (i.e., smooth, sand grain-covered, sand-blasted rods). This separation makes understanding easier, within each group, differences in the bond resistance mechanisms and influences of various parameters on the bond strength.

Typically, two types of tests are conducted to measure the bond strength of reinforcing bars: pullout tests and beam tests, both of which give different values.

Bond strength from beam tests is typically found to be lower than from the pullout tests. This is because in the pullout tests, the splitting of concrete is avoided due to the absence of a local bending on the bar, a higher thickness of the concrete cover, and the confining action of the reaction plate on the concrete specimen (i.e., the concrete surrounding the reinforcing bars is in compression). Alternatively, in the beam tests the concrete surrounding the reinforcing bars is in tension, which varies along the span length and leads to cracking under low stresses and reduction in the bond strength. Thus, the pullout tests give an upper-bound value for the bond-slip performance of FRP bars. However, some researchers consider the beam tests as more realistic than the pullout tests in simulating the real behavior of concrete members in flexure [Tighiourt, Benmokrane, and Gao 1998].

The two types of bond tests, the pull-out and the beam bending, can provide a qualitative understanding of the bond stress distribution from the crack tip, crack spacing, and modes of failure. Note that the bond stress distribution in a bar in a concrete cylinder pull-out test subjects concrete near the contact area of the pull-out specimen with the platen of the machine in compression while the bar acts in tension. However, in a beam bending test both the bar and the surrounding concrete are in tension or in compression, depending on the bar location with respect to the bending load.

## 7.5  ESTIMATION OF BOND STRENGTH

Models representing possible bar slippage from surrounding concrete in a beam and the corresponding stress redistribution are complex. This is attributed to the progression of cracking in a concrete beam under bending due to bar slippage (Figure 7.4). Therefore, the appropriate bond length for a bar in a certain type of concrete must necessarily be developed from first principles, i.e., with the idea of a bar "ideally" attaining full tensile strength before bond failure. Neglecting the resistance offered by other forces and focusing on only the anchorage bond and flexural bond, the development length equation for FRP bars has been derived as follows.

The expression for the development length of an FRP bar can be derived by considering the equilibrium of forces developed at the bar-concrete interface. The optimal bond length can be defined as the minimum length of the bar that must be embedded in concrete to resist the ultimate strength of the FRP bar [Tighiourt, Benmokrane, and Gao 1998]. This bar length is called the "basic development length," $\ell_{bf}$.

Figure 7.7 shows the equilibrium condition of such a bar having a length equal to the basic development length, $\ell_{bf}$. The force in the FRP bar is resisted by the bond

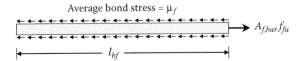

**FIGURE 7.7** Force transfer in the region of development length.

force induced between the concrete and the surface of the bar. The force in the FRP bar can be expressed in terms of its ultimate strength, $f_{pu}$, and its cross-sectional area, $A_{f,bar}$:

$$\text{Force in FRP bar at rupture} = A_{f,bar}f_{pu} \qquad (7.3)$$

The bond resistance developed over the concrete-bar interface area can be expressed in terms of the circumferential area of concrete surrounding the bar and the average bond stress, $\mu_f$, developed over the development length, $\ell_{bf}$:

$$\text{Circumferential area surrounding the bar over the development length} = \pi d_b \ell_{bf}$$
$$\text{Average bond stress} = \mu_f$$
$$\text{Bond resistance} = (\pi d_b \ell_{bf})\mu_f \qquad (7.4)$$

Equating the forces given by Equation 7.3 and Equation 7.4 yields

$$\ell_{bf}\pi d_b \mu_f = A_{f,bar}f_{fu} \qquad (7.5)$$

Rearranging Equation 7.5, the development length of an FRP bar can be expressed as:

$$\ell_{bf} = \frac{A_{f,bar}f_{fu}}{\pi d_b \mu_f} \qquad (7.6)$$

Substitution of $A_{f,bar} = \dfrac{\pi d_b^2}{4}$ in Equation 7.6 yields

$$\ell_{bf} = \frac{d_b f_{fu}}{4\mu_f} \qquad (7.7)$$

Research on development length by Orangun, Jirsa, and Breen [1977] has illustrated that the bond stress of steel bars is a function of the square root of the concrete strength and the bar diameter. A general expression for the average bond stress can be developed as follows:

$$\mu_f = \frac{K_1\sqrt{f_c'}}{d_b} \qquad (7.8)$$

where $K_1$ is a constant. Substitution of the value of $\mu_f$ from Equation 7.8 into Equation 7.7 yields the development length of FRP reinforcement as:

**TABLE 7.1**
**Suggested Values of the Development Length Coefficient, $K_2$, in Equation 7.10**

| Source | Value of $K_2$ for FRP Bars | | Types of Tests |
|---|---|---|---|
| | GFRP | Aramid | |
| Pleimann [1987; 1991] | 1/19.4 | 1/18 | Pullout tests on No. 2, No. 3, and No. 4 FRP bars |
| Faza and GangaRao [1990] | 1/16.7 | | Pullout tests and cantilever beams |
| Ehsani, Sadaatmanesh, and Tao [1996a] | 1/21.3 | | 48 beam specimens and 18 pullout specimens with No. 3, No. 6, and No. 9 bars |
| Tighiouart, Benmokrane, and Gao [1998] | 1/5.6 | | 45 beam specimens using two types of No. 4, No. 5, and No. 8 bars |

$$\ell_{bf} = \frac{d_b f_{fu}}{4\left[\dfrac{K_1\sqrt{f_c'}}{d_b}\right]}$$

$$= \frac{d_b^2 f_{fu}}{4K_1\sqrt{f_c'}} \tag{7.9}$$

For simplicity, Equation 7.9 can be expressed as

$$\ell_{bf} = K_2 \frac{d_b^2 f_{fu}}{\sqrt{f_c'}} \tag{7.10}$$

where $K_2$ is a new constant called the *development length coefficient*. Researchers have conducted many tests to determine a realistic value of $K_2$, which has been found to vary with the size and the type of the bar and the type of test (beam tests versus pullout tests). The many different values of $K_2$ reported in the literature are presented in Table 7.1.

In view of the complexity of Equation 7.10 and the wide variation in the reported values of $K_2$, Ehsani, Saadatmanesh, and Tao [1996a] and Gao, Benmokrane, and Tighiourt [1998] proposed Equation 7.11 for the basic development length of FRP bars based on pullout controlled failure rather than concrete splitting:

$$\ell_{bf} = \frac{d_b f_{fu}}{K_3} \tag{7.11}$$

where $K_3$ had a numerical constant of approximately 2850 [ACI 440.1R-03], so that Equation 7.11 can be expressed as

$$\ell_{bf} = \frac{d_b f_{fu}}{2850} \tag{7.12}$$

Considering all of the aforementioned findings, the ACI Committee 440 has suggested a conservative value of $K_3$ as 2700 for the pullout controlled failures:

$$\ell_{bf} = \frac{d_b f_{fu}}{2700} \tag{7.13}$$

Note that ACI 318-02 (Equation 12-1) accounts for concrete confinement effects from steel shear reinforcement across potential splitting planes through a coefficient $K_{tr}$ as defined earlier under Equation 7.2. In the case of FRP bars, a more general approach to computing the development length has been adopted by the ACI Committee 440 because of the paucity of bond test data on FRP reinforcement utilizing FRP stirrups.

## 7.6  CURRENT RESEARCH FINDINGS

A common conclusion arrived at from all tests is that the bond behavior and failure mechanisms of FRP bars are significantly different from those of steel bars in similar concrete environments. Typically, the chemical bond between concrete and FRP bars is found to be very low, and the mechanical interlock acts as the primary means of force transfer. A summary of test data, with a comprehensive discussion of tests done prior to 1996, has been provided by Cosenza, Manfredi, and Realfonzo [1997]. The authors provide a detailed discussion on the influence of various parameters on the bond strength of FRP bars. Freimanis et al. [1998] have provided a comparison of pullout and tensile behaviors of FRP reinforcement in concrete in terms of pitch and depth (indent) of helical wrappings. Loaded-end slip of FRP bars at the onset of free-end slip was relatively constant for bars with nearly equal ratios of embedment length to bar diameter. The maximum bond stress was found to increase with increasing depth and the indent depth does not seem to depend on indent pitch, within practical limits [Freimis et al. 1998]. However, the tensile strength of bars embedded in concrete revealed an increase in strength with decreasing pitch, possibly due to better fiber confinement and a decrease in strength with increasing indent size. Small-indent was found to be better for bond-strength improvement, but no optimal spacing is suggested [Freimis et al. 1998]. Focacci, Nanni, and Bakis [2000] have developed a model for the local bond-slip relationship to serve as a basis for computing anchorage lengths. A discussion on computational models for the bond between concrete and CFRP bars has been presented by Cox and Cochran [2003].

Research indicates that the bond behavior of straight and deformed FRP bars, in general, is influenced by the following parameters:

1. Confinement pressure
2. Bar diameter

3. Effect of bar position in the cast
4. Top-bar effect
5. Embedment length
6. Temperature change
7. Environmental conditions

However, recognizing that the research findings are based on tests on bars of specific diameters, bars from selected manufacturers, specific embedment lengths, selected range of temperatures, and environmental conditions (typically at room temperature and under freeze-thaw conditions in environmental chamber) is important. Note that not all of these parameters have been investigated sufficiently to arrive at definitive conclusions that can be applied to different types of commercially available FRP bars.

Evolving knowledge about the aforementioned factors notwithstanding, their influence on the bond strength must be accounted. This is done typically, as for steel reinforcing bars, by introducing the concept of "modification factors," which are to be used as multipliers to Equation 7.12. The ACI Guidelines recommend the following modification factors [ACI 440.1R-03].

## 7.6.1 Bar Location Effect or Top-Bar Effect

During the placement of concrete, air, water, and fine particles migrate upward through the poured concrete. This phenomenon can cause a significant drop in the bond strength under the horizontal reinforcement placed near the top of the pour. Tests have shown that the bond strength of top bars is about 66% of that of the bottom bars. "Top bar" is typically a horizontal bar with more than 12 in. of concrete below it at the time of embedment. To reflect the adverse effect of this top casting position, a modification factor (called "bar location modification factor") of 1.3 is recommended by the ACI Guide [ACI 440.1R-03] based on the recommendations of Tighiouast, Benmokrane, and Gao [1998].

## 7.6.2 Effect of Concrete Cover

Concrete cover to a reinforcing bar influences the bond failure mechanism by virtue of its confining effect. This effect is similar to that observed in concrete members reinforced with steel bars. Research has shown the dependence of bond failure on concrete cover in two forms, splitting failure and the pullout failure. The former can occur when the concrete cover is equal to or less than two bar diameters, whereas the latter can occur if the cover exceeds two bar diameters [Ehsani et al. 1996a]. Tests indicated that the ratio of the nominal strength of the FRP bar to the measured strength with concrete covers of $2d_b$ to $d_b$ varied between 1.2 and 1.5 [Ehsani et al. 1996b]. The ACI Guide [ACI 440.1R-03] recommends a minimum cover equal to the diameter of the bar. For concrete cover greater than the bar diameter, the Guide recommends the "concrete cover modification factors" as shown in Table 7.2.

---

**TABLE 7.2**
**Concrete Cover Modification Factors**
**for the Basic Development Lengths of**
**FRP Bars**

| Concrete Cover or Reinforcement Spacing | Modification Factor |
|---|---|
| $d_b$ | 1.5 |
| Between $d_b$ and $2d_b$ | Use linear interpolation between 1.5 and 1.0 |
| Over $2d_b$ | 1.0 |

---

### 7.6.3 DEVELOPMENT LENGTH OF A BENT BAR

Although the concept of the development length of a bent FRP bar is similar to that for a steel reinforcing bar, the corresponding provisions of ACI 318 cannot be used because of major differences in material characteristics. Data available on the bond behavior of hooked bars are rather scant. Based on test data of 36 specimens with hooked GFRP bars, Ehsani, Saadatmanesh, and Tao [1996b] proposed the following expression for the development length of a 90°-hooked FRP bar, $\ell_{bhf}$:

$$\ell_{bhf} = K_4 \frac{d_b}{\sqrt{f_c'}} \tag{7.14}$$

where $K_4$ is a coefficient, the value of which has been found to vary with the ultimate strength of the FRP bar as follows:

$$K_4 = 1820 \text{ for } f_{fu} < 75,000 \text{ psi}$$

$$K_4 = 1820\left(\frac{f_{fu}}{75,000}\right) \text{ for } 75,000 \text{ psi} \le f_{fu} < 150,000 \text{ psi}$$

Equation 7.14 is similar to that used for steel reinforcing bars for which $K_4$ is taken as 1200 [ACI 318 Sec. 12.5.2]. Similar to the case for steel reinforcing bars as provided for in ACI 318-02, another multiplier of 0.7 can be applied when the cover normal to the plane of the hook (i.e., the side cover) is more than 2.5 in. and the cover extension beyond the hook is not less than 2 inches. The two multipliers, 0.7 and $K_4$, are to be used cumulatively when applicable. However, because of the lack of substantial test data, the ACI Guide recommends the following expression for calculating the development length of a hooked bar [ACI 440.1R-03]:

$$\ell_{bhf} = \begin{cases} 2000\dfrac{d_b}{\sqrt{f_c'}} & \text{for } f_{fu} \leq 75{,}000 \text{ psi} \\[2ex] \dfrac{f_{fu}}{37.5}\dfrac{d_b}{\sqrt{f_c'}} & \text{for } 75{,}000 < f_{fu} < 150{,}000 \text{ psi} \\[2ex] 4000\dfrac{d_b}{\sqrt{f_c'}} & \text{for } f_{fu} \geq 150{,}000 \text{ psi} \end{cases}$$

The development length given by Equation 7.14 should not be less than 12 times the bar diameter or 9 in., whichever is greater. The radius of the bend should not be less than three times the bar diameter to preclude the possibility of localized shear failure (i.e., at the end). Furthermore, the straight portion of the bar beyond the bend (called the *tail end*) should not be less than 12 times the bar diameter. This provision is identical to that for steel reinforcing bars in ACI 318-02. Longer tail lengths have been found to be inconsequential as to their influence on the ultimate tensile force and the slippage of the hook.

### 7.6.4 TENSION LAP SPLICE

The splicing of bars — steel or FRP — is unavoidable in reinforced concrete construction due to limitations on available or transportable lengths. FRP bars can be spliced by lapping them in contact or separated. Whether steel or FRP bars, splices in a beam should be located away from the points of maximum bending moments. Also, splices should not be located at the same location in a beam (i.e., they should be staggered).

Spliced ends in reinforcing bars present zones of stress concentration, which can lead to splitting at the early ages of loading unless precautionary steps are taken [Nawy 2002]. Therefore, the overlap distance required for a tension splice is always a matter of concern. In any case, the splice length ought to be longer than the development length of the bar.

Parallel to the provisions of ACI 318 for steel reinforcing bars, tension splices for FRP bars are classified as *Class A* and *Class B*. The requirements for minimum splice lengths for deformed steel bars are shown in Table 7.3 [ACI 318 Sect. 12.15.1].

---

**TABLE 7.3**
**Tension Splices for Steel Reinforcing Bars**

| | |
|---|---|
| Class A splice | $1.0\ell_d$ but not less than 12 in. |
| Class B splice | $1.3\ell_d$ but not less than 12 in. |

*Note:* $\ell_d$ is the development length for the specified yield strength of the bar.

*Source:* ACI 318-05.

---

**TABLE 7.4**
**Tension Lap Splices for Deformed Steel Bars**

| $\dfrac{A_s \ provided *}{A_s \ required}$ | Maximum Percent of $A_s$ Spliced within Required Lap Length | |
|---|---|---|
| | 50% | 100% |
| Equal to or greater than 2 | Class A | Class B |
| Less than 2 | Class B | Class B |

* Ratio of area of reinforcement provided to area of reinforcement required by analysis at the same location.

*Source*: American Concrete Institute (ACI), Guide for the design and construction of concrete reinforced with FRP bars, ACI 440.1 R-03, ACI International, Reported by ACI Committee 440, Farmington Hills, MI: American Concrete Institute, 2001.

**TABLE 7.5**
**Types of Tension Lap Splices Required for FRP Bars**

| $\dfrac{A_{f,provided}}{A_{f,required}}$ | Equivalent to | $\dfrac{f_f}{f_{fu}}$ | Maximum Percentage of $A_f$ Spliced within Required Lap Length | |
|---|---|---|---|---|
| | | | 50% | 100% |
| 2 or more | | 0.5 or less | Class A | Class B |
| Less than 2 | | More than 0.5 | Class B | Class B |

*Source*: American Concrete Institute (ACI), Guide for the design and construction of concrete reinforced with FRP bars, ACI 440.1 R-03, ACI International, Reported by ACI Committee 440, Farmington Hills, MI: American Concrete Institute, 2001.

The requirement for providing a Class A or Class B splice depends on the ratio of the area of reinforcement provided to the area of reinforcement required by analysis at the splice location. The pertinent splice requirements for steel bars specified in ACI 318-02 are summarized in Table 7.4. The Guide [ACI 440.1R-03] suggests a parallel approach by considering the ratio of actual stress in the FRP bar to the ultimate strength of the FRP (instead of the ratio of reinforcement areas), and provides the recommendation shown in Table 7.5.

Note that only limited research data are available on the development lengths for tension splices. According to tests for a Class B splice, the ultimate capacity of the FRP bar was achieved at $1.6\ell_{df}$ [Benmokrane 1997]. For a Class A, a value of $1.3\ell_{df}$ is considered to be conservative becausee the stress level for a Class A splice is not to exceed 50% of the tensile strength of the bar. Accordingly, the Guide recommends a minimum length of lap for tension splices of FRP bars as shown in Table 7.6.

Tests results of limited scope and the wide randomness introduced by many differences in geometrical and mechanical properties of FRP bars make arriving at

TABLE 7.6
Recommended Minimum Lap Lengths
for Tension Lap Splices

| | |
|---|---|
| Class A splice | $1.3\ell_{bhf}$ |
| Class B splice | $1.6\ell_{bhf}$ |

*Source*: American Concrete Institute (ACI), "Guide for the design and construction of concrete reinforced with FRP bars," ACI 440.1 R-03, ACI International, Reported by ACI Committee 440, Farmington Hills, MI: American Concrete Institute, 2001.

any general constitutive laws that can be applied to all types of bars for use in design very difficult. For example, a great deal of uncertainty exists with regard to the influence of confinement pressures and concrete strengths, particularly for deformed bars. Tests to investigate the influence of variation in confinement pressure have been reported by Malvar [1994] who used confinement pressures of 500, 1500, 2500, 3500, and 4500 psi, using four different types of commercially available bars. He found that deformed bars produced by gluing a fiber spiral on the outer surface exhibit a bond behavior similar to that of a smooth bar with no mechanical interlock. Tests have shown that bond failure is due to the detachment of the spiral, with the concrete remaining uncrushed [Malvar 1994]. Because of the specific conditions used in these tests, the dependence of the Malvar model upon the confinement stress is valid for their test specimens but not easily extended to FRP bars under other conditions.

## 7.7 EXAMPLES

Calculations for the development length of FRP bars are presented in Chapter 6 along with the examples on the FRP-internally reinforced concrete beams.

## 7.8 SUMMARY

1. Smooth FRP rebars are unsuitable for use in reinforced concrete structures. The bond-slip curves for smooth FRP rebars show very low values of bond strength due to the activation of a friction mechanism with large damage to rebar surface [Cosenza, Manfredi, and Realfonzo 1997].
2. Sand-covered continuous fiber bars show good bond resistance; however, the interface between sand grains and bars detaches abruptly at limited locations if the curing of resin for the bonding of the sand coat is inadequate, or if the environmental attacks (pH, freeze-thaw, wet-dry cycles) are too severe.
3. Under static loads, helically wrapped bars offer higher bond strength (greater than 2000 psi) compared to those obtained for smooth bars (about 600 psi), provided all other parameters remain identical.

4. In the case of deformed bars, different bond mechanisms have been observed, depending on the mechanical properties of the reinforcements, surface texture of the bars, and compressive strength of concrete [Cosenza, Manfredi, and Realfonzo 1997].

5. Helical wrapping of FRP bars can lower the bond resistance to cyclic loading even for a small load (approximately 20% to 30% of the ultimate strength) due to inadequate adhesion of helical wrap to the FRP bar surface. Special attention must be given to the design and manufacture of bars including curing of the wrapping to improve bond resistance under cyclic loading [Bakis et al. 1998]. Improper and inadequate curing of resins can lead to premature failure of the helical wrap, which can be construed as bond failure being independent of concrete strength [Malvar 1994].

## REFERENCES

Al-Dulaijan, S.U., Nanni, A., Al-Zaharani, M.M., Bakis, C.E., and Boothby, T.E., Bond evaluation of environmentally conditioned GFRP/concrete systems, *Proc. 2nd Intl. Conf. Adv. Comp. Mat. in Bridges and Structures*, El Badry, M. (Ed.), 1996.

Alunno, R.V., Galeota, D., and Giammatteo, M.M., Local bond stress slip relationships of glass fiber reinforced plastic bars embedded in concrete, *Materials and Structures*, 28(180), 1995.

Al-Zaharani, M.M., Bond behavior of fiber reinforced plastic (FRP) reinforcements with concrete, Ph.D. dissertation, University Park, PA: Pennsylvania State University, 1995.

Al-Zaharani, M.M., Al-Dulaijan, S.U., Nanni, A., Bakis, C.E., and Boothby, T.E., Evaluation of bond using FRP rods with axisymmetric deformations, *Construction and Building Materials*, 13, 299–309, 1999.

Al-Zaharani, M.M., Nanni, A., Al-Dulaijan, S.U., and Bakis, C.E., Bond of FRP to concrete in reinforcement rods with axisymmetric deformation, *Proc. 2nd Int. Conf. Adv. Comp. Mat. in Bridges and Structures*, El-Badry, M. (Ed.), 1996.

American Concrete Institute (ACI), Building code requirements for structural concrete (ACI 318-02) and commentary (ACI 318R-02), Farmington Hills, MI: American Concrete Institute, 2002.

American Concrete Institute (ACI), Guide for the design and construction of concrete reinforced with FRP bars, ACI 440.1R-03, ACI International, Reported by ACI Committee 440, Farmington Hills, MI: American Concrete Institute, 2001.

Bakis, C.E., Al-Dulaijan, S.U., Nanni, A., Boothby, T.E., and Al Zahrani, M.M., Effect of cyclic loading in bond behavior of GFRP rods embedded in concrete beams, *J. Composite Tech. Res.*, 20(1), 29–37, 1998.

Bakis, C.E., Uppuluri, V.S., Nanni, A., and Boothby, T.E., Analysis of bonding mechanisms of smooth and lugged FRP rods embedded in concrete, *Composites Sci. Tech.*, 58, 1307–1319, 1983.

Benmokrane, B., Bond strength of FRP rebar splices, *Proc. 3rd Intl. Symp. Non-Metallic (FRP) Reinforcement for Concrete Structures (FRPRCS-3)*, Vol. 2, Japan Concrete Institute, Sapporo, Japan, 1997.

Benmokrane, B., Tighiouart, B., and Chaallal, O., Bond strength and load distribution of composite GFRP reinforcing bars in concrete, *ACI Mat. J.*, 93(3), 1996.

Boothby, T.E., Nanni, A., Bakis, C.E., and Huang, H., Bond of FRP rods embedded in concrete, *Proc. 10th ASCE Eng. Mech. Specialty Conf.*, May 1995.

Chaallal, O. and Benmokrane, B., Pullout and bond of glass-fibre rods embedded in concrete and cement grout, *Materials and Structures*, 26, 1993.

Chambers, R.E., Design standard for pultruded fiber-reinforced-plastic (FRP) structures, *J. Composites for Construction*, ASCE, 1(1), 26–38, 1997.

Ciampi, V., Eligehausen, R., Popov, E.P., and Bertero, V.V., Analytical model for deformed bar bond under generalized excitations, *Trans. IABSE Coll. Adv. Mech. Reinforced Concrete*, Delft, Netherlands, 1981.

Cosenza, E., Manfredi, G., and Realfonzo, R., Behavior and modeling of bond of FRP rebars to concrete, *J. Composites for Construction*, ASCE, 1(2), 40–51, 1997.

Cosenza, E., Manfredi, G., and Realfonzo, R., Bond characteristics and anchorage length of FRP reinforcing bars, *Proc. 2nd Intl. Conf. Adv. Comp. Mat. in Bridges and Structures*, El-Badry, M. (Ed.), 1996.

Cosenza, E., Manfredi, G., and Realfonzo, R., Analytical modeling of bond between FRP reinforcing bars and concrete, *Proc. 2nd Intl. RILEM Symp. (FRPRCS-2)*, Taerwe, L. (Ed.), 1995.

Cox, J.V. and Cochran, K.B., Bond between carbon fiber reinforced polymer bars and concrete, II: computational modeling. *Journal of Composites for Construction,* 7(2): 164–171, May 2003.

Cox, J.V. and Guo, J., Modeling the stress state dependency of the bond behavior of FRP tendon, *4th Intl. Symp. FRP for RC Structures, ACI SP 188-61*, Dolan, C.W., Rizkalla, S.H., and Nanni, A. (Eds.), 1999.

Cox, J.V. and Herrmann, L.R., Confinement stress dependent bond behavior, Part II: A two degree of freedom plasticity model, *Proc. Intl. Conf. Bond in Concrete*, CEB, 3, Riga, Latvia, 1992.

Cox, J.V. and Yu, H.A., Micromechanical analysis of the radial elastic response associated with slender reinforcing elements within a matrix, *J. Composite Materials*, 33(23), 2161–2192, 1999.

Daniali, S., Development length for fibre-reinforced plastic bars, *Proc. 1st Intl. Conf. Adv. Composite Materials in Bridges and Structures*, Neale, K.W. and Labossiere, P. (Eds.), 1992.

Den Uijl, J.A., Bond and fatigue properties of arapree, *Proc. RILEM Intl. Conf. Non-metallic (FRP) Reinforcement for Concrete Structures*, Taerwe, L.E. and Spon, F.N. (Eds.), London, 1995.

Dolan, C.W., Rizkalla, S.H., and Nanni, A. (Eds.), *4th Intl. Symp. Fiber Reinforced Polymer Reinforcement for Reinforced Concrete Structures, ACI SP-188*, Farmington Hill, MI: American Concrete Institute, 1999.

Ehsani, M.R., Saadatmanesh, H., and Tao, S., Bond of GFRP rebars to ordinary-strength concrete, *Proc. Intl. Symp. Fiber Reinforced Plastic Reinforcement for Concrete Structures, ACI SP-138*, Nanni, A. and Dolan, C.W. (Eds.), 1993.

Ehsani, M.R., Saadatmanesh, M., and Tao, S., Design recommendation for bond of GFRP rebars to concrete, *J. Struct. Eng.*, 122 (3), 247-257, 1996a.

Ehsani M.R., Saaddatmanesh, M., and Tao, S., Bond behavior and design recommendations for fiber-glass reinforcing bars, *Proc. ICCI '96*, Saadatmanesh, H. and Ehsani, M.R. (Eds.), Tucson, AZ, January 1996b.

Eligehausen, R., Popov, E.P., and Bertero, V.V., *Local Bond Stress-Slip Relationships of Deformed Bars under Generalized Excitations*, Rep. No. 83/23, Earthquake Eng. Res. Ctr. (EERC), Berkeley, CA: University of California, 1983.

EUROCOMP, *Design Code and Handbook. Structural Design of Polymer Composites*, Clarke, J.L.E. and Spon, F.N. (Ed.), London: Chapman and Hall, 1996.

Faoro, M., Bearing and deformation behavior of structural components with reinforcements comprising resin bounded glass fibre bars and conventional ribbed steel bars, *Proc. Intl. Conf. Bond in Concrete*, CEB, 3, Riga, Latvia, 1992.

Faza, S.S., Bending and bond behavior and design of concrete beams reinforced with fiber reinforced plastic rebars, Ph.D. dissertation, Morgantown, WV: West Virginia State University, 1991.

Faza, S.S. and GangaRao, H.V.S., Bending and bond behavior of concrete beams reinforced with fiber reinforced plastic rebars, *Transportation Res. Rec.* 1290, 185–193, 1990.

Focacci, F., Nanni, A., and Bakis, C.E., Local bond-slip relationship for FRP reinforcement in concrete, *J. Composites for Construction*, ASCE, 4(1): 24–31, February 2000.

Freimanis, A.J., Bakis, C.E., Nanni, A., and Gremel, D., A comparison of pullout tests and tensile behaviors of FRP reinforcement for concrete, *Proc. ICCI '98*, Vol. II, Saadatmanesh, H. and Ehsani, M.R. (Eds.), Tucson, AZ: University of Arizona, 1998.

FRPRCS I, Fiber-reinforced-plastic (FRP) reinforcement for concrete structures: properties and applications, Nanni, A. (Ed.), St. Louis, MO: Elsevier, 1993.

FRPRCS IV, *4th Intl. Symp. Fiber Reinforced Polymer Reinforcement for Reinforced Concrete Structures, ACI, SP-188*, Dolan, C.W., Rizkalla, S.H. and Nanni, A. (Eds.), Farmington Hills, MI: American Concrete Institute, 1999.

GangaRao, H.V.S., Vijay, P.V., Gupta, R.K., and Barbero, E., FRP composites application for non-ductile structural systems: Short and long-term behavior, Submitted to CERL, U.S. Army Corps of Engineers, Morgantown, WV: Constructed Facilities Center, West Virginia University, 2001.

Gao, D., Benmokrane, B., and Tigheourt, B., Bond properties of FRP bars to concrete, Technical report, Department of Civil Engineering, University of Sherbrooke, Quebec, Canada, 1998.

Guo, J. and Cox, J.V., An interface model for the mechanical interaction between FRP bars and concrete, *J. Reinforced Plastics and Composites*, 1999.

Hattori, A., Inoue, S., Miyagawa, T., and Fujii, M.A., Study on bond creep behavior of frp rebars embedded in concrete, *Proc. 2nd Intl. RILEM Symp. (FRPRCS-2)*, Taerwe, L. (Ed.), 1995.

Honma, M. and Maruyama, T., Bond properties of carbon fiber reinforced plastic rods at elevated temperatures, *Proc. Arch. Inst. of Japan Convention*, C (in Japanese), 1989.

Itoh, S., Maruyama, T., and Nishiyama, H., Study of bond characteristics of deformed fiber reinforced plastic rods, *Proc. Japan Concrete Inst.*, 11(1) (in Japanese), 1989.

Japanese Society of Civil Engineers (JSCE), Application of continuous fiber reinforcing materials to concrete structures, Subcommittee on Continuous Fiber Reinforcing Materials, Concrete Library International of JSCE, No. 19, Tokyo, 1992a.

Japanese Society of Civil Engineers (JSCE), Utilization of concrete FRP rods for concrete reinforcement, *Proc. Japanese Society of Civil Engineers*, Tokyo, Japan, 1992b.

Japanese Society of Civil Engineers (JSCE), Recommendations for design and construction for concrete structures using continuous fiber reinforcing materials, Concrete Engineering Series, No. 23, Tokyo, 1997.

Kachlakev, D.I., Bond strength investigations and structural applicability of composite fiber-reinforced polymer (FRP bars), Ph.D. dissertation, Corvallis, OR: Oregon State University, 1997.

Kachlakev, D.I. and Lundy, J.R., Performance of hollow glass fiber-reinforced polymer bars, *J. Composites for Construction*, ASCE, 3(2): 87–91, May 1999.

Kanakubo, T., Yonemaru, K., Fukuyama, H., Fujisawa, M., and Sonobe, Y., Bond performance of concrete members reinforced with FRP bars, *Proc. Intl. Symp. Fiber Reinforced Plastic Reinforcement for Concrete Structures, ACI SP-138*, Nanni, A. and Dolan, C.W. (Eds.), 1993.

Katz, A., Bond to concrete of FRP rebars after cyclic loading, *J. Composites for Construction*, ASCE, 4(3): 137–144, August 2000.

Larralde, J. and Silva Rodriguez, R., Bond and slip of FRP rebars in concrete, *J. Mat. in Civ. Eng.*, ASCE, 5(1), 1993.

Makitani, E., Irisawa, I., and Nishiura, N., Investigation of bond in concrete member with fiber reinforced plastic bars, *Proc. Intl. Symp. Fiber-Reinforced-Plastic Reinforcement for Concrete Structures*, ACI SP-138, Nanni, A. and Dolan, C.W. (Eds.), Vancouver, 1993.

Malvar, L.J., Bond of reinforcement under controlled confinement *ACI Materials J.*, 89(6), 593–601, 1992.

Malvar, L.J., Bond stress-slip characteristics of FRP rebars, Rep. TR-2013 SHR, Port Hueneme, CA: Naval Facilities Engineering Service Center, 1994.

Malvar, L.J., Tensile and bond properties of GFRP reinforcing bars, *ACI Materials J.*, 92(3), 276–285, 1995.

Mashima, M. and Iwamoto, K., Bond characteristics of FRP rod and concrete after freezing and thawing deterioration, *Proc. Intl. Symp. Fiber-Reinforced-Plastic Reinforcement for Concrete Structures*, ACI SP-138, Nanni, A. and Dolan, C.W. (Eds.), 1993.

Nanni, A. (Ed.), Fiber-reinforced-plastic (FRP) reinforcement for concrete structures: Properties and applications, *Developments in Civil Engineering*, Vol. 42, St. Louis, MO: Elsevier.

Nanni, A, Al-Zaharani, M.M., Al-Dulaijan, S.U., Bakis, C.E., and Boothby, T.E., Bond of FRP reinforcement to concrete: experimental results, *Proc. 2nd Intl. RILEM Symp. (FRPRCS-2)*, Taerwe, L. (Ed.), 1995.

Nawy, E.G., *Reinforced Concrete: A Fundamental Approach*, 5th ed., Upper Saddle River, NJ: Prentice Hall, 2002.

Noghabai, K., Splitting of concrete in the anchoring zone of deformed bars, Licentiate Thesis, 26L, Div. of Struct. Eng., Lulea University of Technology, Sweden, 1995.

Ohnstad, T.S. and Kachlakev, D.I., Hollow FRP rebar — Its unique manufacturing process and applications, *Proc. 46th Intl. SAMPE Symp., Vol. 46: A Materials and Processes Odyssey*, Paper code: 46-064, 2001.

Orungun, C., Jirsa J.O., and Breen J.E., A reevaluation of test data on development length and splices, *Proc. ACI J.*, 74, 3, 114–122, March 1977.

Park, R. and Paulay, T., *Reinforced Concrete Structures*, New York: John Wiley & Sons, 1975.

Pleimann, L.G., Tension and bond pullout tests of deformed fiberglass rods, Final Report for Marshall-Vega Corporation, Fayetteville, AK: Civil Engineering Department, University of Arkansas, 1987.

Pleimann, L.G., Strength, modulus of elasticity, and bond of deformed FRP rods, *Proc. Specialty Conf. Adv. Composite Matls. in Civil Eng. Structures*, Materials Engineering Division, American Society of Civil Engineers, 1991.

Rossetti, A.V., Galeota, D., and Giammatteo, M.M., Local bond stress-slip relationships of glass fibre reinforced plastic bars embedded in concrete, *Materials and Structures*, 28(180), 1995.

Soroushian, P., Choi, K.B., Park, G.H., and Aslani, F., Bond of deformed bars to concrete: Effects of confinement and strength of concrete, *ACI Materials J.*, 88(3), 1991.

Taly, N., *Design of Reinforced Masonry Structures*, New York: McGraw-Hill, 2001.

Taly, N. and GangaRao, H.V.S., Bond behavior of FRP reinforcing bars, the-state-of-the-art, *Proc. 45th SAMPE (Society for the Advancement of Materials and Process Engineering) Intl. Conf.*, Long Beach, CA, May 6–10, 2001.

Tepfers, R., Molander, I., and Thalenius, K., Experience from testing of concrete reinforced with carbon fiber and aramid fiber strands, XIV, *Nordic Concrete Cong. and Nordic Concrete Ind. Mtg.*, August 1992.

Tighiouart, B., Benmokrane, B., and Gao, D., Investigation of bond in concrete member with fibre reinforced polymer (FRP) bars, *Construction and Building Materials*, 12, 453–462, 1998.

Tomosawa, F. and Nakatsuji, T., Evaluation of ACM reinforcement durability by exposure test, *Proc. 3rd Intl. Symp. Non-Metallic FRP Reinforcement for Concrete Structures*, Vol. 2, Japan Concrete Institute, Tokyo, 1997.

Uppuluri, V.S., Bakis, C.E., Nanni, A., and Boothby, T.E., Finite element modeling of the bond between fiber reinforced plastic (FRP) reinforcement and concrete, Report CMTC-9602, Composites Manufacturing Technology Center, University Park, PA: Pennsylvania State University, 1996.

Vijay, P.V. and GangaRao, H.V.S., Unified limit state approach using deformability factors in concrete beams reinforced with GFRP bars, *Proc. 4th Matls. Eng. Conf.*, 1999.

Vijay, P.V. and GangaRao, H.V.S., Development of fiber reinforced plastics for highway application: Aging behavior of concrete beams with GFRP bars, CFC-WVU Report No. 99-265 (WVDOT Rep. No. 699-FRP1), Department of Civil and Environmental Engineering, Morgantown, WV: West Virginia University, 1999.

Yonezawa, T., Ohno, S., Kakizawa, T., Inoue, K., Fukata, T., and Okamoto, R., A new three-dimensional FRP reinforcement: Fiber-reinforced-plastic (FRP) for concrete structures: Properties and applications, Nanni, A. (Ed.), St. Louis, MO: Elsevier, 1993.

Yu, H. and Cox, J.V., Radial elastic modulus for the interface between FRP reinforcing bars and concrete, *J. Reinforced Plastics and Composites*, 21(14): 1285–1318, 2002.

# 8 Serviceability: Deflection and Crack Width

## 8.1 INTRODUCTION

A structure must satisfy two basic criteria: strength and serviceability. The first relates to the ability of a structure to carry design loads for its specified service life. The second relates to its ability to satisfactorily operate or perform its intended function during its service life. Strength considerations for concrete beams internally reinforced with steel that are further strengthened with external FRP fabrics and beams internally reinforced with FRP reinforcement were presented, respectively, in Chapter 5 and Chapter 6. Whereas strength considerations are important to ensure the safety of a structure under design loads, its satisfactory performance under service loads is equally important because functional considerations are the reasons for creating a structure. Deflection, crack-width control, creep-rupture stress, and fatigue are related to performance of a concrete structure during its service life. This chapter presents a discussion along with design examples on the serviceability aspects of concrete members internally reinforced with FRP reinforcement.

## 8.2 SERVICEABILITY OF CONCRETE STRUCTURES

Four major considerations affect the serviceability of concrete structures:

1. Deflection
2. Crack width
3. Creep-rupture
4. Fatigue

The above listed four design considerations are discussed in this chapter.

To understand the ramifications of providing FRP reinforcement in concrete structures in lieu of steel reinforcement, consider two identical reinforced concrete sections: one with steel reinforcement and the other with FRP reinforcement of an equal cross-sectional area. Of the two sections, the cracked FRP-reinforced concrete section will exhibit a higher deflection and crack width because of the low stiffness of the FRP reinforcement. Deflection and crack width at service loads can control the overall design process. Generally, a FRP-reinforced concrete beam designed for strength and failing in concrete-crushing failure mode satisfies the serviceability

criteria for deflection and crack width as well as the deformability criteria because
the tensile reinforcement does not rupture.

The model of serviceability criteria for FRP-reinforced beams suggested in ACI
440 [ACI 440.1R-03] is based on the serviceability provisions for conventional
reinforced concrete beams (i.e., beams with steel reinforcing bars) specified in ACI
318. Essentially, for FRP-reinforced beams the serviceability criteria of ACI 318
have been modified to reflect and account for the differences in the properties of
steel and FRP reinforcement in terms of failure strain, including creep-rupture, bond
strength, and corrosion resistance.

## 8.3  DEFLECTIONS

*Deflections* in concrete structures must be controlled (or limited) for a variety of
reasons. The most obvious reason is that an excessive deflection would convey a
sense of impending failure. Excessive deflection in a concrete beam or a slab can
lead to ponding and cracking, which is highly undesirable as these can lead to
structural damage. Even though deflection may not be sufficiently large to cause
distress in a concrete member, it can be large enough to cause distress in the slab-
stiffening elements or even render them nonfunctional.

Two types of deflections should be considered to ensure the serviceability of
concrete members: *immediate* deflection (also called *short-term* or *instantaneous*
deflection) and *long-term* deflection. Immediate deflection occurs during the normal
service life of a member as a result of a percent of the live loads. Long-term deflection
occurs as a result of sustained loads, which is affected by a time-dependent phenomenon called *creep*, and shrinkage through moisture escape from the beam or
slab. The sustained load consists of a dead load and a specified percentage of the
live load. Under sustained loading conditions, stresses in a concrete member remain
in the elastic range. Accordingly, deflection in a beam caused by sustained loads
can be determined based on the elastic properties of the member cross-section. ACI
318 permits the calculations of concrete members based on the effective moment of
inertia, $I_e$, discussed in the next section. An exhaustive discussion on the deflection
of concrete structures can be found in many papers published in ACI Special Publication SP-43 [ACI 1974a].

### 8.3.1  ACI 318 Provisions for Deflection Control

The ACI 318 provisions for deflection control refer to the instantaneous deflections
under service loads and long-term deflections under sustained loads. ACI 318 suggests two methods for the control of deflections of one-way flexural members:

1. Indirect method, which mandates minimum thickness of nonprestressed
   beams (ACI 318, Table 9.5a)
2. Direct method, which prescribes limitations on computed deflections (ACI
   318, Table 9.5b).

**TABLE 8.1**
**Minimum Thickness Requirements of ACI 318 and Suggested Minimum Thickness Requirements for FRP-Reinforced Members**

| | | Minimum thickness, $h$ (in.) | | | |
|---|---|---|---|---|---|
| | | Simply Supported | One End Continuous | Both Ends Continuous | Cantilever |
| Member | | Members not supporting or attached to partitions or other construction likely to be damaged by large deflections. | | | |
| Solid one-way slabs | Steel rein. | $L/20$ | $L/24$ | $L/28$ | $L/10$ |
| | FRP rein.* | $L/16$ | $L/19$ | $L/22$ | $L/8$ |
| Beams or ribbed one-way slabs | Steel rein. | $L/16$ | $L/18.5$ | $L/21$ | $L/8$ |
| | FRP rein.* | $L/13$ | $L/15$ | $L/17$ | $L/6$ |

*Note:* 1. Span $L$ is in inches. 2. For members with steel reinforcement, values shall be used for members cast from normal weight concrete ($w_c$ = 145 lb/ft³) and Grade 60 reinforcement. 3. For members with steel reinforcement having $f_y$ other than 60,000 psi, the tabulated values shall be multiplied by (0.4 + $f_y/100{,}000$). 4. Steel rein. = beams with steel reinforcement 5. FRP rein.* = suggested minimum thickness for beams with FRP reinforcement

### 8.3.1.1 Minimum Thickness for Deflection Control (Indirect Method)

ACI 318 Table 9.5a (see Table 8.1) prescribes the minimum thickness of nonprestressed beams and one-way slabs, and it applies to members not supporting or attached to partitions and other construction that are likely to be damaged due to deflection in the supporting beams. The philosophy behind this prescriptive method is that members having a prescribed minimum thickness would have deflections under service loads small enough to be of any practical concern to members themselves. Because the stiffness of FRP reinforcement is much smaller than that of steel, the FRP-reinforced members will exhibit deflections relatively larger than intended in ACI 318 Table 9.5a. In other words, the prescribed values of the minimum member thickness are not conservative for FRP-reinforced, one-way systems. Stated differently, their thickness should be larger or at least equal to have the deflections intended in ACI 318. Therefore, for preliminary member design the thicknesses prescribed in the ACI 318 table is suggested to be increased by at least 25% (see Table 8.1), followed by verification with allowable or prescribed deflection limits.

### 8.3.1.2 Direct Method of Limiting Deflection Values

Deflections are inversely proportional to the moment of inertia. Whereas determination of the moment of inertia of a homogeneous uncracked section is simple ($I = bh^3/12$ for a rectangular section, where $b$ is the width and $h$ is the overall depth), that of a reinforced concrete section is complex because of the cracking that occurs under service loads. As long as a concrete section remains uncracked, its moment

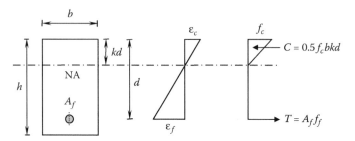

**FIGURE 8.1** Strain and stress distribution in a reinforced concrete beam.

of inertia $I_{unc}$ equals the sum of the moment of inertia of the gross section $I_g$ and the moment of inertia of the transformed area of steel reinforcement taken about the centroidal axis of the transformed uncracked section. When the applied moment $M_a$ exceeds the cracking moment $M_{cr}$, the beam cracks at several locations along the span, resulting in reduced stiffness of the beam as a whole (also called the overall flexural stiffness of the beam). Under this condition, the moment of inertia must be based on the properties of the cracked section, $I_{cr}$. To determine the deflection of a beam intermittently cracked along the span, the *effective moment of inertia $I_e$* must be used, which is a function of $I_{cr}$. Noting that the extent of cracking depends also on the applied moment and the cracking moment, the effective moment of inertia is a complex function of $I_g$, $I_{cr}$, $M_a$, and $M_{cr}$.

Referring to Figure 8.1, the gross moment of inertia of a FRP-reinforced rectangular section can be taken as $I_g = bh^3/12$, whereas $I_{cr}$ is determined using an elastic analysis similar to that used for steel-reinforced concrete. The moment of inertia of the cracked section $I_{cr}$, determined from the parallel axis theorem, consists of two components: the moment of inertia of the solid part, having a depth equal to $kd$, and the contribution of the equivalent area of FRP bars (i.e., the cross-sectional area of FRP bars multiplied by the modular ratio, $n_f$). Thus,

$$I_{cr} = \frac{1}{3}b(kd)^3 + n_fA_f(d - kd)^2 \qquad (8.1)$$

Equation 8.1 can be written in a simplified form as

$$I_{cr} = \frac{bd^3}{3}k^3 + n_fA_fd^2(1 - k)^2 \qquad (8.2)$$

where

$$k = \sqrt{(\rho_fn_f)^2 + 2\rho_fn_f} - \rho_fn_f \qquad (8.3)$$

and $n_f$ is the modular ratio between FRP reinforcement and concrete,

$$n_f = \frac{E_f}{E_c} \tag{8.4}$$

At service load, the effective moment of inertia $I_e$ varies between $I_g$ and $I_{cr}$, depending on the magnitude of the applied moment. Branson [1965; 1977] proposed Equation 8.5 — an empirical relationship — to determine an effective constant moment of inertia $I_e$, which represents a gradual transition from the transformed moment of inertia of the uncracked section $I_{unc}$ to the moment of inertia of a fully cracked section $I_{cr}$. For practical purposes, $I_{unc}$ is usually replaced by the moment of inertia of the gross section $I_g$ ($I_{unc} > I_g$), as the contribution of reinforcement to the moment of inertia is neglected as a conservative approach except for heavily reinforced sections.

$$I_e = \left(\frac{M_{cr}}{M_a}\right)^m I_g + \left[1 - \left(\frac{M_{cr}}{M_a}\right)^m\right] I_{cr} \le I_g \tag{8.5}$$

Because the beam is assumed to be cracked under service loads, the flexural stiffness of the beam varies along the span. An average stiffness for the entire span is obtained by assuming the exponent $m = 3$ in Equation 8.5, which accounts for the variation in the value of the moment of inertia of the cracked beam along the member span. Thus, Equation 8.5 with $m = 3$ has been adopted by ACI 318 to determine the effective moment of inertia $I_e$ of a steel-reinforced section:

$$I_e = \left(\frac{M_{cr}}{M_a}\right)^3 I_g + \left[1 - \left(\frac{M_{cr}}{M_a}\right)^3\right] I_{cr} \le I_g \tag{8.6}$$

Tests have shown that Equation 8.6 does not predict the deflection response of FRP-reinforced beams very well. Because of the lower stiffness of FRP reinforcement and its bond characteristics, Equation 8.6 can overestimate the value of $I_e$ and thus underestimate the deflections [Benmokrane et al. 1996]. Therefore, ACI 440 suggests modification of Equation 8.6 by introducing a bond- and stiffness-dependent factor, $\beta_d$, in Equation 8.6 [Gao et al. 1998]:

$$I_e = \left(\frac{M_{cr}}{M_a}\right)^3 \beta_d I_g + \left[1 - \left(\frac{M_{cr}}{M_a}\right)^3\right] I_{cr} \le I_g \tag{8.7}$$

where

$M_{cr}$ = the cracking moment

$M_a$ = the applied service load moment where deflection is being considered

$$\beta_d = \alpha_b \left[ \frac{E_f}{E_s} + 1 \right]$$  (8.8)

Note that Equation 8.6 is valid only as long as $M_a > M_{cr}$ because when $M_a < M_{cr}$, the beam remains uncracked. The coefficient $\alpha_b$ in Equation 8.8 is a bond-dependent coefficient. ACI 440 suggests the use of $\alpha_b = 0.5$ for GFRP bars. For types of bars for which $\alpha_b$ is not known, ACI 440 recommends the use of $\alpha_b = 0.5$ as a conservative approach. A comprehensive discussion on deflection predictions for concrete beams reinforced with steel and FRP bars has been presented by Bischoff [2005].

## 8.3.2 CALCULATION OF DEFLECTION (DIRECT METHOD)

The immediate deflection of an FRP-reinforced, one-way flexural member can be determined from conventional methods using the effective moment of inertia of the FRP-reinforced beam. Long-term deflections are significant and can be as large as two or three times the short-term deflection. Both short-term and long-term deflections due to service loads must be accounted for in design.

Shrinkage and creep due to sustained loads cause long-term deflections that are in addition to the immediate deflections. Such deflections are influenced by factors such as:

1. Temperature variations
2. Humidity
3. Curing conditions
4. Age at the time of loading
5. Quantity of compression reinforcement
6. Magnitude of sustained loads

Immediate deflections of FRP-reinforced beams are three to four times greater than those of steel-reinforced members having a similar design strength [ACI 440]. Limited data on long-term deflections of FRP reinforced members indicate that creep behavior of FRP-reinforced members is similar to that of steel-reinforced members [Brown 1997; Vijay and GangaRao 1998]. The time versus deflection curves of FRP-reinforced and steel-reinforced members have similar shapes. Because of these similarities, the ACI 318 approach for estimating the long-term deflections of steel-reinforced beams is also used for estimating the long-term deflections of FRP-reinforced beams.

According to ACI 318, Section 9.5.2.5, the long-term deflection due to creep and shrinkage, $\Delta_{cp+sh}$, can be determined by multiplying the immediate deflection caused by sustained loads by a factor as expressed by Equation 8.9 (ACI 440, Equation 8-13a):

$$\Delta_{cp+sh} = \lambda(\Delta_i)_{sus}$$  (8.9)

where $(\Delta_i)_{sus}$ is the immediate deflection under sustained loads. In Equation 8.9, the value of the coefficient $\lambda$ is a function of the time-dependent factor $\xi$ that accounts

**TABLE 8.2**
**Values of Time-Dependent Factor for**
**Various Durations of Loading**

| Duration of Loading | Value of $\xi$ in Equation 8.10 |
|---|---|
| 5 years or more | 2.0 |
| 12 months | 1.4 |
| 6 months | 1.2 |
| 3 months | 1.0 |

for the duration of sustained loads. The relationship between $\lambda$ and $\xi$, developed by Branson [1977], is given by Equation 8.10:

$$\lambda = \frac{\xi}{1 + 50\rho_f'} \tag{8.10}$$

where

$\xi$ = time-dependent factor for sustained loads ($\xi_{FRP}/\xi_{steel}$)
$\rho_f'$ = steel compression reinforcement ratio

As pointed out by Branson, the multiplier on $\xi$ (on the right-hand side of Equation 8.10) accounts for the effect of compression reinforcement in reducing long-term deflections. Because the contribution of compression reinforcement to flexural strengths of FRP-reinforced beams is ignored, $\rho_f' = 0$ for FRP-reinforced beams, and for such beams, $\lambda = \xi$. The value of $\xi$ increases with the duration of sustained loading and reaches a value of 2.0 for a sustained loading duration of 5 or more years [ACI 318]. Values of $\xi$ for steel-reinforced beams under sustained loads for various periods are given in Table 8.2.

To account for the compressive stress in concrete and the influence of FRP on the long-term behavior of FRP-reinforced beams, ACI 440 recommends a further modification of its Equation 8.9 as suggested by Brown [1997]. This change is suggested because these aspects of beam behavior are not addressed in ACI 318, Equation 8.9, which only multiplies the immediate deflection by the time-dependent factor, $\xi$. This additional modification factor (equal to the ratio $\xi_{FRP}/\xi_{steel}$) has been determined from testing. Some test data indicated the value of this additional modification factor varies from 0.46 for AFRP to 0.53 for CFRP [Brown 1997; Kage et al. 1995]. In another study [Vijay and GangaRao 1998], the modification factor for $\xi$ (based on tests on GFRP-reinforced beams failing in concrete crushing mode) varied from 0.75 after 1 year to 0.58 after 5 years. Based on this research, ACI 440 recommends a factor 0.6 (as an average value) as a modifier on $\xi$. Accordingly, Equation 8.9 is modified and expressed as Equation 8.11:

$$\Delta_{(cp+sh)} = 0.6\xi(\Delta_i)_{sus} \tag{8.11}$$

## 8.4  PRACTICAL CONSIDERATIONS FOR THE CONTROL OF DEFLECTIONS

When using FRP reinforcement, deflection control can be a problem because the effective moment of inertia of the beam ($I_e$) may not be large enough. A beam can likely satisfy the strength requirements as well as crack-width limits suggested by ACI 440 or applicable code and still not meet the deflection requirements. Because deflection is inversely proportional to the effective moment of inertia, the deflection, if excessive, can be reduced by increasing the value of $I_e$. The latter can be done by (a) increasing the depth of the beam, (b) the width of the beam, or (c) the amount of FRP reinforcement.

An examination of ACI 440 Equation 8-12a, which gives the value of the effective of moment inertia, shows that the second term (the one containing the multiplier $I_{cr}$) is predominant. It follows that by far the most efficient way to increase $I_e$ is to increase the depth $d$ of the beam (the first option above). However, note that an increased beam depth will result in an increased self-weight of the beam, which will lead to an increase in the value of the applied moment, $M_a$. To illustrate this concept, Example 8.2 has been reworked as Example 8.3 by increasing the overall beam depth $h$ from 18 in. to 21 in., thus increasing the depth, $d$. In situations where an increased beam depth is not practically feasible, the other two alternatives should be tried.


## 8.5  EXAMPLES ON DEFLECTIONS

Several examples are presented in this section to illustrate the procedures for determining deflections (both immediate and long-term) in FRP-reinforced beams. Examples are presented in both U.S. standard units and SI units.

### 8.5.1  EXAMPLES IN U.S. STANDARD UNITS

*Example 8.1: Long-term deflection of beams reinforced with GFRP bars.* A rectangular 14 in. × 24 in. concrete beam is reinforced with eight No. 6 GFRP bars arranged in two rows so that $d = 20.25$ in. (Figure E8.1). The beam spans 20 ft and carries a superimposed dead load of 900 lb/ft and a live load of 600 lb/ft in addition to its own dead weight. The allowable deflection is $l/240$. Check if the beam satisfies the deflection requirements. Assume $f_c' = 4000$ psi, $f_{fu}{}^* = 90$ ksi, and $E_f = 6500$ ksi.

*Solution*: The maximum deflection in a uniformly loaded simple beam is given by

$$\Delta_{D+L} = \frac{5Ml^2}{48E_c I_{e,D+L}}$$

The key term in the above expression is the effective moment of inertia of the beam, $I_e$, which is given by ACI 440 Equation 8-12a:

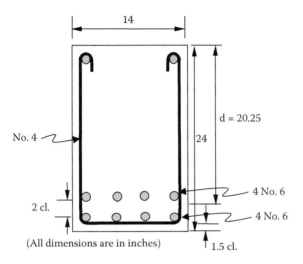

**FIGURE E8.1**  Beam cross section for Example 8.1.

$$(I_e)_{D+L} = \left(\frac{M_{cr}}{M_a}\right)^3 \beta_d I_g + \left[1 - \left(\frac{M_{cr}}{M_a}\right)^3\right] I_{cr} \leq I_g \qquad \text{(ACI 440, Equation 8-12a)}$$

Various quantities in the above expression are calculated step-by-step as follows.

*Step 1.* Calculate the cracking moment of the beam, $M_{cr}$.

$$f_r = 7.5\sqrt{f_c'} = 7.5\sqrt{4000} = 474 \text{ psi}$$

$$M_{cr} = \frac{2 f_r I_g}{h}$$

$$I_g = \frac{bh^3}{12} = \frac{(14)(24)^3}{12} = 16{,}128 \text{ in.}^4$$

$$M_{cr} = \frac{2 f_r I_g}{h} = \frac{2(474)(16{,}128)}{24(12{,}000)} = 53.1 \text{ k-ft}$$

*Step 2.* Calculate the neutral axis factor, $k$, from ACI 440, Equation 8-11.

$$h = 24 \text{ in., } d = 20.25 \text{ in., } A_f = 8(0.31) = 2.48 \text{ in.}^2$$

$$\rho_f = \frac{A_f}{bd} = \frac{2.48}{(14)(20.25)} = 0.0087$$

$$E_c = 57{,}000\sqrt{f_c'} = 57{,}000\sqrt{4000} = 3{,}605{,}000 \text{ psi} = 3605 \text{ ksi}$$

$$n_f = \frac{E_f}{E_c} = \frac{6500}{3605} = 1.8$$

$$k = \sqrt{2\rho_f n_f + (\rho_f n_f)^2} - \rho_f n_f$$

$$= \sqrt{2(0.0087)(1.8) + [(0.0087)(1.8)]^2} - (0.0087)(1.8)$$

$$= 0.162$$

*Step 3.* Calculate the moment of inertia of the cracked section, $I_{cr}$, from ACI 440, Equation 8-10.

$$I_{cr} = \frac{bd^3}{3}k^3 + n_f A_f d^2 (1-k)^2$$

$$= \frac{(14)(20.25)^3}{3}(0.162)^3 + (1.8)(2.48)(20.25)^2(1-0.162)^2$$

$$= 1450 \text{ in.}^4$$

*Step 4.* Calculate $\beta_d$ from ACI 440, Equation 8-12b.

$$\beta_d = \alpha_b \left( \frac{E_f}{E_s} + 1 \right) \qquad \text{(ACI 440, Equation 8-12b)}$$

$$\alpha_b = 0.5 \text{ (by default)}$$

$$\beta_d = 0.5 \left( \frac{6500}{29{,}000} + 1 \right) = 0.61$$

*Step 5.* Calculate $M_a$, the moment due to dead load and service live load.

$$D_{beam} = \frac{(14)(24)}{144}(150) = 350 \text{ lb/ft}$$

$$D_{sup} = 900 \text{ lb/ft}, \ L = 600 \text{ lb/ft}$$

$$w_{ser} = 350 + 600 + 900 = 1850 \text{ k/ft}$$

$$M_a = M_{ser} = M_{D+L} = \frac{w_{ser}l^2}{8} = \frac{(1850)(20)^2}{8(1000)} = 92.5 \text{ k-ft}$$

*Step 6.* Calculate the effective moment of inertia, $I_e$.

$$(I_e)_{D+L} = \left(\frac{M_{cr}}{M_a}\right)^3 \beta_d I_g + \left[1 - \left(\frac{M_{cr}}{M_a}\right)^3\right] I_{cr} \le I_g$$

$$= \left(\frac{53.1}{92.5}\right)^3 (0.61)(16,128) + \left[1 - \left(\frac{53.1}{92.5}\right)^3\right](1450)$$

$$= 3037 \text{ in.}^4 \le I_g = 16,128 \text{ in.}^4, \text{ OK}$$

*Step 7.* Calculate the instantaneous deflection due to the service load.

$$\Delta_{D+L} = \frac{5M_{ser}l^2}{48E_c I_{e,D+L}}$$

$$= \frac{5(92.5)(20)^2(12)^3}{(48)(3605)(3037)} = 0.61 \text{ in.}$$

*Step 8.* Calculate the long-term deflection, $\Delta_{LT}$, which is given by

$$\Delta_{LT} = (\Delta_i)_L + \lambda[(\Delta_i)_D + 0.2(\Delta_i)_L]$$

where $(\Delta_i)_D$ and $(\Delta_i)_L$ are, respectively, the instantaneous deflections due to dead and live load and are, respectively, equal to $\Delta_D$ and $\Delta_L$ calculated earlier.

Calculate instantaneous deflections due to dead load ($\Delta_D$) and live load ($\Delta_L$) separately. This can be done by finding the proportion of dead and live load to the total service load.

$$\Delta_D = \frac{D}{D+L}(\Delta_{D+L}) = \frac{1250}{1850}(0.61) = 0.41 \text{ in.}$$

$$\Delta_L = \frac{L}{D+L}(\Delta_{D+L}) = \frac{600}{1850}(0.61) = 0.20 \text{ in.}$$

(Check: $\Delta_D + \Delta_L = 0.41 + 0.2 = 0.61$ in. $= \Delta_{D+L}$)

$$\lambda = \frac{\xi}{1 + \rho'} \qquad \text{(ACI 440, Equation 8-13b)}$$

For loads sustained for 5 years or more, $\xi = 2.0$. In the present case, the beam does not have any compression reinforcement, so the compression reinforcement ratio $\rho'$ = 0. Therefore,

$$\lambda = 0.6(\xi) = 0.6(2.0) = 1.2 \quad \text{(ACI 440, Equation 8-14)}$$

$$\Delta_{LT,actual} = 0.2 + 1.2[0.41 + 0.2(0.20)] = 0.74 \text{ in.}$$

$$\Delta_{LT,all} = \frac{l}{240} = \frac{(20)(12)}{240} = 1.0 \text{ in.}$$

$$\Delta_{LT,actual} = 0.74 \text{ in.} < \Delta_{LT,allow} = 1.0 \text{ in.}$$

Hence, the deflection requirement is satisfied.

*Example 8.2*: A rectangular 12 in. × 18 in. concrete beam spanning 16 ft is reinforced with five No. 6 GFRP bars arranged as shown in Figure E8.2. It supports a superimposed dead load of 750 lb/ft in addition to its own dead weight and a live of 1200 lb/ft. Check if the beam satisfies the long-term deflection requirements. The allowable long-term deflection is 1/240 of span. Assume that $f_c' = 4$ ksi, $f_{fu}^* = 90$ ksi, and $E_f = 6500$ ksi.

*Solution*: Calculate $(I_e)_{D+L}$ from ACI 440, Equation 8-12a.

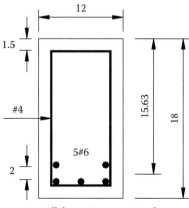

All dimensions are in inches.

**FIGURE E8.2**. Beam cross section for Example 8.2.

$$(I_e)_{D+L} = \left(\frac{M_{cr}}{M_a}\right)^3 \beta_d I_g + \left[1 - \left(\frac{M_{cr}}{M_a}\right)^3\right] I_{cr} \leq I_g \qquad \text{(ACI 440, Equation 8-12a)}$$

Calculate the cracking moment of the beam, $M_{cr}$.

$$f_r = 7.5\sqrt{f_c'} = 7.5\sqrt{4000} = 474 \text{ psi}$$

$$I_g = \frac{bh^3}{12} = \frac{(12)(18)^3}{12} = 5832 \text{ in.}^4$$

$$M_{cr} = \frac{2 f_r I_g}{h} = \frac{2(474)(5832)}{18(12,000)} = 25.6 \text{ k-ft}$$

Calculate the neutral axis factor, $k$. For five No. 6 bars, $A_f = 5(0.44) = 2.20 \text{ in}^2$.

$$d = h - 0.5d_b - d_{bar,shear} - \text{clear cover} = 18 - 0.5(0.75) - 0.5 - 1.5 = 15.63 \text{ in.}$$

$$\rho_f = \frac{A_f}{bd} = \frac{2.2}{(12)(15.63)} = 0.0117$$

$$E_c = 57,000\sqrt{f_c'} = 57,000\sqrt{4000} = 3,605,000 \text{ psi} = 3605 \text{ ksi}$$

$$n_f = \frac{E_f}{E_c} = \frac{6500}{3605} = 1.8$$

$$k = \sqrt{2\rho_f n_f + (\rho_f n_f)^2} - \rho_f n_f$$

$$= \sqrt{2(0.0117)(1.8) + [(0.0117)(1.8)]^2} - (0.0117)(1.8)$$

$$= 0.185$$

Calculate the moment of inertia of the cracked section, $I_{cr}$.

$$I_{cr} = \frac{bd^3}{3}k^3 + n_f A_f d^2 (1-k)^2$$

$$= \frac{(12)(15.63)^3}{3}(0.185)^3 + (1.8)(2.2)(15.63)^2 (1-0.185)^2$$

$$= 739 \text{ in.}^4$$

Calculate $\beta_d$ from ACI 440, Equation 8.12b.

$$\beta_d = \alpha_b\left(\frac{E_f}{E_s}+1\right) \qquad \text{(ACI 440, Equation 8-12b)}$$

$$\alpha_b = 0.5 \text{ (by default)}$$

$$\beta_d = 0.5\left(\frac{6500}{29,000}+1\right)=0.61$$

Calculate $M_a$, the moment due to dead load and service live load.

$$D_{beam} = \frac{(12)(18)}{144}(150)=225 \text{ lb/ft}$$

$$D_{sup} = 750 \text{ lb/ft, } L = 1200 \text{ lb/ft}$$

$$w_{ser} = 225 + 750 + 1200 = 2175 \text{ lb/ft} = 2.175 \text{ k/ft}$$

$$M_a = M_{D+L} = \frac{w_{ser}l^2}{8} = \frac{(2.175)(16)^2}{8} = 69.6 \text{ k-ft}$$

Calculate $(I_e)_{D+L}$ from ACI 440, Equation 8-12a.

$$(I_e)_{D+L} = \left(\frac{M_{cr}}{M_a}\right)^3 \beta_d I_g + \left[1-\left(\frac{M_{cr}}{M_a}\right)^3\right]I_{cr} \le I_g$$

$$= \left(\frac{25.6}{69.6}\right)^3(0.61)(5832)+\left[1-\left(\frac{25.6}{69.6}\right)^3\right](739)$$

$$= 879 \text{ in.}^4 \le I_g = 5832 \text{ in.}^4, \text{ OK}$$

The instantaneous deflection due to service loads is given by

$$\Delta_{D+L} = \frac{5M_{ser}l^2}{48E_cI_{e,D+L}}$$

$$= \frac{5(69.6)(16)^2(12)^3}{(48)(3605)(879)} = 1.01 \text{ in.}$$

Calculate the long-term deflection, $\Delta_{LT}$, which is given by

$$\Delta_{LT} = (\Delta_i)_L + \lambda[(\Delta_i)_D + 0.2(\Delta_i)_L]$$

where $(\Delta_i)_D$ and $(\Delta_i)_L$, respectively, are instantaneous deflections due to dead and live load and are, respectively, equal to $\Delta_D$ and $\Delta_L$ calculated earlier.

Calculate instantaneous deflections due to dead load $(\Delta_D)$ and live load $(\Delta_L)$ separately. This can be done by finding the proportion of dead and live load to the total service load.

$$\Delta_D = \frac{D}{D+L}(\Delta_{D+L}) = \frac{975}{2175}(1.01) = 0.45 \text{ in.}$$

$$\Delta_L = \frac{L}{D+L}(\Delta_{D+L}) = \frac{1200}{2175}(1.01) = 0.56 \text{ in.}$$

(Check: $\Delta_D + \Delta_L = 0.45 + 0.56 = 1.01$ in. $= \Delta_{D+L}$)

$$\lambda = \frac{\xi}{1+\rho'} \qquad \text{(ACI 440, Equation 8-13b)}$$

For loads sustained for 5 years or more, $\xi = 2.0$. In the present case, the beam does not have any compression reinforcement, so the compression reinforcement ratio $\rho' = 0$. Therefore,

$$\lambda = 0.6(\xi) = 0.6(2.0) = 1.2 \quad \text{(ACI 440, Equation 8-14)}$$

$$\Delta_{LT} = 0.56 + 1.2[0.45 + 0.2(0.56)] = 1.23 \text{ in.}$$

$$\Delta_{LT,all} = \frac{l}{240} = \frac{(16)(12)}{240} = 0.8 \text{ in.}$$

$$\Delta_{LT,actual} = 1.23 \text{ in.} > \Delta_{LT,allow} = 0.8 \text{ in.}$$

Hence, the deflection requirement is not satisfied.

*Example 8.3*: Rework Example 8.2 by increasing the depth of the beam to satisfy deflection requirements. All other data remain the same.

*Solution*: As a first trial, the overall depth, $h$, of the beam shown in Figure E8.2 is increased arbitrarily by 3 in., from 18 in. to 21 in., so that the revised depth, $d$ is (see Figure E8.3)

$$d = 15.63 + 3 = 18.63 \text{ in.} \quad (d = 15.63 \text{ in. was determined in Example 8.2})$$

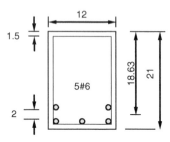

All dimensions are in inches

**FIGURE E8.3** Beam cross section for Example 8.3.

Calculate $M_{cr}$.

$$f_r = 7.5\sqrt{f_c'} = 7.5\sqrt{4000} = 474 \text{ psi}$$

$$I_g = \frac{bh^3}{12} = \frac{(12)(21)^3}{12} = 9261 \text{ in.}^4$$

$$M_{cr} = \frac{2f_4 I_g}{h} = \frac{2(474)(9261)}{(21)(12,000)} = 34.84 \text{ k-ft}$$

Calculate the neutral axis factor, $k$.

$$\rho_f = \frac{A_f}{bd} = \frac{2.2}{(12)(18.63)} = 0.0098$$

$$E_c = 57,000\sqrt{f_c'} = 57,000\sqrt{4000} = 3,605,000 \text{ psi} = 3605 \text{ ksi}$$

$$n_f = \frac{E_f}{E_c} = \frac{6500}{3605} = 1.8$$

$$k = \sqrt{2\rho_f n_f + (\rho_f n_f)^2} - \rho_f n_f$$

$$= \sqrt{2(0.0098)(1.8) + [(0.0098)(1.8)]^2} - (0.0098)(1.8)$$

$$= 0.171$$

Calculate the moment of inertia of the cracked section, $I_{cr}$.

$$I_{cr} = \frac{bd^3}{3}k^3 + n_f A_f d^2 (1-k)^2$$

$$= \frac{(12)(18.63)^3}{3}(0.171)^3 + (1.8)(2.2)(18.63)^2(1-0.171)^2$$

$$= 1074 \text{ in.}^4$$

Calculate $\beta_d$ from ACI 440, Equation 8.12b.

$$\beta_d = \alpha_b \left( \frac{E_f}{E_s} + 1 \right) \qquad \text{(ACI 440, Equation 8-12b)}$$

$$\alpha_b = 0.5 \text{ (by default)}$$

$$\beta_d = 0.5 \left( \frac{6500}{29,000} + 1 \right) = 0.61$$

Calculate $M_a$, the moment due to dead load and service live load moment. Because the beam depth has been increased, $M_a$ should be revised.

$$D_{beam} = \frac{(12)(21)}{144}(150) = 262.5 \text{ lb/ft}$$

$$D_{sup} = 750 \text{ lb/ft}, \ L = 1200 \text{ lb/ft}$$

$$D = 262.5 + 750 = 1012.5 \text{ lb/ft}$$

$$w_{ser} = 262.5 + 750 + 1200 = 2212.5 \text{ lb/ft} = 2.2125 \text{ k/ft}$$

$$M_a = M_{D+L} = \frac{w_{ser}l^2}{8} = \frac{(2.2125)(16)^2}{8} = 70.8 \text{ k-ft}$$

Calculate $(I_e)_{D+L}$ from ACI 440, Equation 8-12a.

$$(I_e)_{D+L} = \left( \frac{M_{cr}}{M_a} \right)^3 \beta_d I_g + \left[ 1 - \left( \frac{M_{cr}}{M_a} \right)^3 \right] I_{cr} \le I_g$$

$$= \left( \frac{34.84}{70.8} \right)^3 (0.61)(9261) + \left[ 1 - \left( \frac{34.84}{70.8} \right)^3 \right](1074)$$

$$= 1619 \text{ in.}^4 \le I_g = 9261 \text{ in.}^4, \text{ OK}$$

The instantaneous deflection due to service loads is given by

$$\Delta_{D+L} = \frac{5M_{ser}l^2}{48E_cI_{e,D+L}}$$

$$= \frac{5(70.8)(16)^2(12)^3}{(48)(3605)(1619)} = 0.56 \text{ in.}$$

Calculate the long-term deflection, $\Delta_{LT}$, which is given by

$$\Delta_{LT} = (\Delta_i)_L + \lambda[(\Delta_i)_D + 0.2(\Delta_i)_L]$$

where $(\Delta_i)_D$ and $(\Delta_i)_L$, respectively, are the instantaneous deflections due to dead and live loads and are, respectively, equal to $\Delta_D$ and $\Delta_L$ calculated earlier.

Calculate the instantaneous deflections due to dead load ($\Delta_D$) and live load ($\Delta_L$) separately. This can be done by finding the proportion of dead and live load to the total service load.

$$\Delta_D = \frac{D}{D+L}(\Delta_{D+L}) = \frac{1012.5}{2212.5}(0.56) = 0.26 \text{ in.}$$

$$\Delta_L = \frac{L}{D+L}(\Delta_{D+L}) = \frac{1200}{2212.5}(0.56) = 0.30 \text{ in.}$$

(Check: $\Delta_D + \Delta_L = 0.26 + 0.30 = 0.56$ in. $= \Delta_{D+L}$)

$$\lambda = \frac{\xi}{1+\rho'} \qquad \text{(ACI 440, Equation 8-13b)}$$

For loads sustained for 5 years or more, $\xi = 2.0$. In the present case, the beam does not have any compression reinforcement, so the compression reinforcement ratio $\rho' = 0$. Therefore,

$$\lambda = 0.6(\xi) = 0.6(2.0) = 1.2 \quad \text{(ACI 440, Equation 8-14)}$$

$$\Delta_{LT,actual} = 0.3 + 1.2[0.26 + 0.2(0.30)] = 0.68 \text{ in.}$$

$$\Delta_{LT,allow} = \frac{l}{240} = \frac{(16)(12)}{240} = 0.8 \text{ in}$$

$$\Delta_{LT,actual} = 0.68 \text{ in.} < \Delta_{LT,allow} = 0.8 \text{ in.}$$

Hence, the deflection requirement is satisfied.

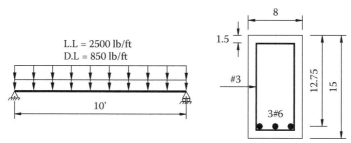

All dimensions are in inches.

**FIGURE E8.4** Loading and beam cross section for Example 8.4.

*Example 8.4*: *Determination of nominal strength of a beam reinforced with CFRP bars*. An 8 in. × 15 in. rectangular concrete beam is reinforced with three No. 6 CFRP bars and No. 3 stirrups as shown in Figure E8.4. The beam carries a super-imposed dead load of 875 lb/ft in addition to its own weight and a floor live load of 2500 lb/ft over a span of 10 ft. Check the adequacy of this beam for flexural strength. Assume $f_c' = 4000$ psi, $f_{fu}^* = 240$ ksi, and $E_f = 20,000$ ksi.

*Solution*: The service loads are given by

$$D_{beam} = \frac{(8)(15)}{144}(150) = 125 \text{ lb/ft}$$

$$D_{super} = 875 \text{ lb/ft}$$

$$D = 125 + 875 = 1000 \text{ lb/ft}$$

$$L = 2500 \text{ lb/ft}$$

$$w_{ser} = 1000 + 2500 = 3500 \text{ lb/ft} = 3.5 \text{ k/ft}$$

The factored loads are given by

$$w_u = 1.2D + 1.6L = 1.2(1000) + 1.6(2500) = 5200 \text{ lb/ft} = 5.2 \text{ k/ft}$$

$$M_u = w_u l^2/8 = (5.2)(10)^2/8 = 65 \text{ k-ft}$$

$$f_{fu} = C_E f_{fu}^* \qquad \text{(ACI 440, Equation 7-1)}$$

For an interior conditioned space, the environmental modification factor, $C_E$, is 1.0 (ACI 440, Table 7.1).

$$f_{fu} = (1.0)(240) = 240.0 \text{ ksi}$$

Calculate the reinforcement ratio, $\rho_f$. First, determine $d$:

$$d = h - \text{clear cover} - \text{diameter of stirrup} - 1/2(\text{diameter of tensile reinforcing bar})$$
$$= 15 - 1.5 - 0.375 - 0.75/2 = 12.75 \text{ in.}$$

$$A_f = 3(0.44) = 1.32 \text{ in}^2$$

$$\rho_f = \frac{A_f}{bd} = \frac{1.32}{(8)(12.75)} = 0.0129$$

Calculate the stress in CFRP reinforcement from ACI 440, Equation 8-4d:

$$f_f = \sqrt{\frac{(E_f\varepsilon_{cu})^2}{4} + \frac{0.85\beta_1 f_c'}{\rho_f} E_f\varepsilon_{cu}} - 0.5E_f\varepsilon_{cu} < f_{fu}$$

$$= \sqrt{\frac{[(20,000)(0.003)]^2}{4} + \frac{0.85(0.85)(4)}{0.0129}(20,000)(0.003)} - 0.5(20,000)(0.003)$$

$$= 89.8 \text{ ksi} < 240 \text{ ksi, OK}$$

Calculate the strength reduction factor $\phi$ from ACI 440, Equation 8-7. First, calculate $\rho_{fb}$ from ACI 440, Equation 8.3.

$$\rho_{fb} = 0.85\beta_1 \frac{f_c'}{f_{fu}} \frac{E_f\varepsilon_u}{E_f\varepsilon_{cu} + f_{fu}} \qquad \text{(ACI 440 Equation 8-3)}$$

For $f_c' = 4000$ psi, $\beta_1 = 0.85$, and $f_{fu} = 240$ ksi. Therefore,

$$\rho_{fb} = (0.85)(0.85)\frac{4}{240}\left[\frac{(20,000)(0.003)}{(20,000)(0.003) + 240}\right] = 0.0024$$

$$1.4\rho_{fb} = 1.4(0.0024) = 0.0034 > 1.4\rho_f = 0.0034$$

Because $(\rho_f = 0.0129) > (1.4\rho_{fb} = 0.0034)$, $\phi = 0.7$ (ACI 440, Equation 8-7).

Calculate the nominal strength of beam from ACI 440, Equation 8-5.

$$M_n = \rho_f f_f \left(1 - 0.59 \frac{\rho_f f_f}{f_c'}\right) bd^2$$

$$= \frac{(0.012)(89.8)}{12}\left[1 - 0.59\frac{(0.0129)(89.8)}{4}\right](8)(12.75)^2 \text{ (ACI 440, Equation 8-5)}$$

$$= 104.1 \text{ k-ft}$$

Calculate $\phi M_n$.

$$\phi M_n = 0.7(104.1) = 72.9 \text{ k-ft} > M_u = 65 \text{ k-ft}$$

The beam has adequate flexural strength.

*Example 8.5: Long-term deflection of a beam reinforced with CFRP bars.* A rectangular 8 in. × 15 in. concrete beam is reinforced with three No. 6 CFRP bars and No. 3 stirrups (Figure E8.5). It supports a superimposed dead load of 875 lb/ft in addition to its own dead weight, and a live load of 2500 lb/ft over span of 10 ft. Calculate the long-term deflection of this beam. The allowable long-term deflection is $L/240$. Assume that $f_c' = 4$ ksi, $f_{fu}^* = 240$ ksi, and $E_f = 20,000$ ksi.

*Solution*: This is the same beam for which the flexural strength was checked in Example 8.4, and from which the following data are obtained: $d = 12.75$ in., $A_f = 1.32$ in$^2$, $\rho_f = 0.0129$, $E_c = 3605$ ksi, $n_f = 5.55$, $k = 0.314$, $j = 0.895$, $D = 1000$ lb/ft, $L = 2500$ lb/ft, and $M_{ser} = M_a = 43.75$ k-ft.

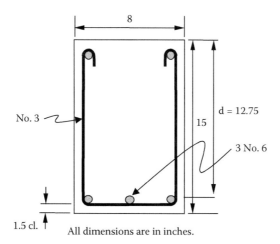

No. 3

1.5 cl.    All dimensions are in inches.

d = 12.75

3 No. 6

**FIGURE E8.5** Beam cross section for Example 8.5.

Calculate $(I_e)_{D+L}$ from ACI 440, Equation 8-12a.

$$(I_e)_{D+L} = \left(\frac{M_{cr}}{M_a}\right)^3 \beta_d I_g + \left[1-\left(\frac{M_{cr}}{M_a}\right)^3\right]I_{cr} \leq I_g$$

Calculate $M_{cr}$.

$$f_r = 7.5\sqrt{f_c'} = 7.5\sqrt{4000} = 474 \text{ psi}$$

$$I_g = \frac{bh^3}{12} = \frac{(8)(15)^3}{12} = 2250 \text{ in.}^4$$

$$M_{cr} = \frac{2 f_r I_g}{h} = \frac{2(474)(2250)}{(15)(12,000)} = 11.85 \text{ k-ft}$$

Calculate the moment of inertia of the cracked section, $I_{cr}$.

$$I_{cr} = \frac{bd^3}{3}k^3 + n_f A_f d^2 (1-k)^2$$

$$= \frac{(8)(12.75)^3}{3}(0.314)^3 + (5.55)(1.32)(12.75)^2(1-0.314)^2$$

$$= 732 \text{ in.}^4$$

Calculate $\beta_d$ from ACI 440, Equation 8.12b.

$$\beta_d = \alpha_b\left(\frac{E_f}{E_s}+1\right) \qquad \text{(ACI 440, Equation 8-12b)}$$

$$\alpha_b = 0.5 \text{ (by default)}$$

$$\beta_d = 0.5\left(\frac{20,000}{29,000}+1\right) = 0.845$$

Calculate the effective moment of inertia, $(I_e)_{D+L}$ from ACI 440, Equation 8-12a.

$$(I_e)_{D+L} = \left(\frac{M_{cr}}{M_a}\right)^3 \beta_d I_g + \left[1 - \left(\frac{M_{cr}}{M_a}\right)^3\right] I_{cr} \leq I_g$$

$$= \left(\frac{11.85}{43.75}\right)^3 (0.845)(2250) + \left[1 - \left(\frac{11.85}{43.75}\right)^3\right](732)$$

$$= 755 \text{ in.}^4 \leq I_g = 2250 \text{ in.}^4, \text{ OK}$$

The instantaneous deflection due to service loads is given by

$$\Delta_{D+L} = \frac{5M_{ser}l^2}{48E_c I_{e,D+L}}$$

$$= \frac{5(43.75)(10)^2(12)^3}{(48)(3605)(755)} = 0.29 \text{ in.}$$

Calculate the long-term deflection, $\Delta_{LT}$, which is given by

$$\Delta_{LT} = (\Delta_i)_L + \lambda[(\Delta_i)_D + 0.2(\Delta_i)_L]$$

where $(\Delta_i)_D$ and $(\Delta_i)_L$, respectively, are the instantaneous deflections due to dead and live loads and are, respectively, equal to $\Delta_D$ and $\Delta_L$ calculated earlier.

Calculate the deflection due to dead load ($\Delta_D$) and live load ($\Delta_L$) separately. This can be done by finding the proportion of dead and live load to the total service load.

$$\Delta_D = \frac{D}{D+L}(\Delta_{D+L}) = \frac{1000}{3500}(0.29) = 0.08 \text{ in.}$$

$$\Delta_L = \frac{L}{D+L}(\Delta_{D+L}) = \frac{2500}{3500}(0.29) = 0.21 \text{ in.}$$

(Check: $\Delta_D + \Delta_L = 0.08 + 0.21 = 0.29 \text{ in.} = \Delta_{D+L}$)

$$\lambda = \frac{\xi}{1+\rho'} \qquad \text{(ACI 440, Equation 8-13b)}$$

For loads sustained for 5 years or more, $\xi = 2.0$. In the present case, the beam does not have any compression reinforcement, so the compression reinforcement ratio $\rho' = 0$. Therefore,

$$\lambda = 0.6(\xi) = 0.6(2.0) = 1.2 \quad \text{(ACI 440, Equation 8-14)}$$

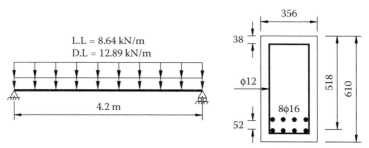

All dimensions are in mm.

**FIGURE E8.6** Beam cross section for Example 8.6.

The long-term deflection of the beam is given by

$$\Delta_{LT} = (\Delta_i)_L + \lambda[(\Delta_i)_D + 0.2(\Delta_i)_L]$$

where $(\Delta_i)_D$ and $(\Delta_i)_L$, respectively, are the instantaneous deflection due to dead and live load and are, respectively, equal to $\Delta_D$ and $\Delta_L$ calculated earlier.

$$\Delta_{LT,actual} = (\Delta_i)_L + \lambda[(\Delta_i)_D + 0.2(\Delta_i)_L]$$
$$= 0.21 + 1.2[0.08 + 0.2(0.21)] = 0.36 \text{ in.}$$

$$\Delta_{LT,allow} = \frac{l}{240} = \frac{(10)(12)}{240} = 0.5 \text{ in.}$$

$$\Delta_{LT,actual} = 0.36 \text{ in.} < \Delta_{LT,allow} = 0.5 \text{ in.}$$

Hence, the deflection requirement is satisfied.

## 8.5.2 Examples in SI Units

The following examples are the same as Example 8.1 through Example 8.5 except that the units have been converted to SI units.

*Example 8.6: Long-term deflection of beams reinforced with GFRP bars.* A rectangular 356 mm × 610 mm concrete beam is reinforced with eight $\phi$-20 mm GFRP bars arranged in two rows so that $d = 518$ mm (Figure E8.6). The beam spans 6.1 m and carries a superimposed dead load of 12.89 kN/m and a live load of 8.64 kN/m in addition to its own weight. The allowable deflection is $l/240$. Check if the beam satisfies the deflection requirements. Assume $f_c' = 27.6$ N/mm², $f_{fu}* = 620.55$ N/mm², and $E_f = 45.5$ kN/mm².

*Solution*: The maximum deflection in a uniformly loaded simple beam is given by

$$\Delta_{D+L} = \frac{5Ml^2}{48E_cI_{e,D+L}}$$

The key term in the above expression is the effective moment of inertia of the beam, $I_e$, which is given by ACI 440, Equation 8-12a:

$$(I_e)_{D+L} = \left(\frac{M_{cr}}{M_a}\right)^3 \beta_d I_g + \left[1 - \left(\frac{M_{cr}}{M_a}\right)^3\right]I_{cr} \leq I_g \qquad \text{(ACI 440, Equation 8-12a)}$$

Various quantities in the above expression are calculated step-by-step as follows.
*Step 1*. Calculate the cracking moment of the beam, $M_{cr}$.

$$f_r = 0.62\sqrt{f_c'} = 0.62\sqrt{27.6} = 3.257 \text{ N/mm}^2$$

$$M_{cr} = \frac{2f_rI_g}{h}$$

$$I_g = \frac{bh^3}{12} = \frac{(356)(610)^3}{12} = 6.73\times10^9 \text{ mm}^4$$

$$M_{cr} = \frac{2f_rI_g}{h} = \frac{2(3.25)(6.73\times10^9)}{610\times10^6} = 71.7 \text{ kN-m}$$

*Step 2*. Calculate the neutral axis factor, $k$, from ACI 440, Equation 8-11.

$$h = 610 \text{ mm}$$

$$d = 548 \text{ mm}$$

$$A_f = 8(201) = 1608 \text{ mm}^2$$

$$\rho_f = \frac{A_f}{bd} = \frac{1608}{(356)(518)} = 0.0087$$

$$E_c = 4750\sqrt{f_c'} = 4750\sqrt{27.6} = 24.95 \text{ kN/mm}^2$$

$$n_f = \frac{E_f}{E_c} = \frac{45.5}{24.95} = 1.8$$

$$k = \sqrt{2\rho_f n_f + (\rho_f n_f)^2} - \rho_f n_f$$

$$= \sqrt{2(0.0087)(1.8) + [(0.0087)(1.8)]^2} - (0.0087)(1.8)$$

$$= 0.162$$

*Step 3*. Calculate the moment of inertia of the cracked section, $I_{cr}$, from ACI 440, Equation 8-10.

$$I_{cr} = \frac{bk^3 d^3}{3} + n_f A_f d^2 (1-k)^2$$

$$= \frac{(356)(0.162)^3(518)^3}{3} + (1.8)(1608)(518)^2(1-0.162)^2$$

$$= 6.15 \times 10^8 \text{ mm}^4$$

*Step 4*. Calculate $\beta_d$ from ACI 440, Equation 8-12b.

$$\beta_d = \alpha_b \left( \frac{E_f}{E_s} + 1 \right) \qquad \text{(ACI 440 Equation 8-12b)}$$

$$\alpha_b = 0.5 \text{ (by default)}$$

$$\beta_d = 0.5 \left( \frac{45.5}{200} + 1 \right) = 0.61$$

*Step 5*. Calculate $M_a$, the moment due to dead load and service live load.

$$D_{beam} = \left( \frac{356}{1000} \right) \left( \frac{610}{1000} \right) (23.56) = 5.12 \text{ kN/m}$$

$$D_{sup} = 12.89 \text{ kN/m}$$

$$L = 8.64 \text{ kN/m}$$

$$w_{ser} = 5.12 + 8.64 + 12.81 = 26.65 \text{ kN/m}$$

$$M_a = M_{ser} = M_{D+L} = \frac{w_{ser} l^2}{8} = \frac{(26.65)(6.1)^2}{8} = 123.95 \text{ kN-m}$$

*Step 6.* Calculate the effective moment of inertia, $I_e$, from ACI 440, Equation 8.12a.

$$I_e = \left(\frac{M_{cr}}{M_a}\right)^3 \beta_D I_g + \left[1 - \left(\frac{M_{cr}}{M_a}\right)^3\right] I_{cr}$$

$$= \left(\frac{71.71}{123.95}\right)^3 (0.61)(6.73 \times 10^9) + \left[1 - \left(\frac{71.71}{123.95}\right)^3\right] (6.15 \times 10^8)$$

$$= 1.29 \times 10^9 \text{ mm}^4 \leq I_g = 6.73 \times 10^9 \text{ mm}^4, \text{ OK}$$

*Step 7.* Calculate the instantaneous deflection due to service load.

$$\Delta_{D+L} = \frac{5 M_{ser} l^2}{48 E_c I_e}$$

$$= \frac{5(123.95)(6.1)^2 (1000)^3}{(48)(24.95)(1.29 \times 10^9)}$$

$$= 14.9 \text{ mm}$$

*Step 8.* Calculate the long-term deflection, $\Delta_{LT}$, which is given by

$$\Delta_{LT} = (\Delta_i)_L + \lambda[(\Delta_i)_D + 0.2(\Delta_i)_L]$$

where $(\Delta_i)_D$ and $(\Delta_i)_L$ are, respectively, the instantaneous deflections due to dead and live load and are, respectively, equal to $\Delta_D$ and $\Delta_L$ calculated earlier. Calculate the instantaneous deflections due to dead load ($\Delta_D$) and live load ($\Delta_L$) separately. This can be done by finding the proportion of dead and live load to the total service load.

$$\Delta_D = \frac{D}{D+L}(\Delta_{D+L}) = \frac{18}{26.46}(14.9) = 10.1 \text{ mm}$$

$$\Delta_L = \frac{L}{D+L}(\Delta_{D+L}) = \frac{8.64}{26.46}(14.9) = 4.8 \text{ mm}$$

(Check: $\Delta_D + \Delta_L = 10.1 + 4.8 = 14.9 \text{ mm} = \Delta_{D+L}$)

$$\lambda = \frac{\xi}{1+\rho'} \qquad \text{(ACI 440, Equation 8-13b)}$$

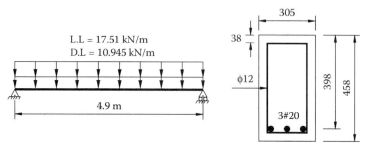

All dimensions are in mm.

**FIGURE E8.7** Beam cross section for Example 8.7.

For loads sustained for 5 years or more, $\xi = 2.0$. In the present case, the beam does not have any compression reinforcement, so the compression reinforcement ratio $\rho' = 0$. Therefore,

$$\lambda = 0.6(\xi) = 0.6(2.0) = 1.2 \quad \text{(ACI 440, Equation 8-14)}$$

$$\Delta_{LT,actual} = 4.8 + 1.2[10.1 + 0.2(4.8)] = 18.1 \text{ mm}$$

$$\Delta_{LT,allow} = \frac{l}{240} = \frac{6100}{240} = 25.4 \text{ mm}$$

$$\Delta_{LT,actual} = 18.1 \text{ mm} < \Delta_{LT,allow} = 25.4 \text{ mm}$$

Hence, the deflection requirement is satisfied.

*Example 8.7*: A rectangular 305 mm × 458 mm concrete beam spanning 4.9 m is reinforced with five φ-20 mm GFRP bars arranged as shown in Figure E8.7. It supports a superimposed dead load of 10.945 kN/m in addition to its own dead weight and a live of 17.513 kN/m. Check if the beam satisfies the long-term deflection requirements. The allowable long-term deflection is $l/240$. Assume that $f_c' = 27.6$ N/mm², $f_{fu}* = 620.55$ N/mm², and $E_f = 45.5$ kN/mm².
*Solution*: Calculate $(I_e)_{D+L}$ from ACI 440, Equation 8-12a.

$$I_e = \left(\frac{M_{cr}}{M_a}\right)^3 \beta_d I_g + \left[1 - \left(\frac{M_{cr}}{M_a}\right)^3\right] I_{cr} \leq I_g \qquad \text{(ACI 440, Equation 8-12a)}$$

Calculate $M_{cr}$.

$$f_r = 0.62\sqrt{f_c'} = 0.62\sqrt{27.6} = 3.257 \text{ N/mm}^2$$

$$M_{cr} = \frac{2 f_r I_g}{h}$$

$$I_g = \frac{bh^3}{12} = \frac{(305)(458)^3}{12} = 2.442 \times 10^9 \text{ mm}^4$$

$$M_{cr} = \frac{2 f_r I_g}{h} = \frac{2(3.257)(2.442 \times 10^9)}{458 \times 10^6} = 34.73 \text{ kN-m}$$

Calculate the neutral axis factor, $k$. For five No. 6 bars, $A_f = 5(314) = 1570 \text{ mm}^2$.

$d = h\ 0.5d_b - d_{bar,shear} - \text{clear cover} = 458 - 0.5(20) - 12 - 38 = 398 \text{ mm}$

$$\rho_f = \frac{A_f}{bd} = \frac{1570}{(305)(398)} = 0.0129$$

$$E_c = 4750\sqrt{f_c'} = 4750\sqrt{27.6} = 24.95 \text{ kN/mm}^2$$

$$n_f = \frac{E_f}{E_c} = \frac{45.5}{24.95} = 1.8$$

$$k = \sqrt{2\rho_f n_f + (\rho_f n_f)^2} - \rho_f n_f$$

$$= \sqrt{2(0.0129)(1.8) + [(0.0129)(1.8)]^2} - (0.0129)(1.8)$$

$$= 0.1935$$

Calculate the moment of inertia of the cracked section, $I_{cr}$.

$$I_{cr} = \frac{bk^3 d^3}{3} + n_f A_f d^2 (1-k)^2$$

$$= \frac{(305)(0.1935)^3(398)^3}{3} + (1.8)(1570)(398)^2(1-0.1935)^2$$

$$= 3.376 \times 10^8 \text{ mm}^4$$

Calculate $\beta_d$ from ACI 440, Equation 8.12b.

$$\beta_d = \alpha_b \left( \frac{E_f}{E_s} + 1 \right) \qquad \text{(ACI 440, Equation 8-12b)}$$

$$\alpha_b = 0.5 \text{ (by default)}$$

$$\beta_d = 0.5 \left( \frac{45.5}{200} + 1 \right) = 0.61$$

Calculate $M_a$, the moment due to dead load and service live load.

$$D_{beam} = \left( \frac{305}{1000} \right) \left( \frac{458}{1000} \right) (23.56) = 3.29 \text{ kN/m}$$

$$D_{sup} = 10.945 \text{ kN/m}$$

$$L = 17.513 \text{ kN/m}$$

$$w_{ser} = 3.29 + 10.945 + 17.513 = 31.748 \text{ kN/m}$$

$$M_a = M_{ser} = M_{D+L} = \frac{w_{ser} l^2}{8} = \frac{(31.748)(4.9)^2}{8} = 95.283 \text{ kN-m}$$

Calculate $I_e$ from ACI 440, Equation 8-12a.

$$I_e = \left( \frac{M_{cr}}{M_a} \right)^3 \beta_d I_g + \left[ 1 - \left( \frac{M_{cr}}{M_a} \right)^3 \right] I_{cr}$$

$$= \left( \frac{34.73}{95.283} \right)^3 (0.61)(2.442 \times 10^9) + \left[ 1 - \left( \frac{34.73}{95.283} \right)^3 \right] (3.376 \times 10^8)$$

$$= 3.93 \times 10^8 \text{ mm}^4 \le I_g = 2.442 \times 10^9 \text{ mm}^4, \text{ OK}$$

The instantaneous deflection due to service loads is given by

$$\Delta_{D+L} = \frac{5 M_{ser} l^2}{48 E_c I_e}$$

$$= \frac{5(95.283)(4.9)^2 (1000)^3}{(48)(24.95)(3.93 \times 10^8)} = 24.3 \text{ mm}$$

Calculate the long-term deflection, $\Delta_{LT}$, which is given by

$$\Delta_{LT} = (\Delta_i)_L + \lambda[(\Delta_i)_D + 0.2(\Delta_i)_L]$$

where $(\Delta_i)_D$ and $(\Delta_i)_L$, respectively, are the instantaneous deflections due to dead and live loads and are, respectively, equal to $\Delta_D$ and $\Delta_L$ calculated earlier. Calculate the instantaneous deflections due to dead load ($\Delta_D$) and live load ($\Delta_L$) separately. This can be done by finding the proportion of dead and live loads to the total service load.

$$\Delta_D = \frac{D}{D+L}(\Delta_{D+L}) = \frac{14.23}{31.74}(24.3) = 10.9 \text{ mm}$$

$$\Delta_L = \frac{L}{D+L}(\Delta_{D+L}) = \frac{17.513}{31.74}(24.3) = 13.4 \text{ mm}$$

(Check: $\Delta_D + \Delta_L = 10.9 + 13.4 = 24.3 \text{ mm} = \Delta_{D+L}$)

$$\lambda = \frac{\xi}{1+\rho'} \qquad \text{(ACI 440, Equation 8-13b)}$$

For loads sustained for 5 years or more, $\xi = 2.0$. In the present case, the beam does not have any compression reinforcement, so the compression reinforcement ratio $\rho' = 0$. Therefore,

$$\lambda = 0.6(\xi) = 0.6(2.0) = 1.2 \quad \text{(ACI 440, Equation 8-14)}$$

$$\Delta_{LT} = 13.41 + 1.2[10.9 + 0.2(13.41)] = 29.7 \text{ mm}$$

$$\Delta_{LT,allow} = \frac{l}{240} = \frac{4900}{240} = 20.42 \text{ mm}$$

$$\Delta_{LT,actual} = 29.7 \text{ mm} > \Delta_{LT,allow} = 20.42 \text{ mm}$$

Hence, the deflection requirement is not satisfied.

*Example 8.8*: Rework Example 8.7 by increasing the depth of the beam to satisfy deflection requirements. All other data remain the same.

*Solution*: As a first trial, the overall depth, $h$, of the beam is increased arbitrarily by 86 mm, from 458 mm to 534 mm, so that the revised depth, $d$ is

$$d = 398 + 86 = 478 \text{ mm} \ (d = 398 \text{ mm was determined in Example 8.7)}$$

Calculate the cracking moment of the beam, $M_{cr}$.

$$f_r = 0.62\sqrt{f_c'} = 0.62\sqrt{27.6} = 3.257 \text{ N/mm}^2$$

$$M_{cr} = \frac{2f_r I_g}{h}$$

$$I_g = \frac{bh^3}{12} = \frac{(305)(534)^3}{12} = 3.87 \times 10^9 \text{ mm}^4$$

$$M_{cr} = \frac{2f_r I_g}{h} = \frac{2(3.257)(3.87 \times 10^9)}{534 \times 10^6} = 47.21 \text{ kN-m}$$

Calculate the neutral axis factor, $k$.

$$\rho_f = \frac{A_f}{bd} = \frac{1570}{(305)(478)} = 0.0108$$

$$E_c = 4750\sqrt{f_c'} = 4750\sqrt{27.6} = 24.95 \text{ kN/mm}^2$$

$$n_f = \frac{E_f}{E_c} = \frac{45.5}{24.95} = 1.8$$

$$k = \sqrt{2\rho_f n_f + (\rho_f n_f)^2} - \rho_f n_f$$
$$= \sqrt{2(0.0108)(1.8) + [(0.0108)(1.8)]^2} - (0.0108)(1.8)$$
$$= 0.178$$

Calculate the moment of inertia of the cracked section, $I_{cr}$.

$$I_{cr} = \frac{bk^3 d^3}{3} + n_f A_f d^2 (1-k)^2$$
$$= \frac{(305)(0.178)^3(478)^3}{3} + (1.8)(1570)(478)^2(1-0.178)^2$$
$$= 4.989 \times 10^8 \text{ mm}^4$$

Calculate $\beta_d$ from ACI 440, Equation 8.12b.

$$\beta_d = \alpha_b \left( \frac{E_f}{E_s} + 1 \right) \qquad \text{(ACI 440, Equation 8-12b)}$$

$$\alpha_b = 0.5 \text{ (by default)}$$

$$\beta_d = 0.5 \left( \frac{45.4}{200} + 1 \right) = 0.61$$

Calculate $M_a$, the moment due to dead load and service live load moment. Because the beam depth has been increased, $M_a$ should be revised.

$$D_{beam} = \left( \frac{305}{1000} \right) \left( \frac{534}{1000} \right) (23.56) = 3.84 \text{ kN/m}$$

$$D_{sup} = 10.945 \text{ kN/m}$$

$$L = 17.512 \text{ kN/m}$$

$$D = 3.84 + 10.945 = 14.785 \text{ kN/m}$$

$$w_{ser} = 3.84 + 10.945 + 17.512 = 32.297 \text{ kN/m}$$

$$M_a = M_{ser} = M_{D+L} = \frac{w_{ser} l^2}{8} = \frac{(32.297)(4.9)^2}{8} = 96.93 \text{ kN-m}$$

Calculate $I_e$ from ACI 440, Equation 8-12a.

$$I_e = \left( \frac{M_{cr}}{M_a} \right)^3 \beta_d I_g + \left[ 1 - \left( \frac{M_{cr}}{M_a} \right)^3 \right] I_{cr}$$

$$= \left( \frac{47.21}{96.93} \right)^3 (0.61)(3.87 \times 10^9) + \left[ 1 - \left( \frac{47.21}{96.93} \right)^3 \right] (4.989 \times 10^8)$$

$$= 7.14 \times 10^8 \text{ mm}^4 \le I_g \, 3.87 \times 10^9 \text{ mm}^4, \text{ OK}$$

The instantaneous deflection due to service loads is given by

$$\Delta_{D+L} = \frac{5 M_{ser} l^2}{48 E_c I_e}$$

$$= \frac{5(96.93)(4.9)^2 (1000)^3}{(48)(24.95)(7.14 \times 10^8)} = 13.6 \text{ mm}$$

Calculate the long-term deflection, $\Delta_{LT}$, which is given by

$$\Delta_{LT} = (\Delta_i)_L + \lambda[(\Delta_i)_D + 0.2(\Delta_i)_L]$$

where $(\Delta_i)_D$ and $(\Delta_i)_L$, respectively, are the instantaneous deflections due to dead and live load and are, respectively, equal to $\Delta_D$ and $\Delta_L$ calculated earlier. Calculate the instantaneous deflections due to dead load ($\Delta_D$) and live load ($\Delta_L$) separately. This can be done by finding the proportion of dead and live load to the total service load.

$$\Delta_D = \frac{D}{D+L}(\Delta_{D+L}) = \frac{14.785}{32.297}(13.6) = 6.2 \text{ mm}$$

$$\Delta_L = \frac{L}{D+L}(\Delta_{D+L}) = \frac{17.512}{32.297}(13.6) = 7.4 \text{ mm}$$

(Check: $\Delta_D + \Delta_L = 6.2 + 7.4 = 13.6$ mm $= \Delta_{D+L}$)

$$\lambda = \frac{\xi}{1+\rho'} \qquad \text{(ACI 440, Equation 8-13b)}$$

For loads sustained for 5 years or more, $\xi = 2.0$. In the present case, the beam does not have any compression reinforcement, so the compression reinforcement ratio $\rho' = 0$. Therefore,

$$\lambda = 0.6(\xi) = 0.6(2.0) = 1.2 \quad \text{(ACI 440, Equation 8-14)}$$

$$\Delta_{LT} = 7.4 + 1.2[6.2 + 0.2(7.4)] = 16.6 \text{ mm}$$

$$\Delta_{LT,allow} = \frac{l}{240} = \frac{4900}{240} = 20.42 \text{ mm}$$

$$\Delta_{LT,actual} = 16.6 \text{ mm} < \Delta_{LT,allow} = 20.42 \text{ mm}$$

Hence, the deflection requirement is satisfied.

*Example 8.9: Long-term deflection of a beam reinforced with CFRP bars.* A rectangular 204 mm × 380 mm concrete beam is reinforced with three φ-20 mm CFRP bars and φ-10 mm stirrups (Figure E8.9). It supports a superimposed dead load of 12.775 kN/m in addition to its own dead weight, and a live load of 36.5 kN/m over a span of 3 m. The allowable long-term deflection is $l/240$. Calculate the long-term deflection of this beam. Assume that $f_c' = 27.6$ N/mm², $f_{fu}{}^* = 1654.8$ N/mm², and $E_f = 137.9$ kN/mm².

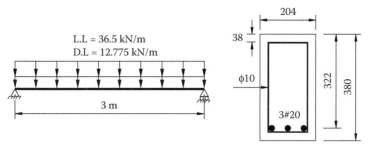

All dimensions are in mm.

**FIGURE E8.9** Beam cross section for Example 8.9.

*Solution*: This is the same beam for which the strength and crack width requirements are checked, respectively, in Example 8.15 and Example 8.16 presented in
Section 8.6. The following data are obtained from Example 8.16: $d = 322$ mm, $A_f$
$= 942$ mm², $\rho_f = 0.0143$, $E_c = 24.95$ kN/mm², $n_f = 5.53$, $k = 0.3264$, $j = 0.8912$, $D$
$= 14.6$ kN/m, $L = 36.5$ kN/m, and $M_{ser} = M_a = 57.48$ kN-m.

Calculate $(I_e)_{D+L}$ from ACI 440, Equation 8-12a.

$$(I_e)_{D+L} = \left(\frac{M_{cr}}{M_a}\right)^3 \beta_d I_g + \left[1 - \left(\frac{M_{cr}}{M_a}\right)^3\right] I_{cr} \leq I_g$$

Calculate the cracking moment of the beam, $M_{cr}$.

$$M_{cr} = \frac{2 f_r I_g}{h}$$

$$I_g = \frac{bh^3}{12} = \frac{(204)(380)^3}{12} = 9.328 \times 10^8 \text{ mm}^4$$

$$M_{cr} = \frac{2 f_r I_g}{h} = \frac{2(3.257)(9.328 \times 10^9)}{380 \times 10^6} = 15.991 \text{ kN-m}$$

Calculate the moment of inertia of the cracked section, $I_{cr}$.

$$I_{cr} = \frac{bk^3 d^3}{3} + n_f A_f d^2 (1-k)^2$$

$$= \frac{(204)(0.3264)^3 (322)^3}{3} + (5.53)(942)(322)^2 (1-0.3264)^2$$

$$= 3.24 \times 10^8 \text{ mm}^4$$

Calculate $\beta_d$ from ACI 440, Equation 8.12b.

$$\beta_d = \alpha_b\left(\frac{E_f}{E_s}+1\right)$$    (ACI 440, Equation 8-12b)

$$\alpha_b = 0.5 \text{ (by default)}$$

$$\beta_d = 0.5\left(\frac{137.9}{200}+1\right)=0.845$$

Calculate the effective moment of inertia, $I_e$, from ACI 440, Equation 8-12a.

$$I_e = \left(\frac{M_{cr}}{M_a}\right)^3\beta_d I_g+\left[1-\left(\frac{M_{cr}}{M_a}\right)^3\right]I_{cr}$$

$$=\left(\frac{15.99}{57.48}\right)^3(0.845)(9.328\times10^8)+\left[1-\left(\frac{15.99}{57.48}\right)^3\right](3.24\times10^8)$$

$$= 3.34\times10^8 \text{ mm}^4 \le I_g = 9.328\times10^8 \text{ mm}^4, \text{ OK}$$

The instantaneous deflection due to service loads is given by

$$\Delta_{D+L}=\frac{5M_{ser}l^2}{48E_cI_e}$$

$$=\frac{5(57.48)(3.0)^2(1000)^3}{(48)(24.95)(3.34\times10^8)}=6.5 \text{ mm}$$

Calculate the long-term deflection, $\Delta_{LT}$, which is given by

$$\Delta_{LT} = (\Delta_i)_L + \lambda[(\Delta_i)_D + 0.2(\Delta_i)_L]$$

where $(\Delta_i)_D$ and $(\Delta_i)_L$, respectively, are the instantaneous deflections due to dead and live load and are, respectively, equal to $\Delta_D$ and $\Delta_L$ calculated earlier. Calculate the instantaneous deflections due to dead load $(\Delta_D)$ and live load $(\Delta_L)$ separately. This can be done by finding the proportion of dead and live load to the total service load.

$$\Delta_D = \frac{D}{D+L}(\Delta_{D+L})=\frac{14.6}{51.1}(6.5)=1.9 \text{ mm}$$

$$\Delta_L = \frac{L}{D+L}(\Delta_{D+L}) = \frac{36.6}{51.1}(6.5) = 4.6 \text{ mm}$$

(Check: $\Delta_D + \Delta_L = 1.9 + 4.6 = 6.5 \text{ mm} = \Delta_{D+L}$)

$$\lambda = \frac{\xi}{1+\rho'} \qquad \text{(ACI 440, Equation 8-13b)}$$

For loads sustained for 5 years or more, $\xi = 2.0$. In the present case, the beam does not have any compression reinforcement, so the compression reinforcement ratio $\rho' = 0$. Therefore,

$$\lambda = 0.6(\xi) = 0.6(2.0) = 1.2 \quad \text{(ACI 440, Equation 8-14)}$$

$$\Delta_{LT} = 4.6 + 1.2[1.9 + 0.2(4.6)] = 8.0 \text{ mm}$$

$$\Delta_{LT,allow} = \frac{l}{240} = \frac{3000}{240} = 12.5 \text{ mm}$$

$$\Delta_{LT,actual} = 8.0 \text{ mm} < \Delta_{LT,allow} = 12.5 \text{ mm}$$

Hence, the deflection limit is satisfied.

## 8.6 CRACK WIDTHS

### 8.6.1 ACI 318 Provisions for Crack Widths

The ACI 318 limits on crack-widths in steel-reinforced concrete structures are based on considerations of steel corrosion, aesthetics, and psychological effects (i.e., a crack indicating a weak structure or failure). Because FRP reinforcements are corrosion resistant, the crack-width limits of ACI 318 for steel-reinforced beams can be relaxed if corrosion is the primary reason for crack-width limitation. If steel is also used in conjunction with FRP reinforcement (e.g., steel stirrups for shear reinforcement), the use of ACI 318 provisions for crack-width limits is suggested for FRP-reinforced members also.

The ACI 318 provisions for crack-width limits in steel-reinforced structures are 0.013 in. (0.3 mm) for exterior exposure and 0.016 in. (0.4 mm) for interior exposure. The Canadian Highways Bridge Design Code allows crack widths of 0.020 in. (0.5 mm) for exterior exposure and 0.028 in. (0.7 mm) for interior exposure when FRP reinforcement is used [CSA 1996]. ACI 440 recommends that based on the function, type of exposure, and aesthetics considerations, an appropriate crack-width limit from either of the two sets of crack-width limits may be used for FRP-reinforced concrete structures.

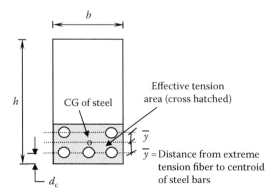

**FIGURE 8.2** Effective tension area $A$ to be used in Equation 8.12.

Crack widths exhibited in FRP-reinforced members are typically larger than those in steel-reinforced members. Research on crack width has shown that the widely-used Gergely-Lutz crack-width equation used for steel-reinforced members can be suitably modified to predict crack widths in FRP-reinforced members [Faza and GangaRao 1993; Masmoudi et al. 1996]. Referring to Figure 8.2, the Gergely-Lutz equation is expressed by Equation 8.12:

$$w = 0.076\beta f_s \sqrt[3]{d_c A} \qquad (8.12)$$

where

    $w$ = the crack width (mils, or $10^{-3}$ in.)

    $\beta$ = the ratio of the distance from the neutral axis to the extreme tension fiber to the distance from neutral axis to the centroid of tensile reinforcement

$$= \frac{h - kd}{d(1 - k)}$$

    $f_s$ = the stress in steel reinforcement due to service loads (ksi)

    $d_c$ = the thickness of the concrete cover measured from the extreme tension fiber to the center of a bar or the closest wire location (in.)

    $A$ = the effective tension area of concrete having the same centroid as that of tensile reinforcement, divided by the number of bars (in²)

Research shows that the crack width is proportional to strain rather than stress in the tensile reinforcement. Accordingly, the value of $f_s$ in Equation 8.12 can be expressed in the form of the strain in steel reinforcement based on Hooke's law ($f_s = E_s\varepsilon_s$) resulting in Equation 8.13:

$$w = 0.076\beta(E_s\varepsilon_s)\sqrt[3]{d_c A} \qquad (8.13)$$

where $E_s$ is the modulus of elasticity of steel in ksi. For predicting the crack width of FRP-reinforced flexural members, Equation 8.13 is modified by replacing the steel strain $\varepsilon_s$ with the FRP strain, $\varepsilon_f = f_f/E_f$. The resulting equation is

$$w = 0.076\beta \frac{E_s}{E_f} f_f \sqrt[3]{d_c A} \qquad (8.14)$$

Substitution of $E_s = 29{,}000$ ksi in Equation 8.14 yields

$$w = \frac{2200}{E_f} \beta f_f \sqrt[3]{d_c A} \qquad (8.15)$$

The stress in the FRP bars due to service load, $f_f$, can be determined from Equation 8.16:

$$f_{f,ser} = \frac{M_{ser}}{A_s jd} \qquad (8.16)$$

Equation 8.15 has been found to predict the crack width reasonably well when FRP bars having a bond strength similar to that of steel are used [Faza and GangaRao 1993]. Crack widths can be overestimated if Equation 8.15 is applied when FRP bars having a bond strength greater than that of steel are used. Similarly, Equation 8.15 may underestimate crack widths if FRP bars having a bond strength lower than that of steel are used. To overcome these difficulties and to reflect the quality of bond, ACI 440 suggests a corrective bond coefficient, $k_b$, to be used as a modifier on the right-hand side of Equation 8.15. Thus, Equation 8.15 is modified to read

$$w = \frac{2200}{E_f} \beta k_b f_f \sqrt[3]{d_c A} \qquad (8.17)$$

For SI units, Equation 8.17 can be expressed as

$$w = \frac{2.2}{E_f} \beta k_b f_f \sqrt[3]{d_c A} \qquad (8.18)$$

where $f_f$ and $E_f$ are in MPa, $d_c$ is in mm, and the bar cross-sectional area $A$ is in mm$^2$.

The corrective bond coefficient $k_b$, introduced in Equation 8.17 and Equation 8.18, accounts for the degree of bond between a FRP bar and the surrounding concrete. ACI 440 suggests the values of $k_b$ as shown in Table 8.3.

**TABLE 8.3**
**ACI 440 Recommended Values of $k_b$**

| Bond Behavior of FRP Bars | ACI 440 Recommended Value of $k_b$ |
|---|:---:|
| Bond behavior similar to that of steel bars | 1.0 |
| Bond behavior inferior to that of steel bars | >1.0 |
| Bond behavior superior to that of steel bars | <1.0 |
| Bond behavior unknown | 1.2 |

## 8.7  EXAMPLES ON CRACK WIDTHS

Several examples are presented in this section to illustrate the calculation procedures for determining crack widths in FRP-reinforced beams. Calculations are presented in a step-by-step format to develop an easy understanding and the confidence level of readers. Example 8.10 presents the calculations for a beam that is one of the several beams supporting a floor system of a building. This is a special (combination) example in that it presents calculations for loads and the nominal strength of an FRP-reinforced beam as well as calculations for checking the crack width of the same beam. In design practice, beams must be first designed for flexure and then checked for serviceability (including crack width). Accordingly, Example 8.11 and Example 8.12 present calculations for the nominal strength of beams followed by calculations for crack widths.

### 8.7.1  EXAMPLES IN U.S. STANDARD UNITS

*Example 8.10: To check the adequacy of a given beam for flexural strength and crack width.* A rectangular 12 in. × 21 in. beam is used to support a 6 in. thick concrete floor having a live load of 50 lb/ft². The beams are spaced at 12 ft on centers and span 20 ft as shown in Figure E8.10. The tensile reinforcement consists of eight No. 5 GFRP bars arranged in two rows. Check the adequacy of a typical beam for moment strength and crack width. Assume $f_c' = 4000$ psi, $f_{fu}^* = 90$ ksi, and $E_f = 6500$ ksi.
*Solution*:
*Step 1.* Calculate service loads.

$$\text{Tributary width of beam} = 12 \text{ ft}$$

$$D_{slab} = (6/12)(1)(12)(150) = 900 \text{ lb/ft}$$

$$D_{beam} = (12)(21)(150)/144 = 262.5 \text{ lb/ft}$$

$$D = D_{slab} + D_{beam} = 1162.5 \text{ lb/ft}$$

$$L = (50)(12) = 600 \text{ lb/ft}$$

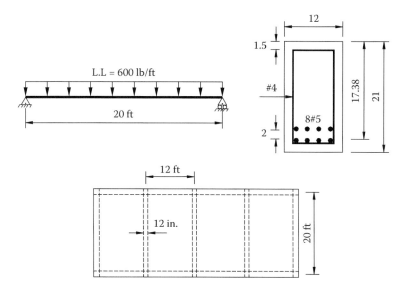

**FIGURE E8.10** Floor plan and beam cross-section for Example 8.10.

$$w_{ser} = D + L = 1162.5 + 600 = 1762.5 \text{ lb/ft}$$

*Step 2.* Calculate the factored loads.

$$w_u = 1.2D + 1.6L = 1.2(1162.5) + 1.6(600) = 2355 \text{ lb/ft} \approx 2.36 \text{ k/ft}$$

$$M_u = w_u l^2/8 = (2.36)(20)2/8 = 118 \text{ k-ft}$$

*Step 3.* Determine the tensile design strength of GFRP reinforcement, $f_{fu}$.

$$f_{fu} = C_E f_{fu}^* \qquad \text{(ACI 440, Equation 7-1)}$$

For an interior conditioned space, the environmental modification factor, $C_E$, is 0.8 (ACI 440, Table 7.1).

$$f_{fu} = (0.8)(90) = 72 \text{ ksi}$$

*Step 4.* Calculate the reinforcement ratio, $\rho_f$. First, determine $d$. For a beam with two layers of tension reinforcement,

$$d = h - \text{clear cover} - \text{diameter of stirrup} - \text{diameter of tensile reinforcing bar} -$$
$$1/2(2 \text{ in. clear distance between the two layers of bars})$$
$$= 21 - 1.5 - 0.5 - 0.625 - 1/2(2) = 17.375 \text{ in.} \approx 17.38 \text{ in.}$$

$$A_f = 8(0.31) = 2.48 \text{ in}^2$$

$$\rho_f = \frac{A_f}{bd} = \frac{2.48}{(12)(17.38)} = 0.012$$

*Step 5.* Calculate the stress in GFRP reinforcement from ACI 440, Equation 8-4d.

$$f_f = \sqrt{\frac{(E_f\varepsilon_{cu})^2}{4} + \frac{0.85\beta_1 f_c'}{\rho_f} E_f\varepsilon_{cu}} - 0.5E_f\varepsilon_{cu} < f_{fu}$$

$$= \sqrt{\frac{[(6500)(0.003)]^2}{4} + \frac{0.85(0.85)(4)}{0.012}(6500)(0.003)} - 0.5(6500)(0.003)$$

$$= 59.5 \text{ ksi} < 72 \text{ ksi, OK}$$

*Step 6.* Calculate the strength reduction factor $\phi$ from ACI 440, Equation 8-7. First, calculate $\rho_{fb}$ from ACI 440, Equation 8.3.

$$\rho_{fb} = 0.85\beta_1 \frac{f_c'}{f_{fu}} \frac{E_f\varepsilon_u}{E_f\varepsilon_{cu} + f_{fu}} \qquad \text{(ACI 440, Equation 8-3)}$$

For $f_c' = 4000$ psi, $\beta_1 = 0.85$, and $f_{fu} = 72$ ksi (calculated earlier). Therefore,

$$\rho_{fb} = (0.85)(0.85)\frac{4}{72}\left[\frac{(6500)(0.003)}{(6500)(0.003)+72}\right] = 0.0086$$

$$1.4\rho_{fb} = 1.4(0.0086) = 0.012$$

Because $(\rho_f = 0.012) \le (1.4\rho_{fb} = 0.012)$, $\phi = 0.7$ (ACI 440, Equation 8-7)

*Step 7.* Calculate the nominal strength of beam from ACI 440, Equation 8-5.

$$M_n = \rho_f f_f \left(1 - 0.59\frac{\rho_f f_f}{f_c'}\right) bd^2$$

$$= \frac{(0.012)(59.5)}{12}\left[1 - 0.59\frac{(0.012)(59.5)}{4}\right](12)(17.38)^2 \quad \text{(ACI 440, Equation 8-5)}$$

$$= 193 \text{ k-ft}$$

*Step 8.* Calculate $\phi M_n$.

$$\phi M_n = 0.7(193) = 135 \text{ k-ft} > M_u = 118 \text{ k-ft}$$

*Step 9.* Calculate the crack width, $w$, from ACI 440, Equation 8-9c.

$$w = \frac{2200}{E_f} \beta k_b f_f (d_c A)^{1/3} \qquad \text{(ACI 440, Equation 8-9c)}$$

To calculate $\beta$ and $f_f$, first calculate the neutral axis factor, $k$. The modular ratio is

$$n_f = \frac{E_f}{E_c} = \frac{6,500,000}{57,000\sqrt{4000}} = 1.8$$

$$\rho_f = 0.012 \text{ (calculated earlier)}$$

$$k = \sqrt{2\rho_f n_f + (\rho_f n_f)^2} - \rho_f n_f = \sqrt{2(0.012)(1.8) + [(0.012)(1.8)]^2}$$
$$- (0.012)(1.8) = 0.187$$

The lever arm factor, $j = 1 - k/3 = 1 - 0.187/3 = 0.938$.

To calculate the service load stress in the GFRP bars, first calculate the service load moment.

$$M_{ser} = (w_{ser} l^2)/8 = (1762.5)(20)2/8 = 88.1 \text{ k-ft}$$

$$d = 17.38 \text{ in. (as before)}$$

The stress in FRP bars due to service load is given by

$$f_{f,ser} = \frac{M_{ser}}{A_f jd} = \frac{(88.1)(12)}{(2.48)(0.938)(17.38)} = 26.15 \text{ ksi}$$

$$\beta = \frac{h - kd}{d(1-k)} = \frac{21 - (0.187)(17.38)}{17.38(1 - 0.187)} = 1.256$$

$$d_c = h - d = 21 - 17.38 = 3.62 \text{ in.}$$

$$A = \frac{2d_c b}{\text{Number of bars}} = \frac{2(3.62)(12)}{8} = 10.86 \text{ in.}^2$$

The bond-dependent factor is $k_b = 1.2$ (default value, ACI 440, Section 8.3.1).

$$w = \frac{2200}{E_f} \beta k_b f_b (d_c A)^{1/3}$$

$$= \frac{2200}{6500} (1.256)(1.2)(26.15)[(3.62)(10.86)]^{1/3}$$

$$= 45 \text{ mils} > w_{allow} = 28 \text{ mils, N.G.}$$

Conclusions:

1. The beam satisfies the strength requirements: $\phi M_n = 135$ k-ft $> M_u = 118$ k-ft.
2. The beam does not satisfy the crack width limits of ACI 440: $w = 45$ mils $> w_{allow} = 28$ mils.

Therefore, the beam design is not acceptable.

*Example 8.11: Determination of the nominal strength of a beam and crack width when the strength reduction factor $\phi$ is less than 0.7.* A 12 in. × 24 in. rectangular concrete beam is reinforced with eight No. 5 GFRP bars arranged in two layers as shown in Figure E8.11. The beam carries a superimposed dead load of 900 lb/ft in addition to its own weight and a floor live load of 600 lb/ft. Check the adequacy of this beam for a simple span of 20 ft for strength and crack width. Assume $f_c' = 4000$ psi, $f_{fu}* = 90$ ksi, and $E_f = 6500$ ksi.
*Solution*:
*Step 1*. Calculate the service loads.

$$D_{beam} = \frac{(12)(24)}{144}(150) = 300 \text{ lb/ft}$$

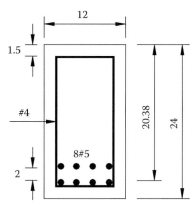

All dimensions are in inches.

**FIGURE E8.11** Beam cross-section for Example 8.11.

$$D_{super} = 900 \text{ lb/ft}$$

$$D = 900 + 300 = 1200 \text{ lb/ft}$$

$$L = 600 \text{ lb/ft}$$

$$w_{ser} = D + L = 1200 + 600 = 1800 \text{ lb/ft} = 1.8 \text{ k/ft}$$

*Step 2*. Calculate the factored loads.

$$w_u = 1.2D + 1.6L = 1.2(1200) + 1.6(600) = 2400 \text{ lb/ft} = 2.4 \text{ k/ft}$$

$$M_u = \frac{w_u l^2}{8} = \frac{2.4(20)^2}{8} = 120 \text{ k-ft}$$

*Step 3*. Determine the tensile design strength of GFRP reinforcement, $f_{fu}$.

$$f_{fu} = C_E f_{fu}^* \qquad \text{(ACI 440, Equation 7-1)}$$

For an interior conditioned space, the environmental modification (reduction) factor, $C_E$, is 0.8 (ACI 440, Table 7.1).

$$f_{fu} = (0.8)(90) = 72 \text{ ksi}$$

*Step 4*. Calculate the reinforcement ratio, $\rho_f$. First, determine $d$. For a beam with two layers of tension reinforcement,

$$d = h - \text{clear cover} - \text{diameter of stirrup} - \text{diameter of tensile reinforcing bar} -$$
$$1/2(2 \text{ in. clear distance between the two layers of bars})$$
$$= 24 - 1.5 - 0.5 - 0.625 - 1/2(1) = 20.375 \text{ in.} \approx 20.38 \text{ in.}$$

$$A_f = 8(0.31) = 2.48 \text{ in.}^2$$

$$\rho_f = \frac{A_f}{bd} = \frac{2.48}{(12)(20.38)} = 0.01$$

*Step 5*. Calculate the stress in GFRP reinforcement from ACI 440, Equation 8-4d.

$$f_f = \sqrt{\frac{(E_f \varepsilon_{cu})^2}{4} + \frac{0.85\beta_1 f_c'}{\rho_f} E_f \varepsilon_{cu}} - 0.5 E_f \varepsilon_{cu} < f_{fu}$$

$$= \sqrt{\frac{[(6500)(0.003)]^2}{4} + \frac{0.85(0.85)(4)}{0.010}(6500)(0.003)} - 0.5(6500)(0.003)$$

$$= 66 \text{ ksi} < 72 \text{ ksi, OK}$$

*Step 6.* Calculate the strength reduction factor $\phi$ from ACI 440, Equation 8-7. First, calculate $\rho_{fb}$ from ACI 440, Equation 8.3.

$$\rho_{fb} = 0.85\beta_1 \frac{f'_c}{f_{fu}} \frac{E_f\varepsilon_{cu}}{E_f\varepsilon_{cu} + f_{fu}} \qquad \text{(ACI 440, Equation 8-3)}$$

For $f'_c = 4000$ psi, $\beta_1 = 0.85$, and $f_{fu} = 72$ ksi (calculated earlier). Therefore,

$$\rho_{fb} = (0.85)(0.85)\frac{4}{72}\left[\frac{(6500)(0.003)}{(6500)(0.003) + 72}\right] = 0.0086$$

$$1.4\rho_{fb} = 1.4(0.0086) = 0.012 > \rho_f = 0.010.$$

Because $(\rho_{fb} = 0.0086) < (\rho_f = 0.010) < (1.4\rho_{fb} = 0.012)$,

$$\phi = \frac{\rho_f}{2\rho_{fb}} = \frac{0.01}{2(0.0086)} = 0.58 \quad \text{(ACI 440, Equation 8-7)}$$

*Step 7.* Calculate the nominal strength of the beam from ACI 440, Equation 8-5.

$$M_n = \rho_f f_f\left(1 - 0.59\frac{\rho_f f_f}{f'_c}\right)bd^2$$

$$= \frac{(0.010)(66)}{12}\left[1 - 0.59\frac{(0.010)(66)}{4}\right](12)(20.38)^2 \quad \text{(ACI 440, Equation 8-5)}$$

$$= 247.4 \text{ k-ft}$$

*Step 8.* Calculate $\phi M_n$.

$$\phi M_n = 0.58(247.4) = 143.5 \text{ k-ft} > M_u = 120 \text{ k-ft}$$

*Step 9.* Calculate the crack width, $w$, from ACI 440, Equation 8-9c.

$$w = \frac{2200}{E_f}\beta k_b f_f(d_c A)^{1/3} \qquad \text{(ACI 440, Equation 8-9c)}$$

To calculate $\beta$ and $f_f$, first calculate the neutral axis factor, $k$. The modular ratio is

$$n_f = \frac{E_f}{E_c} = \frac{6,500,000}{57,000\sqrt{4000}} = 1.8$$

$$\rho_f = 0.010 \text{ (calculated earlier)}$$

$$k = \sqrt{2\rho_f n_f + (\rho_f n_f)^2} - \rho_f n_f = \sqrt{2(0.01)(1.8)+[(0.01)(1.8)]^2}$$
$$- (0.01)(1.8) = 0.173$$

The lever arm factor is $j = 1 - k/3 = 1 - 0.173/3 = 0.942$.

To calculate the service load stress in GFRP bars, first calculate the service load moment.

$$M_{ser} = \frac{w_{ser}l^2}{8} = \frac{1.8(20)^2}{8} = 90 \text{ k-ft}$$

$$d = 17.38 \text{ in. (as before)}$$

The stress in FRP bars due to service loads is given by

$$f_{f,ser} = \frac{M_{ser}}{A_f jd} = \frac{(90)(12)}{(2.48)(0.942)(20.38)} = 22.68 \text{ ksi}$$

$$\beta = \frac{h-kd}{d(1-k)} = \frac{24-(0.173)(20.38)}{20.38(1-0.173)} = 1.215$$

$$d_c = h - d = 24 - 20.38 = 3.62 \text{ in.}$$

$$A = \frac{2d_c b}{\text{Number of bars}} = \frac{2(3.62)(12)}{8} = 10.86 \text{ in.}^2$$

The bond-dependent factor is $k_b = 1.2$ (default value, ACI 440, Section 8.3.1).

$$w = \frac{2200}{E_f} \beta k_b f_f (d_c A)^{1/3}$$

$$= \frac{2200}{6500}(1.215)(1.2)(22.68)[(3.62)(10.86)]^{1/3}$$

$$= 38 \text{ mils} > w_{allow} = 28 \text{ mils, n.g.}$$

Conclusions:

1. The beam satisfies the strength requirements: $\phi M_n = 143.5$ k-ft $> M_u = 120$ k-ft.
2. The beam does not satisfy the crack width limits of ACI 440: $w = 38$ mils $> w_{allow} = 28$ mils.

Therefore, the beam design is not acceptable.

*Example 8.12: Determination of crack width in a beam reinforced with GFRP bars.* A rectangular 14 in. × 24 in. concrete beam is reinforced with 8 No. 5 GFRP bars arranged in two rows as shown in Figure E8.12. The beam spans 14 ft and carries a superimposed load of 900 lb/ft in addition to its own weight and a live load of 600 lb/ft. Check if the beam satisfies the crack width requirements. Assume $f_c' = 4000$ psi, $f_{fu}{}^* = 90$ ksi, and $E_f = 6500$ ksi.

*Solution*: Note that the calculations for crack width are similar to Step 9 of Example 8.10 and Example 8.11. The crack width, $w$, is determined from ACI 440, Equation 8-9c.

$$w = \frac{2200}{E_f}\beta k_b f_f (d_c A)^{1/3} \qquad \text{(ACI 440, Equation 8-9c)}$$

Calculate the service loads.

$$D_{beam} = \frac{(14)(24)}{144}(150) = 350 \text{ lb/ft}$$

$$D_{super} = 900 \text{ lb/ft}$$

$$D = 900 + 350 = 1250 \text{ lb/ft}$$

$$L = 600 \text{ lb/ft}$$

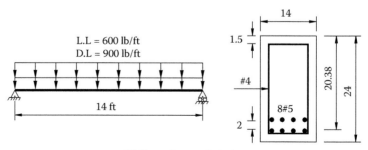

All dimensions are in inches.

**FIGURE E8.12** Beam cross-section for Example 8.12.

$$w_{ser} = D + L = 1250 + 600 = 1850 \text{ lb/ft} = 1.85 \text{ k/ft}$$

$$M_{ser} = \frac{w_{ser}l^2}{8} = \frac{(1.85)(14)^2}{8} = 45.33 \text{ k-ft}$$

To determine $\beta$ and $f_f$, first calculate the neutral axis factor, $k$. The modular ratio is

$$n_f = \frac{E_f}{E_c} = \frac{6,500,000}{57,000\sqrt{4000}} = 1.8$$

$$A_f = 8(0.31) = 2.48 \text{ in.}^2$$

$d = h -$ clear cover $-$ diameter of stirrup $-$ diameter of tensile reinforcing bar $-$
1/2(2 in. clear distance between the two layers of bars)
$= 24 - 1.5 - 0.5 - 0.625 - 1/2(2) = 20.375 \text{ in.} \approx 20.38 \text{ in.}$

$$\rho_f = \frac{A_f}{bd} = \frac{2.48}{(14)(20.38)} = 0.0087$$

$$k = \sqrt{2\rho_f n_f + (\rho_f n_f)^2} - \rho_f n_f = \sqrt{2(0.0087)(1.8) + [(0.0087)(1.8)]^2}$$
$$- (0.0087)(1.8) = 0.162$$

The lever arm factor is $j = 1 - k/3 = 1 - 0.162/3 = 0.946$.

The stress in FRP bars due to service loads is given by

$$f_{f,ser} = \frac{M_{ser}}{A_f jd} = \frac{(45.33)(12)}{(2.48)(0.946)(20.38)} = 11.38 \text{ ksi}$$

$$\beta = \frac{h - kd}{d(1-k)} = \frac{24 - (0.162)(20.38)}{20.38(1 - 0.162)} = 1.21$$

$$d_c = h - d = 24 - 20.38 = 3.62 \text{ in.}$$

$$A = \frac{2d_c b}{\text{Number of bars}} = \frac{2(3.62)(14)}{8} = 12.67 \text{ in.}^2$$

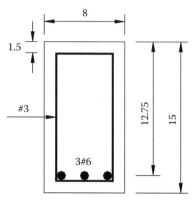

All dimensions are in inches.

**FIGURE E8.13** Beam cross-section for Example 8.13.

The bond-dependent factor is $k_b = 1.2$ (default value, ACI 440, Section 8.3.1).

$$w = \frac{2200}{E_f} \beta k_b f_f (d_c A)^{1/3}$$

$$= \frac{2200}{6500} (1.21)(1.2)(11.38)[(3.62)(12.67)]^{1/3}$$

$$= 20 \text{ mils} < w_{allow} = 28 \text{ mils, OK}$$

Thus, the beam satisfies the crack width limits of ACI 440.

*Example 8.13: Determination of crack width in a beam reinforced with CFRP bars.* An 8 in. × 15 in. rectangular concrete beam is reinforced with three No. 6 CFRP bars and No. 3 stirrups as shown in Figure E8.13. The beam carries a superimposed dead load of 875 lb/ft in addition to its own weight and a floor live load of 2500 lb/ft over a span of 10 ft. Determine if this beam satisfies the crack width requirements. Assume $f_c' = 4000$ psi, $f_{fu}{}^* = 240$ ksi, and $E_f = 20,000$ ksi.

*Solution*: Note that this is the same beam for which the flexural strength was determined in Example 7.4. The crack width, $w$, is determined from ACI 440, Equation 8-9c.

$$w = \frac{2200}{E_f} \beta k_b f_f (d_c A)^{1/3} \qquad \text{(ACI 440, Equation 8-9c)}$$

Calculate the service loads, obtained from Example 8.4:

$$D = 1000 \text{ lb/ft}$$

$$L = 2500 \text{ lb/ft}$$

$$w_{ser} = D + L = 1000 + 2500 = 3500 \text{ lb/ft} = 3.5 \text{ k/ft}$$

$$M_{ser} = \frac{w_{ser}l^2}{8} = \frac{(3.5)(10)^2}{8} = 43.75 \text{ k-ft}$$

To determine $\beta$ and $f_f$, first calculate the neutral axis factor, $k$. The modular ratio is

$$n_f = \frac{E}{E_c} = \frac{20,000,000}{57,000\sqrt{4000}} = 5.55$$

$A_f = 1.32 \text{ in}^2$ and $\rho_f = 0.0129$ (from Example 8.4)

$$k = \sqrt{2\rho_f n_f + (\rho_f n_f)^2} - \rho_f n_f = \sqrt{2(0.0129)(5.55) + [(0.0129)(5.55)]^2}$$
$$- (0.0129)(5.55) = 0.314$$

The lever arm factor is $j = 1 - k/3 = 1 - 0.314/3 = 0.895$.

The stress in the CFRP bars due to service load is given by

$$f_{f,ser} = \frac{M_{ser}}{A_f jd} = \frac{(43.75)(12)}{(1.32)(0.895)(12.75)} = 34.85 \text{ ksi}$$

$$\beta = \frac{h - kd}{d(1-k)} = \frac{15 - (0.314)(12.75)}{12.75(1 - 0.314)} = 1.257$$

$$d_c = h - d = 15 - 12.75 = 2.25 \text{ in.}$$

$$A = \frac{2d_c b}{\text{Number of bars}} = \frac{2(2.25)(8)}{3} = 12 \text{ in.}^2$$

The bond-dependent factor is $k_b = 1.2$ (default value, ACI 440, Section 8.3.1).

$$w = \frac{2200}{E_f} \beta k_b f_f (d_c A)^{1/3}$$

$$= \frac{2200}{20,000} (1.257)(1.2)(34.85)[(2.25)(12)]^{1/3}$$

$$= 17 \text{ mils} < w_{allow} = 28 \text{ mils, OK}$$

Thus, the beam satisfies the crack width limits of ACI 440.

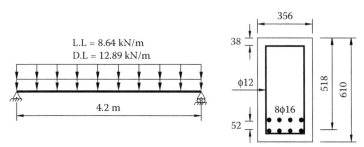

All dimensions are in mm.

**FIGURE E8.14** Beam cross-section for Example 8.14.

## 8.7.2 EXAMPLES IN SI UNITS

The following examples present calculations for the above examples in SI units. Example 8.14 shows all calculations for checking the crack width of a FRP-reinforced beam in a step-by-step format. Example 8.15 and Example 8.16 are special in that Example 8.15 presents calculations determining the nominal strength of an FRP-reinforced beam, whereas Example 8.16 presents calculations for checking the crack width for the same beam as described in Example 8.15.

*Example 8.14: Determination of crack width in a beam reinforced with GFRP bars.* A rectangular 356 mm × 610 mm concrete beam is reinforced with eight φ-16 mm GFRP bars arranged in two rows as shown in Figure E8.14. The beam spans 4.2 m and carries a superimposed load of 12.89 kN/m in addition to its own weight and a live load of 8.64 kN/m. Check if the beam satisfies the crack width requirements. Assume $f_c' = 27.6$ N/mm$^2$, $f_{fu}{}^* = 620.55$ N/mm$^2$, and $E_f = 45.5$ kN/mm$^2$.

*Solution*: Note that the calculations for crack width are similar to Step 9 of Example 8.10 and Example 8.11. The crack width, $w$, is determined from ACI 440, Equation 8-9c.

$$w = \frac{2.2}{E_f}\beta k_b f_f \sqrt[3]{d_c A} \qquad \text{(ACI 440, Equation 8-9c)}$$

Calculate the service loads.

$$D_{beam} = \left(\frac{356}{1000}\right)\left(\frac{610}{1000}\right)(23.56) = 5.11 \text{ kN/m}$$

$$D_{super} = 12.89 \text{ kN/m}$$

$$D = 12.89 + 5.11 = 18 \text{ kN/m}$$

$$L = 8.64 \text{ kN/m}$$

$$w_{ser} = D + L = 18 + 8.4 = 26.64 \text{ kN/m}$$

$$M_{ser} = \frac{w_{ser}l^2}{8} = \frac{(26.64)(4.2)^2}{8} = 58.76 \text{ kN/m}$$

To determine $\beta$ and $f_f$, first calculate the neutral axis factor, $k$.

$$A_f = 8(200.96) = 1607.68 \text{ mm}^2$$

$d = h$ – clear cover – diameter of stirrup – diameter of tensile reinforcing bar – 1/2(52 mm clear distance between the two layers of bars)
$$= 610 - 38 - 12 - 16 - 1/2(52) = 518 \text{ mm}$$

$$\rho_f = \frac{A_f}{bd} = \frac{1607.68}{(365)(518)} = 0.0087$$

$$E_c = 4750\sqrt{f_c'} = 4750\sqrt{27.6} = 24.95 \text{ kN/mm}^2$$

$$n_f = \frac{E_f}{E_c} = \frac{45.5}{24.95} = 1.8$$

$$k = \sqrt{2\rho_f n_f + (\rho_f n_f)^2} - \rho_f n_f$$
$$= \sqrt{2(0.0087)(1.8) + [(0.0087)(1.8)]^2} - (0.0087)(1.8)$$
$$= 0.162$$

The lever arm factor is $j = 1 - k/3 = 1 - 0.162/3 = 0.946$.

Calculate $\beta$.

$$\beta = \frac{h - kd}{d(1 - k)} = \frac{610 - (0.162)(518)}{518(1 - 0.162)} = 1.215$$

Calculate the stress in the FRP bars due to service loads.

$$f_{f,ser} = \frac{M_{ser}}{A_s jd} = \frac{(58.76)(1000)}{(1607.68)(0.946)(518)} = 0.0746 \text{ kN/mm}^2$$

$$d_c = h - d = 610 - 518 = 92 \text{ mm}$$

$$A = \frac{2d_cb}{\text{Number of bars}} = \frac{2(92)(356)}{8} = 8188 \text{ mm}^2$$

The bond-dependent factor is $k_b = 1.2$ (default value, ACI 440, Section 8.3.1).

$$w = \frac{2.2}{E_f}\beta k_b f_f \sqrt[3]{d_cA}$$

$$= \frac{2.2}{45,500}(1.215)(1.2)(7.46)\sqrt[3]{(92)(8188)}$$

$$= 0.5 \leq w_{allow} = 0.71 \text{ mm, OK}$$

Thus, the beam satisfies the crack width limits of ACI 440.

*Example 8.15: Determination of nominal strength of a beam reinforced with CFRP bars.* A 204 mm × 380 mm rectangular concrete beam is reinforced with three φ-20 mm CFRP bars and No. 3 stirrups as shown in Figure E8.15. The beam carries a superimposed dead load of 12.775 kN/m in addition to its own weight and a floor live load of 36.5 kN/m over a span of 3 m. Check the adequacy of this beam for flexural strength. Assume $f_c' = 27.6$ N/mm$^2$, $f_{fu}* = 1654.8$ N/mm$^2$, and $E_f = 137.9$ kN/mm$^2$.
*Solution*: Calculate the service loads.

$$D_{beam} = \frac{(204)(380)(23.56)}{(1000)(1000)} = 1.82 \text{ kN/m}$$

$$D_{super} = 12.775 \text{ kN/m}$$

$$D = 1.82 + 12.775 = 14.6 \text{ kN/m}$$

$$L = 36.5 \text{ kN/m}$$

$$W_{ser} = 14.6 + 36.5 = 51.1 \text{ kN/m}$$

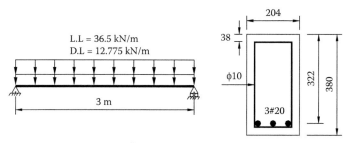

All dimensions are in mm.

**FIGURE E8.15** Beam cross-section for Example 8.15.

Calculate the factored loads.

$$w_u = 1.2D + 1.6L = 1.2(14.6) + 1.6(36.5) = 75.92 \text{ kN/m}$$

$$M_{ser} = \frac{w_{ser}l^2}{8} = \frac{(75.92)(3.0)^2}{8} = 85.41 \text{ kN/m}$$

$$f_{fu} = C_E f_{fu}^* \qquad \text{(ACI 440, Equation 7-1)}$$

For an interior conditioned space, the environmental modification (reduction) factor, $C_E$, is 1.0 (ACI 440, Table 7.1).

$$f_{fu} = (1.0)(1654.8) = 1654.8 \text{ N/mm}^2$$

Calculate the reinforcement ratio, $\rho_f$. First, determine $d$.

$$d = h - \text{clear cover} - \text{diameter of stirrup} - 1/2 \text{ (diameter of tensile reinforcing bar)}$$
$$= 380 - 38 - 10 - 20/2 = 322 \text{ mm}.$$

$$A_f = 3(314) = 942 \text{ mm}^2$$

$$\rho_f = \frac{A_f}{bd} = \frac{942}{(204)(322)} = 0.0143$$

Calculate the stress in the CFRP reinforcement from ACI 440, Equation 8-4d.

$$f_f = \sqrt{\frac{(E_f \varepsilon_{cu})^2}{4} + \frac{0.85\beta_1 f'_c}{\rho_f} E_f \varepsilon_{cu}} - 0.5 E_f \varepsilon_{cu} < f_{fu}$$

$$= \sqrt{\frac{[(137.9)(0.003)]^2}{4} + \frac{0.85(0.8692)(0.0276)}{0.010}(137.9)(0.003)} - 0.5(137.9)(0.003)$$

$$= 0.5885 \text{ kN/mm}^2 \le 1.654 \text{ kN/mm}^2, \text{ OK}$$

Calculate the strength reduction factor $\phi$ from ACI 440, Equation 8-7. First, calculate $\rho_{fb}$ from ACI 440, Equation 8.3.

$$\rho_{fb} = 0.85\beta_1 \frac{f'_c}{f_{fu}} \frac{E_f \varepsilon_{cu}}{E_f \varepsilon_{cu} + f_{fu}} \qquad \text{(ACI 440, Equation 8-3)}$$

For $f'_c = 27.6 \text{ N/mm}^2$, $\beta_1 = 0.8692$, and $f_{fu} = 1654.8 \text{ N/mm}^2$. Therefore,

$$\rho_{fb} = (0.85)(0.8692)\frac{27.6}{1654.8}\left[\frac{(137,900)(0.003)}{(137,900)(0.003)+1654.8}\right] = 0.0025$$

$$1.4\rho_{fb} = 1.4(0.0025) = 0.0034 > 1.4\rho_f = 0.002$$

Because ($\rho_f = 0.0129$) > ($1.4\rho_{fb} = 0.0034$), $\phi = 0.7$ (ACI 440, Equation 8-7). Calculate the nominal strength of the beam from ACI 440, Equation 8-5.

$$M_n = \rho_f f_f\left(1 - 0.59\frac{\rho_f f_f}{f_c'}\right)bd^2$$

$$= (0.0143)(588.5)\left[1 - 0.59\frac{(0.0143)(588.5)}{27.6}\right](204)(380)^2$$

$$= 203.3 \text{ kN-m}$$

Calculate $\phi M_n$.

$$\phi M_n = 0.7(203.3) = 142.3 \text{ kN-m} > M_u = 85.41 \text{ kN-m}.$$

Thus, the beam has adequate flexural strength.

*Example 8.16: Determination of crack width in a beam reinforced with CFRP bars.* A 204 mm × 380 mm rectangular concrete beam is reinforced with three $\phi$-20 mm CFRP bars and $\phi$-10 mm stirrups as shown in Figure E8.16. The beam carries a superimposed dead load of 12.775 kN/m in addition to its own weight and a floor live load of 36.5 kN/m over a span of 3 m. Determine if this beam satisfies the crack width requirements. Assume $f_c' = 27.6$ N/mm², $f_{fu}^* = 1654.8$ N/mm², and $E_f = 137.9$ kN/mm².

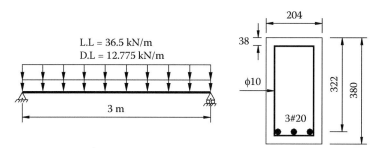

All dimensions are in mm.

**FIGURE E8.16** Beam cross-section for Example 8.16.

*Solution*: Note that this is the same beam for which the flexural strength was determined in Example 8.15. The crack width, $w$, is determined from ACI 440, Equation 8-9c.

$$w = \frac{2.2}{E_f}\beta k_b f_f \sqrt[3]{d_c A}$$

Calculate the service loads, obtained from Example 8.9.

$$D = 14.6 \text{ kN/m}$$

$$L = 36.5 \text{ kN/m}$$

$$w_{ser} = D + L = 14.6 + 36.5 = 51.1 \text{ kN/m}$$

$$M_{ser} = \frac{w_{ser}l^2}{8} = \frac{(51.12)(3.0)^2}{8} = 57.48 \text{ kN/m}$$

To determine $\beta$ and $f_f$, first calculate the neutral axis factor, $k$.

$$n_f = \frac{E_f}{E_c} = \frac{137,900}{4750\sqrt{27.6}} = 5.53$$

$A_f = 942 \text{ mm}^2$ and $\rho_f = 0.0143$ (from Example 8.9).

$$k = \sqrt{2\rho_f n_f + (\rho_f n_f)^2} - \rho_f n_f$$

$$= \sqrt{2(0.0143)(5.53) + [(0.0143)(5.53)]^2} - (0.0143)(5.53)$$

$$= 0.3264$$

The lever arm factor is $j = 1 - k/3 = 1 - 0.3264/3 = 0.8912$.

Calculate the stress in the FRP bars due to service loads.

$$f_{f,ser} = \frac{M_{ser}}{A_s jd} = \frac{(57.48)(1000)}{(942)(0.8912)(322)} = 0.2126 \text{ kN/mm}^2$$

Calculate $\beta$.

$$\beta = \frac{h - kd}{d(1 - k)} = \frac{380 - (0.3264)(322)}{322(1 - 0.3264)} = 1.2674$$

$$d_c = h - d = 380 - 322 = 58 \text{ mm}$$

$$A = \frac{2d_c b}{\text{Number of bars}} = \frac{2(58)(204)}{3} = 7888 \text{ mm}^2$$

The bond-dependent factor is $k_b = 1.2$ (default value, ACI 440, Section 8.3.1).

$$w = \frac{2.2}{E_f} \beta k_b f_f \sqrt[3]{d_c A}$$

$$= \frac{2.2}{137,900}(1.2674)(1.2)(212.6)\sqrt[3]{58(7888)}$$

$$= 0.4 \leq w_{allow} = 0.71 \text{ mm, OK}$$

Thus, the beam satisfies the crack width limits of ACI 440.

## 8.8  CREEP-RUPTURE

### 8.8.1  CONCEPT OF CREEP

*Creep* is a behavior wherein the strain or deformation in a member undergoes changes while the load remains constant. It is defined as a time-dependent deformation of a structural member under constant load. This is the opposite of *relaxation*, which is a time-dependent load response to a constant deformation.

Creep is a function of stress, time, and temperature. A material is likely to creep more (at a higher rate) at higher stress levels or after long periods of time even if the temperature is not elevated. Most engineering materials, particularly metals, hardly creep at ordinary temperatures. However, notable exceptions are found. For example, a mechanical engineer must consider creep problems in the engineering design of things such as tubes of high-pressure boilers, steam-turbine blades, jet-engine blades, and other applications where high temperatures must be sustained for a long periods of time. A chemical engineer must often design containers or vessels for a process that takes place at very high pressures and temperatures.

### 8.8.2  CONCEPT OF CREEP-RUPTURE

What is creep-rupture? Among other factors, the failure of a material depends on the rate of loading. Our understanding of the strengths of materials is based on conventional strength tests in which loads are increased from zero to the failure load, typically at a loading rate at which failure will occur in a few minutes. However, some materials (e.g., glass, concrete, wood, and many ceramics) have long been known to fail under sustained stresses that are significantly less than those required to bring about failure in conventional strength tests in a few minutes. This will occur if the loads (or stresses caused by these loads) are sustained for a sufficiently long time for a micro-crack to grow to the critical (failure) size as a result of creep. An engineer must design a part or a member so that stresses under sustained loads are low enough to limit creep-induced stress to an allowable value. This maximum stress is often referred to as the *creep strength* or *creep limit*. Thus, creep strength can be defined as the highest stress that a material can endure for a specified time without excessive deformation or failure. The *creep-rupture strength* (sometimes also referred to as *rupture strength*) is the highest sustained stress a material can stand without rupture. Creep-rupture is often referred to as *static fatigue*.

FRP reinforcing bars subjected to constant loads over a period of time can suddenly fail after a time period (called the endurance time) has elapsed. Fortunately, creep-rupture is not a matter of concern for steel reinforcing bars in concrete except in extremely high temperatures such as those encountered in a fire. A discussion on creep rupture can be found in the ACI 440 Guidelines [ACI 440.1-R03].

To preclude the possibility of creep-rupture of the FRP reinforcement under sustained, cyclic/fatigue loads, the stresses in the FRP reinforcement must be limited. Because both sustained loads and fatigue loads correspond to working loads, these stresses will be within the elastic range of the member. Therefore, elastic analysis can be used for determining these stresses.

## 8.9  CREEP-RUPTURE STRESS LIMITS

Based on the conventional strength of materials approach used for homogeneous beams subjected to flexure, stresses caused by moment due to sustained loading can be calculated from the basic expression (of the form $f = M_c/I$) used for determining flexural stresses in a beam. The stress in concrete at the level of FRP bars can be expressed by Equation 8.19:

$$f_{c,s} = M_s \frac{d(1-k)}{I_{cr}}$$ (8.19)

where

$f_{c,s}$ = the maximum compressive stress in concrete due to sustained loading
$M_s$ = the moment due to sustained loads (dead load plus prescribed portion of live load)
$k$ = the ratio of the depth of elastic neutral axis from the compression face to the depth of beam
$d$ = the depth of the beam (from the extreme compression face to the centroid of the FRP reinforcement)
$I_{cr}$ = the moment of inertia of the cracked section

The corresponding stress in FRP bars due to sustained loads, $f_{f,s}$, can be obtained by multiplying the right-hand side of Equation 8.19 by the modular ratio $n_f$, resulting in Equation 8.20:

$$f_{f,s} = M_s \frac{n_f d(1-k)}{I_{cr}}$$ (8.20)

**TABLE 8.4**
**Creep-Rupture Stress Limits in FRP Reinforcement**

| Fiber Type | Creep-Rupture Stress Limit ($f_{f,s}$) | Factor of Safety | ACI 440 Threshold Values |
|---|---|---|---|
| Glass FRP | $0.29f_{fu}$ | 1.67 | $0.20f_{fu}$ |
| Aramid FRP | $0.47f_{fu}$ | 1.67 | $0.30f_{fu}$ |
| Carbon FRP | $0.93f_{fu}$ | 1.67 | $0.55f_{fu}$ |

*Source:* ACI 440.1R-03, Table 8.2.

Alternatively, the stress in FRP bars due to sustained loads can be determined from Equation 8.21, which is simpler than Equation 8.20 (because the calculation of $I_{cr}$ is time-consuming):

$$f_{f,s} = \frac{M_s}{A_s jd} \tag{8.21}$$

where $j$ is the lever arm factor, $j = (1 - k/3)$. Note that the form of Equation 8.21 is the same as that of Equation 8.16, except that in Equation 8.16, $M_{ser}$ is the moment due to service loads, whereas in Equation 8.21, $M_s$ is the moment due to sustained loads ($M_{ser} > M_s$).

The degree of susceptibility of FRP bars to creep-rupture varies for different fiber types. Carbon bars are the least susceptible to creep-rupture and glass bars have a relatively higher susceptibility to creep-rupture. The values of maximum permissible creep-rupture stresses, $f_{f,s}$, for different of types FRP bars are expressed as fractions (or percentages) of their tensile strength, $f_{fu}$. Research on the creep-rupture of FRP bars indicates that the creep-rupture threshold for AFRP and GFRP bars is nearly the same even though the ACI 440.1R-03 Table 8.2 states that the creep-rupture threshold of GFRP bars is lower than that of AFRP bars as shown in Table 8.4. These stress values are based on the creep-rupture stress limits of 0.29, 0.47, and 0.93, respectively, for GFRP, AFRP, and CFRP, and have an imposed safety factor of $1/0.60 = 5/3 \approx 1.67$. For example, the creep-rupture stress limit of $0.20f_{fu}$ for glass FRP is obtained as $f_{f,s} = (0.29f_{fu})/1.67 \approx 0.2f_{fu}$, and so on.

## 8.10 EXAMPLES OF CREEP-RUPTURE STRESS LIMITS

The following examples illustrate the ACI 440 philosophy for checking the creep-rupture stress limits for FRP-reinforced beams. Examples are presented in both U.S. standard units and SI units.

### 8.10.1 EXAMPLES IN U.S. STANDARD UNITS

*Example 8.17: Determination of creep-rupture stress in a beam reinforced with CFRP bars.* An 8 in. × 15 in. rectangular concrete beam is reinforced with three

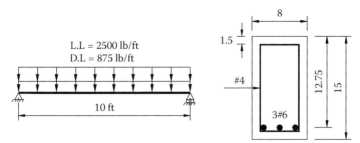

All dimensions are in inches.

**FIGURE E8.17** Beam cross-section for Example 8.17.

No. 6 CFRP bars and No. 3 stirrups as shown in Figure E8.17. The beam carries a superimposed dead load of 875 lb/ft in addition to its own weight and a floor live load of 2500 lb/ft over a span of 10 ft. Assume that the sustained load is 20% of the live load. Check the creep-rupture stress limits for this beam. Assume $f_c' = 4000$ psi, $f_{fu}* = 240$ ksi, and $E_f = 20{,}000$ ksi.

*Solution*: The creep-rupture stress limits are to be determined in accordance with ACI 440, Section 8.4.1. Values of safe sustained stress levels or creep-rupture stress limits are given in ACI 440, Table 8.3.

Calculate the moment due to the sustained load. Calculations for the nominal strength of the beam in this example were presented in Example 7.4, wherein the dead load was determined to be 1000 lb/ft. Thus, $D = 1000$ lb/ft. Therefore, the sustained load is

$$w_s = D + 0.2L = 1000 + 0.2(2500) = 1500 \text{ lb/ft}$$

The moment due to the sustained load is

$$M_s = \frac{w_s l^2}{8} = \frac{(1.5)(10)^2}{8} = 18.75 \text{ k-ft}$$

From Example 6.5, $d = 12.75$ in., $j = 0.895$, and $A_f = 1.32$ in². Therefore, the stress in CFRP bars due to the sustained load is given by

$$f_{f,s} = \frac{M_s}{A_f jd} = \frac{(18.75)(12)}{(1.32)(0.895)(12.75)} = 14.94 \text{ ksi}$$

From Example 6.4, $f_{fu} = 240$ ksi.

$$0.2f_{fu} = 0.2(240) = 48 \text{ ksi}$$

$$f_{f,s} = 14.94 \text{ ksi} < 0.2 \, f_{fu} = 48 \text{ ksi}$$

Hence, the creep-rupture stress is within the allowable limits permitted by ACI 440.

## 8.10.2 Examples in SI Units

The following example is the same as Example 8.17 except that units have been changed to SI units.

*Example 8.18: Determination of creep rupture stress in a beam reinforced with CFRP bars.* A 204 mm × 380 mm rectangular concrete beam is reinforced with three $\phi$-20 mm CFRP bars and $\phi$-10 stirrups as shown in Figure E8.15. The beam carries a superimposed dead load of 12.775 kN/m in addition to its own weight and a floor live load of 36.5 kN/m over span of 3 m. Check the creep rupture stress limits for the beam. Assume that the sustained load is 20% of the live load, and $f_c'$ = 27.6 N/mm², $f_{fu}^*$ = 1654.8 N/mm², and $E_f$ = 137.9 kN/mm².

*Solution*: Note that the creep-rupture stress limits are to be determined in accordance with ACI 440, Section 8.4.1. Values of safe sustained stress levels or creep-rupture stress limits are given in ACI 440, Table 8.3.

Calculate the moment due to the sustained load. The data in this example are same as in Example 8.15, from which the following data are obtained:

$$D = 14.6 \text{ kN/m}$$

$$L = 36.5 \text{ kN/m}$$

Therefore, the sustained load is

$$w_s = D + 0.2L = 14.6 + 0.2(36.5) = 21.9 \text{ kN/m}$$

The moment due to the sustained load is

$$M_s = \frac{w_s l^2}{8} = \frac{(21.9)(3.0)^2}{8} = 24.63 \text{ kN-m}$$

From Example 8.16, $d$ = 322 mm, $j$ = 0.8912, and $A_f$ = 942 mm². Therefore, the stress in CFRP bars due to the sustained load is given by

$$f_{f,s} = \frac{M_s}{A_f jd} = \frac{(24.63)(1000)}{(942)(0.8912)(322)} = 0.0911 \text{ kN/mm}^2$$

From Example 8.15, $f_{fu}$ = 1654.8 N/mm²

$$0.2f_{fu} = 0.2(1654.8) = 330.96 \text{ N/mm}^2$$

$$f_{f,s} = 91.1 \text{ N/mm}^2 < 0.2f_{fu} = 330.96 \text{ N/mm}^2$$

Hence, the creep-rupture stress is within the allowable limits permitted by ACI 440.

## 8.11  FATIGUE STRESS LIMITS

Fatigue stress limits for a structural member must be considered for loading situations that involve fluctuations in stress levels. Concrete is essentially a composite material, and so is the FRP reinforcement. Therefore, the behavior of concrete structures under fatigue loading has three components to consider: concrete, FRP reinforcement, and the combination of the two.

Considerable research has been conducted on steel-reinforced concrete members, which has been reported in the literature [ACI 1974b; 1974c; 1982]. Brief summaries have been presented in ACI [1977] and in Nilson and Winter [1991]. Tests on steel-reinforced concrete beams show that

1. The fatigue strength at which two million or more cycles can be applied without failure is independent of the grade of steel.
2. The sustainable stress range $(f_{max} - f_{min})$ without causing failure depends on the magnitude of $f_{min}$.
3. In deformed bars, the degree of stress concentration at the junction of the rib and the cylindrical body of the bar play a significant role in reducing the safe stress range.

Fatigue failure of concrete requires (1) a cyclic loading generally in excess of 1 million load cycles, and (2) a change of reinforcement stress in each cycle of about 20 ksi. Because in most concrete structures dead load stresses (i.e., $f_{min}$) account for a significant portion of the service load stresses, Case 2 is infrequent (i.e., the stress range $f_{max} - f_{min}$ is small). Consequently, reinforced concrete rarely fails in fatigue.

A substantial amount of data related to fatigue behavior and life prediction of stand-alone FRP materials generated in the last 30 years is reported in the literature [National Research Council 1991]. Different types of fibers have different fatigue characteristics. Of all types of FRP composites currently used for infrastructure applications, CFRP is generally thought to be the least prone to fatigue failure (ACI 440).

Test data specifically on fatigue failure of FRP-reinforced concrete beams and concrete beams wrapped with FRP fabric (for strengthening) is sparse. However, based on the available test data, the following general observations are made with regard to fatigue failure of FRP-reinforced beams:

1. Beams reinforced with glass fiber reinforced polymer (GFRP) bars and designed for compression failure have a better fatigue life as compared to those designed for tension failure. This phenomenon is attributed to a larger depth of compression block under compression failure, which provides adequate ductility before concrete crushing.
2. In contrast to beams designed for compression failure, beams having tension failure collapse into two pieces due to the rupture of FRP bars as well as shear compression failure of concrete because of their very shallow depth (approximately 8 to 17% of the total beam/slab depth, depending on the type of the FRP bars) of compression block at the time of FRP bar

rupture [Vijay and GangaRao 1999]. This is unacceptable from the viewpoint of catastrophic failure.

Fatigue tests on 8-in.-thick FRP-reinforced concrete slabs stiffened with steel stringers have revealed the following trends [Kumar and GangaRao 1998]:

1. The fatigue response of FRP-reinforced concrete decks and those reinforced with steel rebars are nearly identical in terms of the loss of deck stiffness with the number of cycles up to a stress range of about 40% of the ultimate strength of concrete. The primary failure mode is the shear fatigue in both steel and FRP-reinforced concrete decks, which resulted in the punching shear of the slab.
2. The fatigue response of concrete decks is observed to be in three continuous stages, where the first stage is defined through crack initiation (approximately 10% of the fatigue life), whereas the second stage corresponds to gradual crack propagation (approximately 70% to 75% of the fatigue life), leading to the third stage consisting of deck instability and failure (approximately 15% of the fatigue life).
3. Gradual degradation under fatigue in concrete decks prevailed until approximately 80% to 85% of the total fatigue life, which was found to be at least 2 million cycles.
4. The above trends were observed in FRP-reinforced concrete decks subjected to a maximum stress of about 50% of concrete's ultimate strength with a stress range of about 30%. FRP bars were subjected to a stress of about 20% of their ultimate strength with a stress range of about 10%.

ACI 440 suggests limiting FRP stress to those shown in the Table 8.4. The FRP stress range can be determined using elastic analysis. However, $M_s$ is taken as being equal to the moment due to all sustained loads plus the maximum moment induced under a fatigue loading cycle.

## REFERENCES

American Concrete Institute (ACI), *Guide for the Design and Construction of Concrete Reinforced with FRP Bars*, ACI 440.1R-03, ACI Committee 440, Farmington Hills, MI: American Concrete Institute, 2003.

American Concrete Institute (ACI), *Deflection of Concrete Structures, ACI Special Publication SP-43 (ACI SP-43)*, ACI Committee 440, Farmington Hills, MI: American Concrete Institute, 1974a.

American Concrete Institute (ACI), Consideration for design of concrete structures subjected to fatigue loading, Report by ACI Committee 215, *ACI J.*, 71, 97–121, March 1974b.

American Concrete Institute (ACI), Abeles symposium: fatigue of concrete, *ACI Special Publication SP-41*, Farmington Hills, MI: American Concrete Institute, 1974c.

American Concrete Institute (ACI), Analysis of concrete bridges, *ACI Report 343R-77*, ACI Committee 343, Farmington Hills, MI: American Concrete Institute, 1977.

American Concrete Institute (ACI), Fatigue of concrete structures, *ACI Special Publication SP-75*, Farmington Hills, MI: American Concrete Institute, 1982.

Benmokrane, B., Challal, O., and Masmoudi, R., Flexural response of concrete beams reinforced with FRP bars, *ACI Struct. J.*, 93, 1, 46–55, January-February 1996.

Bischoff, P.H., Reevaluation of deflection prediction for concrete beams reinforced with steel and fiber reinforced polymer bars, *J. Struct. Eng.*, 131, 5, 752–767, May 2005.

Branson, D.E., Instantaneous and time-dependent deflections of simple and continuous reinforced concrete beams, *HPR Report No. 7, Part I*, Alabama Highway Department, Bureau of Public Roads, Auburn, AL: Department of Civil Engineering and Auburn Research Foundation, Auburn University, 1965.

Branson, D.E., *Deformation of Concrete Structures*, New York: McGraw-Hill, 1977.

Brown, V., Sustained load deflections in GFRP reinforced beams, *Proc. 3rd Intl. Symp. Non-Metallic (FRP) Reinforcement for Concrete Structures (FRPRCS-3)*, v.2, Japan Concrete Institute, Sapporo, Japan, 1997.

CSA 1996. Canadian Highway Bridge Design Code-Section 16: Fiber Reinforced Structures, Canadian Standard Association.

Gao, D., Benmokrane, B., and Masmoudi, R., A calculating method of flexural properties of FRP-reinforced concrete beam: Part 1: Crack width and deflection, Technical Report, Department of Civil Engineering, University of Sherbrook, Sherbrook, Quebec, Canada, 1998.

Faza, S.S. and GangaRao, H.V.S., Theoretical and experimental correlation of behavior of concrete beams reinforced with fiber reinforced plastic rebars, *Fiber-Reinforced-Plastic Reinforcement for Concrete Structures*, SP-138, Nanni, A. and Dolan, C.W. (Eds.), Farmington Hills, MI: American Concrete Institute, 1993.

Kage, T., Masuda, Y., Tanano, Y., and Sato, K., Long-term deflection of continuous fiber reinforced concrete beams, *Proc. 2nd Intl. RILEM Symp. Non-Metallic (FRP) Reinforcement for Concrete Structures*, Ghent, Belgium, 1995.

Kumar, S., and GangaRao, H.V.S., Fatigue response of concrete decks reinforced with FRP rebars, *J. Struct. Eng.*, ASCE, 124, 1, 11–16, January 1998.

Masmoudi, R., Benmokrane, B., and Challal, O., Cracking behavior of concrete beams reinforced with FRP rebars, *Proc. ICCI '96*, Tucson, AZ, 1996.

National Research Council, Life prediction methodologies for composite materials, NMAB-460, Washington, DC: National Materials Advisory Board, 1991.

Nilson, A. H. and Winter, G., *Design of Concrete Structures*, New York: John Wiley & Sons, 1991.

Vijay, P.V. and GangaRao, H.V.S., Creep behavior of concrete beams reinforced with GFRP bars, *Proc. CDCC '98*, Benmokrane, B. and Rahman, H. (Eds.), Sherbrooke, Quebec, Canada, August 5–7, 1998.

Vijay, P.V. and GangaRao, H.V.S., Development of fiber reinforced plastics for highway application: Aging behavior of concrete beams reinforced with GFRP bars, *CFC-WVU Report No. 99-265, WVDOH RP No. T-699-FRP1*, Morgantown, WV: Department of Civil and Environmental Engineering, West Virginia University, 1999.

# Glossary

Composite materials are uniquely different from the conventional materials, such as steel, reinforced concrete, aluminum, wood, and others. Many terms and acronyms used in the context of composite materials are quite uncommon and not used when discussing other materials. These terms may be unfamiliar to many bridge engineers, and their meanings are also not self-evident. Therefore, a few selected terms are provided in this section. These and many other terms and acronyms pertaining to composites are listed in handbooks on composite materials such as Lubin [1969; 1982], ASM International [1987], and Schwartz [1992].

**A stage** — An early stage in the polymerization reaction of certain thermosetting resins (especially phenolic) in which the material, after application to the reinforcement, is still soluble in certain liquids and is fusible (see also *B stage* and *C stage*).

**ABS** — Acrylonitrile-butadiene-styrene.

**Accelerated aging test** — A test performed under laboratory conditions during a short period (around 12 to 18 months) that simulates the long-term effects of approximately 50 years of real weather aging. For GFRP reinforcing bars, such tests include exposing those bars from –30° to 140°F and a high pH value (to simulate the high alkalinity present in concrete) of about 13.5 for about 12 to 18 months.

**Acrylic resin** — A fast-curing thermoset polymer crosslinked with styrene.

**Adhesiveness** — The property defined by the adhesion stress $\sigma_A = F/A_{gl}$, where $F$ is the perpendicular force to the glue line and $A_{gl}$ is the glue line contact area.

**Advanced composites** — Composites traditionally used in aerospace applications. These composites have fiber reinforcements of a thin diameter in a matrix material, such as epoxy and aluminum. Examples include graphite/epoxy, Kevlar/epoxy, and boron/aluminum.

**AFRP** — Aramid fiber reinforced polymer.

**Aging** — The process or the effect on materials of exposure to an environment over a period of time.

**Anisotropic** — Exhibiting different thermomechanical properties in different directions under applied or environmental loads.

**Aramid** — Aromatic polyamide fibers (a type of organic fibers) characterized by excellent high-temperature performance, flame resistance, and nonconductive properties. Aramid fibers are used to achieve high-strength, high-modulus reinforcement in polymer composites. More usually found as polyaramid — a synthetic fiber (see also *Kevlar*).

**Aspect ratio** — In the case of fibers, the ratio of length to the diameter of a fiber. In the case of plates (as in the theory of flat plates), the ratio of the larger side to the smaller side.

**Attenuation** — The process of making thin and slender, as applied to the formation of fiber from molten glass (reduction of force or intensity as in the decrease of an electrical signal).

**Autoclave** — A closed vessel for conducting a chemical reaction or other operation under pressure and heat.

**B stage** — The intermediate stage in the polymerization (curing) process of thermosetting resins, following which the material will soften with heat and become plastic and fusible. The resin of an uncured prepreg or premix is usually in B stage (see also *A stage* and *C stage*).

**Bar** — An uniaxial structural reinforcement formed into a linear shape.

**Barcol hardness test** — A test to determine the degree of cure by measuring resin hardness (ASTM D 2583).

**Binder** — A compound (resin or cementing constituent) that holds fibers together in some manner (analogous to cement in concrete, which holds aggregate, sand, and reinforcement together).

**Braided string or rope** — A string or rope made by braiding continuous fibers together.

**Braiding** — The intertwining of fibers in an organized fashion.

**Bidirectional laminate** — A reinforced plastic laminate with the fibers oriented in two directions in the plane of the laminate; a cross laminate (see also *unidirectional laminate*).

**BMC** — Bulk molding compound.

**Boron fiber** — A fiber usually made of a tungsten-filament core with elemental boron vapor deposited on it to impart strength and stiffness.

**C-glass** — A glass having a soda-lime-borosilicate composition used for its chemical stability in corrosive environments. Often used in composites that contact or contain acidic materials for corrosion-resistant service in the chemical processing industry.

**C stage** — The final stage in the reaction of certain thermosetting resins in which the material is relatively insoluble and infusible. The resin in a fully cured thermoset molding is in this stage (see also *A stage* and *B stage*).

**Carbon-carbon** — A composite of carbon fiber in a carbon matrix.

**Carbon fiber** — (used interchangeably with *graphite fiber*) An important reinforcing fiber known for its light weight, high strength, and high stiffness that is produced by pyrolysis of an organic fiber in an inert atmosphere at temperatures above 1800°F (980°C). The material can also be graphitized by heat treatment above 3000°F (1650°C).

**Catalyst** — An organic peroxide used to activate polymerization (curing).

**CFRP** — Carbon fiber reinforced polymer. Also includes graphite fiber reinforced polymer.

**Compatibility** — The ability of two or more constituents combined with each other to form a composition with properties better than the constituents.

**Composite material** — Any substance made from two or more materials that are distinct in their physical and chemical properties, each having a recognizable region. The materials are bonded and an interface exists between the materials, often a region such as the surface treatment used on fibers (sizing) to improve matrix adhesion and other performance parameters.

**Compression molding** — A manufacturing process for composites. A predetermined amount of thermosetting resin, filler, and other ingredients are placed in heated (female) mold die having a cavity of the desired shape. The final product is obtained by compressing the contents by a heated male die.

**Compressive modulus ($E_c$)** — The ratio of compressive stress to compressive strain below the proportional limit. Theoretically equal to Young's modulus determined from tensile tests.

**Continuous fibers** — Aligned fibers whose individual lengths are significantly greater than 15 times the critical length. Fibers commonly used are glass, carbon, and aramid.

**Continuous filament** — Individual fiber of small diameter and great or indefinite length.

**Continuous roving** — Parallel filaments coated with sizing, drawn together into single or multiple strands, and wound into a spool.

**Copolymer** — See *polymer*.

**Creep** — (also *plastic flow*) A time-dependent deformation that occurs when a material is subjected to load for a prolonged period of time. Ordinarily, for a given load a fixed deflection will result. But if the material creeps, the deflection will continue to increase even if the load remains the same. Creep at room temperature is called *cold flow*.

**Creep rupture** — The sudden rupture of material under a sustained load. Usually associated with glass reinforced composites.

**Crimp** — The waviness of a fiber, a measure of the difference between the length of an unstraightened fiber and its straight length.

**Critical length** — The minimum length of a fiber necessary to develop its full tensile strength in a matrix.

**Crosslinking** — The development of links between longitudinal molecular chains during curing.

**Cure** — A change of resin state from liquid to solid by polymerization. Usually irreversible and encouraged by initiators and (often) heat.

**D-glass** — A high-boron-content glass made especially for laminates requiring a precisely controlled dielectric constant.

**Deformability** — The ratio of deflection (curvature) at failure to deflection at the service limit state. The term is used in the context of beams using nonductile material, such as glass or aramid fiber reinforced polymers.

**Denier** — A measure of fiber diameter, taken as weight in grams of 9000 meters of fiber.

**Doff** — A roving package.

**Drape** — The ability of a fabric or a prepreg to conform to the shape of a contoured surface.

**Ductility** — The ability of a structure, its components, or of the materials to offer resistance in the inelastic domain of response (i.e., to deform without fracture). Mathematically, ductility ($\mu$) can be defined as the ratio of total imposed displacements at any instant to that of the onset of displacement at yield $\Delta_y$, i.e., $\mu = \Delta/\Delta_y$. In reinforced concrete beams, ductility is defined as the ratio of strain in the reinforcement at failure of the beam to strain in the reinforcement at yield.

**Durability** — The ability of a material to maintain desired properties (physical, chemical, mechanical, thermal, electrical, or others) with time.

**E-Glass** — A lime-alumina-borosilicate glass developed specifically for electrical applications. It is the type most used for glass fibers for reinforced plastics, and has become known as the standard textile glass; suitable for electrical laminates because of its high resistivity (nonconductivity). Also called *electric glass*, as the "E" stands for electrical grade.

**Elasticity** — The ability of a loaded material to return to its original shape upon unloading.

**Elastic modulus** — See *modulus of elasticity.*

**Elastomers** — Polymers that are capable of undergoing large elastic strains, e.g., synthetic and natural rubbers.

**Epoxy plastics** — Plastics based on thermoset resins made by the reaction of epoxides or oxiranes with other materials such as amines, alcohols, phenols, carboxylic acids, acid anhydrides, and unsaturated compounds.

**Epoxy resin** — A thermoset resin formed by the chemical reaction of epoxide groups with amines, alcohols, phenols, and others, that can be cured by the addition of various hardeners.

**Equivalent diameter** — The diameter of a circular cross-section having the same cross-sectional area as that of a noncircular cross-section.

**Esters** — A new molecule that results when the hydrogen atom of the hydroxyl portion of carboxylic acid is replaced by a group containing carbon.

**Exothermic** — The liberation of heat during polymerization.

**Extrusion** — (also *extrusion molding*) A manufacturing process that involves forcing molten resin through a die of a desired shape.

**Fabric** — An arrangement of fibers held together in two or three dimensions. A fabric can be woven, nonwoven, stitched, or knitted.

**Fabric, nonwoven** — A material formed from fibers or yarns without interlacing, This can be stitched, knit, or bonded.

**Fabric, woven** — A material constructed of interlaced yarns, fibers, or filaments.

**Fatigue** — A phenomenon of reduced material strength due to repetitive loading involving fluctuating stress or strain.

**Fatigue life** — The number of stress cycles that will cause failure for a maximum stress or strain and specified stress or strain ratio.

**Fatigue limit** — (also *endurance limit*) The limiting value of the fatigue strength as the fatigue life becomes very large.

**Fatigue strength** — The value of the maximum stress (tensile or compressive) that will cause failure at a specified number of stress or strain cycles for a given stress or strain ratio.

**FEM** — Finite element model.

**Fiber, short** — Relatively short lengths of very small sections of various materials made by chopping filaments; also called filament, thread, or bristle. Typical examples of fibers are glass, carbon, aramid, and steel.

**Fiber blooming** — (used in conjunction with failures of conditioned fiber glass reinforcement). The failure of surface reinforcement of a composite, e.g., a reinforcing bar.

**Fiber glass** — An individual filament made by attenuating molten glass (see also *continuous filament, staple fibers*).

**Fiber reinforced composite (FRC)** — A general term for material consisting of two or more discrete physical phases in which a fibrous phase is dispersed in a continuous matrix phase. The fibrous phase can be macro-, micro-, or submicroscopic, but it must retain its physical identity so that it could conceivably be removed from the matrix intact.

**Fiber reinforced polymer (FRP)** — (also *fiber reinforced plastic*) Any type of polymer-reinforced cloth, mat, strands, or any other form of fibrous glass. A general term used for composite materials but mostly used to denote glass fiber reinforced polymers (GFRP).

**Fiber reinforcement** — Reinforcement consisting mainly of fibers.

**Fiber volume fraction** — The ratio of the volume of fibers to the volume of the fiber reinforced composite. The total volume of a FRC, $v$, consists of two parts: the volume of the fibers ($v_f$) and the volume of the matrix ($v_m$). Thus, $v = v_f + v_m$.

**Filament** — A single continuous fiber. Used interchangeably with *fiber.*

**Fillers** — Substances used to add to thermosetting or thermoplastic polymers to reduce resin cost, control shrinkage, improve mechanical properties, and impart a degree of fire retardancy.

**Flexural modulus** — The ratio of stress to strain, measured within the elastic limit (20% to 80% of ultimate values) in the outermost fibers in a flexural test (see also *modulus of elasticity*).

**Four-point bending test** — A beam-bending test in which two equal loads equidistant from the center line of the beam are applied to cause regions of constant moment (between the two applied loads) and constant shear (between the supports and the adjacent loads). The four points referred to are the two load points on the beam and the two support points.

**Fracture** — A rupture of the surface without the complete separation of laminate.

**FRAT** — Fiber-reinforced advanced titanium.

**FRC** — Fiber reinforced composite.

**FRP** — Fiber reinforced polymer.

**GFRP** — Glass fiber reinforced polymer.

**Gage length** — The length of a portion of a specimen over which deformation is measured.

**Glass** — An inorganic product of fusion that has cooled to a rigid condition without crystallizing. Glass is typically hard, relatively brittle, and has conchoidal* fracture. Commercially, glass is available as *C-glass*, *D-glass*, *E-glass*, *M-glass*, and *S-glass*.

**Glass fiber** — A fiber drawn from an inorganic product of fusion that has cooled without crystallizing. Also, a glass filament that has been cut to a measurable length.

**Glass filament** — A form of glass that has been drawn to a small diameter and extreme length. Most filaments are less than 0.005 in. (0.13 mm) in diameter.

**Graphite fiber** — A fiber containing more than 99% elemental carbon made from a precurser by oxidation.

**Grating** — A prefabricated planar assembly consisting of continuous filaments (or bars), usually in two axial directions.

**Grid** — A prefabricated planar assembly consisting of continuous filaments (or bars) in an orthogonal arrangement or in three axial directions.

**Hand lay-up** — A fabrication method in which reinforcement layers, pre-impregnated or coated afterwards, are placed in a mold by hand, then cured to the formed shape.

**Hardener** — A chemical additive that is added to a thermoset resin to accelerate the curing reaction.

**Hybrid** — The result of attaching a composite body to another material, such as aluminum, steel, and so on.

**Hybrid composite** — A composite with two or more different types of reinforcing fibers.

**Hygroscopic** — Capable of adsorbing and retaining atmospheric moisture.

**Impregnate** — To saturate the voids or interstices of a reinforcement (e.g., cloth, mat, or filament) with resin.

**Impregnated fabric** — Fabric that has been impregnated with resin in flat form.

**Injection molding** — A manufacturing process for thermoplastic polymers.

**Inhibitor** — A substance that retards a chemical reaction. Used in certain type of monomers and resins to prolong shelf life.

**Initiator** — See *catalyst*.

**Interface** — The surface between two adjacent materials.

**Inhibitor** — Any chemical additive that delays or slows the curing cycle.

**Injection molding** — The process of forming a plastic to the desired shape by forcibly injecting the polymer into a mold.

**Interlaminar** — Between two or more adjacent laminae.

**Interlaminar shear** — Shear that produces displacement or deformation along the plane of their interface.

**Isopthalic polyester resin** — A polyster resin based on isopthalic acid, generally higher in properties (such as thermal, mechanical, and chemical resistance) than a general purpose or orthophthalic polyster resin.

---

* A fracture shape whose surface resembles the inside of a clamshell.

**Isotropic** — A material characteristic implying uniform properties in all directions, independent of the direction of loading.

**Jet molding** — A manufacturing process for thermosetting polymers.

**Kevlar™** — A strong, lightweight aramid fiber (a type of organic fiber) used as reinforcement, trademarked by E.I. du Pont de Nemours & Co.

**Knitwork** — A fiber/fabric architecture made by knitting.

**Laminate** — A product resulting from bonding (in a resin matrix) multiple plies of reinforcing fiber or fabric. It is the most common form of fiber-reinforced composites used in structural applications.

**Limit states** — Refers to the condition of a structure under loads. The two commonly recognized limit states are *ultimate limit state* and *serviceability limit state*.

**Load factor** — A factor (or coefficient) used to augment design loads and to account for uncertainties in the calculation of loads, stresses, and deformation.

**Long-term (duration)** — A duration of at least 1 year.

**M-glass** — A high-beryllia-content glass designed especially for its high modulus of elasticity.

**Macromechanics** — A study that deals with the gross behavior of composite materials and structures (in this case, lamina, laminate, and structure). Once the characteristics of a fibrous lamina are determined, its detailed microstructural nature can be ignored and treated simply as a homogenous orthotropic sheet.

**Mandrel** — (1) The core around which resin-impregnated fibers or rovings, such as glass, are wound to form axisymmetric shapes, such as pipes, tubes, or vessels. (2) In extrusion, the central piece of a pipe or a tubing die.

**Mat** — A fibrous reinforcing material available in blankets of various widths, thickness, lengths, and weights. It is composed of chopped filaments (for chopped-strands mat) or swirled filaments (for continuous-strand mat) with a binder applied to maintain form.

**Matrix** — A binder material in which the reinforcing material of a composite is embedded. The matrix can be polymer, metal, or ceramic (analogous to cement in concrete).

**Mer** — A suffix (as in monomer, polymer) used to describe the molecular structure of an organic compound.

**Mesh** — A construction in two or three axial directions made up of continuous filaments.

**Metallic fiber** — Manufactured fiber composed of metal, plastic-coated metal, metal-coated plastic, or a core completely covered by metal.

**Micromechanics** — A study dealing with the mechanical behavior of constituent materials such as fiber and matrix materials, the interaction of these constituents, and the resulting behavior of the basic composite (single lamina in this case). It involves the study of relationships between the effective composite properties and the effective constituent properties.

**Micron ($\mu$m)** — A unit of measurement representing one-millionth of a meter (1 $\mu$m = 0.001 mm = 0.00003937 in.).

**Mil** — A unit of measurement representing one-thousandth of an inch (1 mil = 0.001 in.).

**Modulus\*** — A number that expresses a measure of some property of a material, e.g., modulus of elasticity, modulus of rigidity (or shear modulus), flexural modulus, modulus of rupture, and so on.

**Modulus of elasticity** — The ratio of stress to strain in a material within the elastic limit.

**Molding** — The shaping of a plastic composition in or on a mold, normally accomplished under heat and pressure. Sometimes used to denote the finished part.

**Monomer** — (1) A simple molecule capable of reacting with like or unlike molecules to form a polymer. (2) The smallest repeating structure of a polymer, also called a *mer*.

**Multifilament** — A yarn consisting of many continuous filaments.

**Nonaging** — A term used to describe a material that is unaffected by the process or the effect of exposure to an environment for an interval of time.

**Nonhygroscopic** — Not absorbing or retaining an appreciable quantity of moisture from the air (water vapor).

**Nonwoven roving** — A reinforcement composed of continuous loose fiber strands.

**Nylon** — A low-modulus resin.

**Organic** — Comprising hydrocarbons or their derivatives. Also, material of biological (plant or animal) origin.

**Organic polymers** — Polymers having carbon as a common element in their make-up.

**Orthotropic** — Having three mutually perpendicular planes of elastic symmetry.

**PAN** — See *polyacrylonitrile*.

**PAN carbon fiber** — Carbon fiber made from polyacrylonitrile (PAN).

**PET** — Polyethylene terephthalate, a thermoplastic polyester resin.

**pH** — The measure of the acidity or alkalinity of a substance, neutrally being at pH 7. Acid solutions are less than 7, and alkaline solutions are more than 7.

**Phase** — A homogeneous mass having uniform physical and chemical properties. Solutions, pure elements, and compounds in a single state (i.e., either solid or liquid) and gases are phases.

**Phenolic resin** — A thermosetting resin produced by condensation of an aromatic alcohol with an aldehyde, particularly phenol with formaldehyde. Possesses better thermal resistance than other thermoset resins.

**Pitch carbon fiber** — Carbon fiber made from petroleum pitch.

**Planar/three-dimensional reinforcement** — Biaxial and triaxial reinforcement made of continuous fibers; can be grid-shaped, mesh-shaped, knitted, or woven.

**Plastic** — (also *polymeric material*) A thermosetting or thermoplastic (organic) material composed of very large molecules formed by polymerization. It contains as an essential ingredient an organic substance of high molecular

---

\* Using "modulus" alone without modifying terms is confusing and should be discouraged.

weight, is solid in its finished state, and at some stage in its manufacture or processing into finished articles can be shaped by flow; made of plastic. A major advantage of plastics is that they can significantly deform without rupturing. A *rigid plastic* is one with a stiffness or apparent modulus of elasticity greater than 100 kips/in$^2$ (690 MPa) at 73.4°F (23°C). A *semi-rigid plastic* has a stiffness or apparent modulus of elasticity between 10 and 100 kips/in$^2$ (69 and 690 MPa) at 73.4°F (23°C). Various types of palstics are designated by ASTM D 4000.

**Plastic strain** — (also *permanent* or *inelastic strain*) The deformation remaining after the removal of a load.

**Plasticize** — To make a material moldable by softening it with heat or a plasticizer.

**Plasticizers** — Substances added to polymers and some casting materials to reduce brittleness.

**Ply** — (1) One of the layers that makes up a laminate. (2) The number of single yarns twisted together to form a plied yarn.

**Polyacrylonitrile (PAN)** — A product used as a base material in the manufacture of certain carbon fibers.

**Polyamide** — A polymer in which the structural units are linked by amide or thioamide groupings. Many polyamides are fiber-forming.

**Polyesters** — (also *polyester resins*) Thermosetting resins produced by dissolving unsaturated, generally linear alkyd resins in a vinyl active monomer, e.g., styrene, methyl styrene, or diallyl phthalate.

**Polyimide** — A polymer produced by heating polyamic acid, a highly heat-resistant resin (over 600°F) suitable for use as a binder or an adhesive.

**Polymer** — A high-molecular-weight organic compound, natural or synthetic, whose structure can be represented by a repeated small unit (mer), e.g., polyethylene, rubber, cellulose. Synthetic polymers are formed by the addition or condensation polymerization of monomers. Some polymers are elastomers; some are plastics. When two or more monomers are involved, the product is called a *copolymer*.

**Polymeric material** — (also *polymers*) Materials, commonly called plastic, that are continuously deformable at some stage during their manufacture (see *plastic*).

**Polymerization** — A chemical reaction in which the molecules of a monomer (e.g., ethylene, $C_2H_4$) are linked together to form large molecules (e.g., polyethylene), whose molecular weight is a multiple of that of the original substance. When two or more monomers are involved, the process is called *copolymerization* or *heteropolymerization*.

**Pot life** — The time for which a reactive resin system retains a sufficiently low viscosity to be used in a process.

**Prepreg** — Ready-to-mold (or semi-hardened) material in sheet form, which may be cloth, mat, or paper impregnated with resin, and stored for use. The resin is partially cured to a B stage and supplied to the fabricator, who lays up the finished shape and completes the cure with heat and pressure.

**Proportional limit** — The maximum stress up to which stress and strain remain proportional (or linear).

**Prototype** — A model suitable for use in complete evaluation of form, design, and performance.

**Pultrusion** — A continuous manufacturing process in which fibers and resin are pulled through a prismatic die to form sections similar to metal or other thermoplastic extrusions.

**Pyrolysis** — The chemical transformation (or decomposition) of a compound caused by heat.

**Quasi-isotropic** — Approximating isotropy by orienting reinforcement (or plies) in several directions.

**Reinforcement** — The key element in a composite material, in the form of short or continuous fibers in complex textile forms, embedded in matrix, that imparts strength and stiffness to the end product.

**Relaxation** — The loss of stress in a material held at constant length under sustained loads. The same basic phenomenon is known as *creep* when defined in terms of change in deformation under constant stress.

**Resin** — A polymeric material used to bind fibers together in a composite.

**Resin system** — A mixture of resin and initiators plus any fillers or additives required for processing or performance considerations.

**Rope** — An assembly of bundled continuous fibers.

**Roving** — A group of bundles of continuous filaments either as untwisted strands or as twisted yarn.

**Rupture strength** — A measure of strength obtained by dividing the load at rupture by the area of cross-section of the specimen. When the actual area of the cross-section at the time of rupture is used, the strength is called the *true rupture strength*. When the original cross-sectional area of the specimen is used, the strength is called the *apparent strength* (which is usually intended, unless otherwise specified).

**S-glass** — A magnesia-alumina-silicate glass, specially designed to provide filaments with very high tensile strength and Young's modulus (about 33% and 20% higher, respectively, than D-glass) used in aircraft and military components.

**SCRIMP** — Seeman composite resin infusion molding process. A vacuum process used for comolding composite skins and core in one piece without the need of oven autoclave (i.e., in an open mold).

**Serviceability limit state** — The stage in the loading of a structure deemed undesirable in its loaded condition because of the cracks, deformation, and so on, generated beyond a certain limit.

**Shear span** — The length of a beam in which shear is constant. The length between the support and the adjacent load in a four-point bending test.

**Shear modulus ($G$)** — (also *modulus of rigidity*) The ratio of elastic shear stress to elastic shear strain.

**Shelf life** — (also *storage life*) The period for which a material can be stored under specified conditions without adverse effects on its properties.

**Short-term (duration)** — A duration of 48 hours.

**Sizing** — (1) Applying a material on a surface to fill pores and thus reduce the absorption of the subsequently applied adhesive or coating. (2) To modify the surface properties of the substrate to improve adhesion. (3) The material used for this purpose, also called *size*.

**SMC** — Sheet molding compound.

**Specific modulus** — Stiffness-to-weight ratio: specific modulus = $E/\rho$, where $E$ is Young's modulus of elasticity, and $\rho$ is the density of the material. Also expressed as $E/\rho g$, where $g$ is the acceleration due to gravity, 32.2 ft/sec$^2$.

**Specific strength** — Strength-to-weight ratio: specific strength = $f_u/\rho$, where $f_u$ is the ultimate strength of the material and $\rho$ is the density of the material. Also expressed as $f_u/\rho g$, where $g$ is the acceleration due to gravity, 32.2 ft/sec$^2$.

**Spray-up** — A method of contact molding wherein resin and chopped strands of continuous filament roving are deposited on the mold directly from a chopper gun.

**Staple fibers** — Fibers of spinnable length manufactured directly or by cutting continuous filaments to relatively short lengths, generally less than 17 in. (432 mm).

**Static fatigue** — The failure of a part under continued static load. Analogous to creep-rupture failure in metals testing but often the result of aging accelerated by stress.

**Storage life** — See *shelf life*.

**Strain hardening** — An increase of strength and hardness of metallic materials resulting from inelastic deformation.

**Strand** — A linear component (such as a fiber or a filament) that constitutes either a part or the whole element.

**Styrene** — A monomer used as a crosslinking agent in unsaturated polyester and vynil ester resins.

**Thermal fatigue** — The fracture of materials resulting from repeated thermal cycles that produce damaging fluctuating stresses.

**Thermal shock** — The damaging fracture or distortion as a result of sudden temperature changes.

**Thermoplastic** — A plastics (or polymeric) material that can be softened by heating and hardened by cooling in a repeatable, reversible process.

**Thermoset** — A plastics (polymeric) material that is hardened by an irreversible chemical reaction.

**Three-point bending test** — A beam bending test in which the load is applied at the center of the beam. The three points referred to are the load point on the beam and the two supports.

**Toughness** — The ability to absorb a large amount of energy prior to fracture.

**Tow** — A large bundle of continuous filaments, generally 10,000 or more, not twisted, usually designated by a number followed by "K," indicating multiplication by 1000. For example, 12K tow has 12,000 filaments.

**Transfer molding** — A method of molding thermosetting materials in which the plastic is first softened by heating and pressure in a transfer chamber and then forced by high pressure through suitable sprues, runners, and gates into the closed mold for final curing.

**Triaxial reinforcement** — Fiber reinforcements in three orthogonal directions. The idea is to produce as much near-isotropic characteristics in fiber reinforced composite as possible.

**UDC** — Unidirectional composites.

**Ultimate limit state** — The state (or condition) of a loaded structure such that structural members are deemed to have reached the limit of their ability to withstand external or internal forces.

**Uncrimped** — Fibers without any crimp.

**Under-cure** — A state wherein the degree of polymerization is insufficient to withstand demolding, handling, and subsequent processing stages due to inadequate time or temperatures in the mold.

**Unidirectional laminate** — A reinforced plastic laminate in which substantially all the fibers are oriented in the same direction.

**Unsaturated polyester** — A family of thermosetting resins produced by dissolving alkyd resins in a reactive monomer such as styrene. Curing is encouraged using organic peroxides in combination with accelerators or heat. The product of a condensation reaction between dysfunctional acids and alcohols, one of which — generally the acid — contributes an olefinic reaction.

**Vinyl esters** — (also *vinyl ester resins*) A class of thermoset resins produced from esters of acrylic acids, often dissolved in styrene.
Resin characterized by reactive unsaturation located primarily in terminal positions that can be compounded with styrol monomers to give highly crosslinked thermoset copolymers.

**Viscosity** — The property of resistance to flow exhibited within the body of a material expressed in terms of the relationship between the applied shearing stress and the resulting rate of strain in shear.

**Viscoelastic** — A term used to describe materials that exhibit characteristics of both viscous fluids and elastic solids. Polymeric materials, which are known to be viscoelastic, may behave like solids depending on the time scale or the temperature.

**V-RTM** — (also *VA-RTM*) Vacuum resin transfer molding, a vacuum process to combine resin and reinforcement in an open mold.

**Whisker** — A very short fiber form of reinforcement, usually crystalline.

**Wet-out** — The condition of an impregnated roving or yarn wherein substantially all voids between the sized strands and filaments are filled with resin.

**Wet-out rate** — The time required for a plastic to fill the interstices of a reinforcement material and wet the surface of the reinforcement fibers. Usually determined by optical or light-transmission means.

**Yarn** — A group of fibers held together to form a string or rope.

**Yield point** — The stress below the tensile strength at which a marked increase in deformation first occurs without corresponding increase in load during a tensile test. It is commonly observed in mild steels and occasionally in few other alloys.

**Yield strength** — The stress corresponding to some arbitrarily selected permanent deformation.

**Young's modulus (*E*)** — The modulus of elasticity in tension (ratio of stress to strain within the elastic limit).

## REFERENCES

ASM International, *Composites*, Engineered Materials Hand Book, Vol. 1, Metals Park, OH: ASM International, 1987.

Chawla, K.K., *Composite Materials*, New York: Springer Verlag, 1997.

Gibson, R.F., *Principles of Composite Material Mechanics*, New York: McGraw-Hill, 1994.

Keyser, C.A., *Materials Science in Engineering*, 4th Ed., Columbus, OH: Charles Merrill Publishing Co., 1986.

Lubin, G. (Ed.), *Handbook of Composites*, New York: Van Nostrand Reinhold Co., 1982.

Lubin, G. (Ed.), *Handbook of Fiberglass and Advanced Composites*, New York: Van Nostrand Reinhold Co., 1969.

Mallick, P.K., *Fiber-Reinforced Composites — Materials, Manufacturing, and Design*, Mechanical Engineering Series, Vol. 62, New York: Marcel Dekker, 1988.

Owen, M.J., Middleton, V., and Jones, I.A., (Eds.), *Integrated Design and Manufacture Using Fibre-Reinforced Polymeric Composites*, Boca Raton, FL: CRC Press, 2000.

Schwartz, M., *Composite Materials Handbook*, 2nd Ed., New York: McGraw-Hill, 1992.

Tsai, S.W., Hahn, H.T., *Introduction to Composite Materials*, Lancaster, PA: Technomic Publishing Co., 1980.

# Appendix A: Tables

## TABLE A1
### Cross-Sectional Areas of FRP Reinforcing Bars

| Bar Size Designation | | Nominal diameter in. (mm) | Area in.² (mm²) |
|---|---|---|---|
| Standard | Metric Conversion | | |
| No. 2 | No. 6 | 0.250 (6.4) | 0.05 (31.6) |
| No. 3 | No. 10 | 0.375 (9.5) | 0.11 (71) |
| No. 4 | No. 13 | 0.500 (12.7) | 0.20 (129) |
| No. 5 | No. 16 | 0.625 (15.9) | 0.31 (199) |
| No. 6 | No. 19 | 0.750 (19.1) | 0.44 (284) |
| No. 7 | No. 22 | 0.875 (22.2) | 0.60 (387) |
| No. 8 | No. 25 | 1.000 (25.4) | 0.79 (510) |
| No. 9 | No. 29 | 1.128 (28.7) | 1.00 (645) |
| No. 10 | No. 32 | 1.270 (32.3) | 1.27 (819) |
| No. 11 | No. 36 | 1.410 (35.8) | 1.56 (1006) |
| No. 14 | No. 43 | 1.693 (43.0) | 2.25 (1452) |
| No. 18 | No. 57 | 2.257 (57.3) | 4.00 (2581) |

## TABLE A2
### Area of Groups of Standard Bars

| Bar No. | Number of Bars | | | | | | | | |
|---|---|---|---|---|---|---|---|---|---|
| | 2 | 3 | 4 | 5 | 6 | 7 | 8 | 9 | 10 |
| 4 | 0.39 | 0.53 | 0.78 | 0.98 | 1.18 | 1.37 | 1.57 | 1.77 | 1.96 |
| 5 | 0.61 | 0.91 | 1.23 | 1.53 | 1.84 | 2.15 | 2.45 | 2.76 | 3.07 |
| 6 | 0.88 | 1.32 | 1.77 | 2.21 | 2.65 | 3.09 | 3.53 | 3.98 | 4.42 |
| 7 | 1.20 | 1.80 | 2.41 | 3.01 | 3.61 | 4.21 | 4.81 | 5.41 | 6.01 |
| 8 | 1.57 | 2.35 | 3.14 | 3.93 | 4.71 | 5.50 | 6.28 | 7.07 | 7.85 |
| 9 | 2.00 | 3.00 | 4.00 | 5.00 | 6.00 | 7.00 | 8.00 | 9.00 | 10.00 |
| 10 | 2.53 | 3.79 | 5.06 | 6.33 | 7.59 | 8.86 | 10.12 | 11.39 | 12.66 |
| 11 | 3.12 | 4.68 | 6.25 | 7.81 | 9.37 | 10.94 | 12.50 | 14.06 | 15.62 |
| 14 | 4.50 | 6.75 | 9.00 | 11.25 | 13.50 | 15.75 | 18.00 | 20.25 | 22.50 |
| 18 | 8.00 | 12.00 | 16.00 | 20.00 | 24.00 | 28.00 | 32.00 | 36.00 | 40.00 |

*Continued.*

363

**TABLE A2** *(Continued)*
**Area of Groups of Standard Bars**

| | Number of Bars | | | | | | | | | |
|---|---|---|---|---|---|---|---|---|---|---|
| Bar No. | 11 | 12 | 13 | 14 | 15 | 16 | 17 | 18 | 19 | 20 |
| 4 | 2.16 | 2.36 | 2.55 | 2.75 | 2.95 | 3.14 | 3.34 | 3.53 | 3.73 | 3.93 |
| 5 | 3.37 | 3.68 | 3.99 | 4.30 | 4.60 | 4.91 | 5.22 | 5.52 | 5.83 | 6.14 |
| 6 | 4.86 | 5.30 | 5.74 | 6.19 | 6.63 | 7.07 | 7.51 | 7.95 | 8.39 | 8.84 |
| 7 | 6.61 | 7.22 | 7.82 | 8.42 | 9.02 | 9.62 | 10.22 | 10.82 | 11.43 | 12.03 |
| 8 | 8.64 | 9.43 | 10.21 | 11.00 | 11.78 | 12.57 | 13.35 | 14.14 | 14.92 | 15.71 |
| 9 | 11.00 | 12.00 | 13.00 | 14.00 | 15.00 | 16.00 | 17.00 | 18.00 | 19.00 | 20.00 |
| 10 | 13.92 | 15.19 | 16.45 | 17.72 | 18.98 | 20.25 | 21.52 | 22.78 | 24.05 | 25.31 |
| 11 | 17.19 | 18.75 | 20.31 | 21.87 | 23.44 | 25.00 | 26.56 | 28.13 | 29.69 | 31.25 |
| 14 | 24.75 | 27.00 | 29.25 | 31.50 | 33.75 | 36.00 | 38.25 | 40.50 | 42.75 | 45.00 |
| 18 | 44.00 | 48.00 | 52.00 | 56.00 | 60.00 | 64.00 | 68.00 | 72.00 | 76.00 | 80.00 |

**TABLE A3**
**Minimum Modulus of Elasticity,**
**by Fiber Type, for Reinforcing Bars**
**(ACI 440.1R-03)**

| Bar Type | Modulus grade $\times 10^3$ ksi (GPa) |
|---|---|
| GFRP bars | 5.7 (39.3) |
| AFRP bars | 10.0 (68.9) |
| CFRP bars | 16.0 (110.3) |

**TABLE A4**
**Strength Properties of Steel and FRP Bars**
**(ACI 440.1R-03)**

| Bar Type | Tensile strength $f_y$ or $f_{pu}$, ksi (MPa) | Modulus of Elasticity ksi (GPa) |
|---|---|---|
| Steel | 60 (414) | 29,000 (200) |
| GFRP bars | 80 (552) | 6,000 (41.4) |
| AFRP bars | 170 (1172) | 12,000 (82.7) |
| CFRP bars | 300 (2070) | 22,000 (152) |

## TABLE A5
## Usual Tensile Properties of Steel and FRP Reinforcing Bars*
## (ACI 440.1R-03)

| Property | Steel | GFRP | CFRP | AFRP |
|---|---|---|---|---|
| Nominal yield stress, ksi (MPa) | 40–75 | — | — | — |
| | (276–517) | | | |
| Tensile strength, ksi (MPa) | 70–100 | 70–230 | 87–535 | 250–368 |
| | (483–690) | (483–1600) | (600–3690) | (1720–2540) |
| Elastic modulus, ksi (GPa) | 29.0 | 5.1–7.4 | 15.9–84.0 | 6.0–18.2 |
| | (200) | (35.0–51.0) | (120.0–580.0) | (41.0–125.0) |
| Yield strain, % | 1.4–2.5 | — | — | — |
| Rupture strain, % | 6.0–12.0 | 1.2–3.1 | 0.5–1.7 | 1.9–4.4 |

* Typical values for fiber volume fraction ranging from 0.5 to 0.7.

## TABLE A6
## Typical Densities of FRP Reinforcing Bars,
## lb/ft³ (g/cm³) (ACI 440.1R-03)*

| Steel | GFRP | CFRP | AFRP |
|---|---|---|---|
| 490.00 | 77.8–131.00 | 93.3–100.00 | 77.8–88.10 |
| (7.90) | (1.25–2.10) | (1.5–1.6) | (1.25–1.40) |

* Typical values for fiber volume fraction ranging from 0.5 to 0.7.

## TABLE A7
## Coefficients of Thermal Expansion (CTE) for FRP
## Reinforcing Bars (ACI 440.1R-03)*

| Direction of Expansion | Coefficient of Thermal Expansion $\times 10^{-6}/°F$ ($\times 10^{-6}/°C$) | | | |
|---|---|---|---|---|
| | Steel | GFRP | CFRP | AFRP |
| Longitudinal, $\alpha_L$ | 6.5 | 3.3–5.6 | –4.0 to 0.0 | –3.3 to –1.1 |
| | (11.7) | (6.0–10.0) | (9.0 to 0.0) | (–6 to –2) |
| Transverse, $\alpha_T$ | 6.5 | 11.7–12.8 | 41–58 | 33.3–44.4 |
| | (11.7) | (21.0–23.0) | (74.0–104.0) | (60.0–80.0) |

* Typical values for fiber volume fraction ranging from 0.5 to 0.7.

## TABLE A8

**Values of Coefficient** $\beta_d = \alpha_b\left(\dfrac{E_f}{E_s}+1\right)$

## ACI 440.1R-03 Eq. 8-12b*

| $E_f$ (ksi) | $\beta_d$ | $E_f$ (ksi) | $\beta_d$ | $E_f$ (ksi) | $\beta_d$ |
|---|---|---|---|---|---|
| 5,000 | 0.5862 | 9,500 | 0.6638 | 18,000 | 0.8103 |
| 5,500 | 0.5948 | 10,000 | 0.6724 | 19,000 | 0.9276 |
| 6,000 | 0.6034 | 11,000 | 0.6897 | 20,000 | 0.8448 |
| 6,500 | 0.6121 | 12,000 | 0.7069 | 22,000 | 0.8793 |
| 7,000 | 0.6207 | 13,000 | 0.7241 | 24,000 | 0.9138 |
| 7,500 | 0.6293 | 14,000 | 0.7413 | 26,000 | 0.9483 |
| 8,000 | 0.6379 | 15,000 | 0.7586 | 28,000 | 0.9828 |
| 8,500 | 0.6466 | 16,000 | 0.7759 | 29,000 | 1.0000 (max) |
| 9,000 | 0.6551 | 17,000 | 0.7931 | — | — |

* Interpolation may be used for values not shown in the table.

## TABLE A9
## Conversion Factors: U.S. Customary Units to SI Units

|  | Multiply |  | by |  | to obtain |
|---|---|---|---|---|---|
| Length | inches | × | 25.4 | = | millimeters |
|  | feet | × | 0.3048 | = | meters |
|  | yards | × | 0.9144 | = | meters |
|  | miles (statute) | × | 1.609 | = | kilometers |
| Area | square inches | × | 645.2 | = | square millimeters |
|  | square feet | × | 0.0929 | = | square meters |
|  | square yards | × | 0.8361 | = | square meters |
| Volume | cubic inches | × | 16,387 | = | cubic millimeters |
|  | cubic feet | × | 0.02832 | = | cubic meters |
|  | cubic yards | × | 0.7646 | = | cubic meters |
|  | gallons | × | 0.003785 | = | cubic meters |
|  | (U.S. liquid) |  |  |  |  |
| Force | pounds | × | 4.448 | = | newtons |
|  | kips | × | 4448 | = | newtons |
| Force per unit length | pounds per foot | × | 14.594 | = | newtons per meter |
|  | kips per foot | × | 14,594 | = | newtons per meter |
| Load per unit volume | pounds per cubic foot | × | 0.15714 | = | kilonewtons per cubic meter |
| Bending moment or | inch-pounds | × | 0.1130 | = | newton meters |
| torque | foot-pounds | × | 1.356 | = | newton meters |
|  | inch-kips | × | 113.0 | = | newton meters |
|  | foot-kips | × | 1356 | = | newton meters |
|  | inch-kips | × | 0.1130 | = | kilonewton meters |
|  | foot-kips | × | 1.356 | = | kilonewton meters |
| Stress, pressure, loading | pounds per sq inch | × | 6895 | = | pascals |
| (force) per unit area) | pounds per sq inch | × | 6.895 | = | kilopascals |
|  | pounds per sq inch | × | 0.006895 | = | megapascals |
|  | kips per sq inch | × | 6.895 | = | megapascals |
|  | pounds per sq foot | × | 47.88 | = | pascals |
|  | pounds per sq foot | × | 0.04788 | = | kilopascals |
|  | kips per sq foot | × | 47.88 | = | kilopascals |
|  | kips per sq foot | × | 0.04788 | = | megapascals |
| Mass | pounds | × | 0.454 | = | kilograms |
| Mass per unit volume | pounds per cubic foot | × | 16.02 | = | kilograms per cubic meter |
| (density) | pounds per cubic yard | × | 0.5933 | = | kilograms per cubic meter |
| Moment of inertia | inches | × | 416.231 | = | millimeters |
| Mass per unit length | pounds per foot length | × | 1.488 | = | kilograms per meter |
| Mass per unit area | pounds per square foot | × | 4.882 | = | kilograms per square meter |

# Appendix B:
# GFRP Reinforced Concrete Beams: Theoretical and Experimental Results Comparison

Comparison of experimental and theoretical values of maximum moment values in GFRP bar reinforced concrete beams tested to failure by different researchers is provided in Table B.1 (Vijay and GangaRao, 2001). Dimensions of the beams, GFRP reinforcement details, concrete properties, ratio of experimental to theoretical values, actual failure modes, and explanation of notations used in Table B.1 are provided.

## NOTATIONS

| | |
|---|---|
| $a$ | depth of equivalent rectangular stress block as defined by ACI 318 |
| $A_f$ | total area of longitudinal FRP reinforcement in tension |
| $b$ | width of the rectangular concrete member |
| $c$ | distance from extreme compression fiber to neutral axis |
| $d$ | distance from extreme compression fiber to centroid of tension reinforcement |
| $(c/d)_{bal}$ | c/d for producing balanced strain conditions in concrete and FRP |
| $E_f$ | tensile modulus of elasticity of FRP reinforcement |
| $f_c'$ | specified compressive strength of concrete |
| $f_f$ | nominal strength of FRP reinforcement |
| $M_{th}$ | theoretical moment capacity |
| $M_{exptl.}$ | experimental moment capacity |
| $\beta_1$ | factor between 0.85-0.65 as defined in ACI 318 based on $f_c'$ |
| $l$ | test span of the GFRP reinforced beam |
| $l/d$ | ratio of test span to effective depth |
| $\sigma$ | standard deviation for ultimate moment capacity of concrete beams reinforced with GFRP bars and designed for compression failure. |

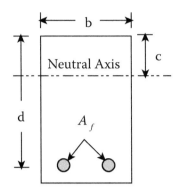

**FIGURE B.1** Rectangular concrete section with FRP bars.

## REFERENCE

Vijay P.V. and Hota V.S. GangaRao, Bending behavior and deformability of glass fiber-reinforced polymer reinforced concrete members, *ACI Structural Journal,* Nov.–Dec. 98 (6), 2001, pp. 834–842.

| Beam | $f'_c$ psi | $\beta_1$ | $b$ (in) | $d$ (in) | $l/d$ | $A_f$ | $E_f$ ($\times 10^6$ psi) | $f_i$ ksi | $(c/d)bal$ | Moment Calculations for Comp. Failure | | | | Actual Failure |
|---|---|---|---|---|---|---|---|---|---|---|---|---|---|---|
| | | | | | | | | | | (expt.) k-ft | $c/d$ | $M_{th}$ lb-ft | $M_{expt}/M_{th}$ | |
| **GangaRao, H.V.S. and Faza, S.S., 1991** | | | | | | | | | | | | | | |
| C4 | 4200 | 0.840 | 6.00 | 10.50 | 10.29 | 1.571 | 6.61 | 80.00 | 0.199 | 40.000 | 0.332 | 47.223 | 0.847 | C/S |
| C8 | 5000 | 0.800 | 6.00 | 10.56 | 10.22 | 1.203 | 7.35 | 80.00 | 0.216 | 41.630 | 0.295 | 49.302 | 0.844 | C |
| C-H5 | 6500 | 0.725 | 6.00 | 10.50 | 10.29 | 1.571 | 6.61 | 80.00 | 0.199 | 54.750 | 0.295 | 58.181 | 0.941 | C/S |
| CC | 7500 | 0.675 | 6.00 | 10.50 | 10.29 | 1.571 | 6.61 | 80.00 | 0.199 | 60.000 | 0.286 | 61.370 | 0.978 | C/S |
| EH2 | 6500 | 0.725 | 6.00 | 10.75 | 10.05 | 0.589 | 7.00 | 107.00 | 0.164 | 31.130 | 0.196 | 42.183 | 0.738 | C |
| EH4 | 6500 | 0.725 | 6.00 | 10.60 | 10.19 | 0.552 | 6.92 | 130.00 | 0.138 | 37.500 | 0.191 | 39.978 | 0.938 | T/C |
| **Nawy, E.W., Neuwerth, A.M. and Philips, C.J., 1971** | | | | | | | | | | | | | | |
| 1 | 4810 | 0.810 | 3.50 | 6.50 | 11.08 | 0.045 | 7.30 | 154.80 | 0.124 | 3.900 | 0.108 | 4.225 | 0.923 | C |
| 2 | 4100 | 0.845 | 3.50 | 6.50 | 11.08 | 0.045 | 7.30 | 154.80 | 0.124 | 2.950 | 0.114 | 3.953 | 0.746 | C |
| 5 | 5030 | 0.799 | 3.50 | 6.30 | 11.43 | 0.057 | 7.30 | 154.80 | 0.124 | 4.200 | 0.120 | 4.527 | 0.928 | C |
| 6 | 5030 | 0.799 | 3.50 | 6.30 | 11.43 | 0.057 | 7.30 | 154.80 | 0.124 | 4.200 | 0.120 | 4.527 | 0.928 | C |
| 9 | 4740 | 0.813 | 3.50 | 6.25 | 11.52 | 0.068 | 7.30 | 154.80 | 0.124 | 5.000 | 0.134 | 4.730 | 1.057 | C |
| 10 | 4530 | 0.824 | 3.50 | 6.25 | 11.52 | 0.068 | 7.30 | 154.80 | 0.124 | 2.900 | 0.136 | 4.641 | 0.625 | C |
| 13 | 4490 | 0.826 | 3.50 | 6.29 | 11.45 | 0.079 | 7.30 | 154.80 | 0.124 | 3.600 | 0.146 | 4.992 | 0.721 | C |
| 14 | 4970 | 0.802 | 3.50 | 6.29 | 11.45 | 0.079 | 7.30 | 154.80 | 0.124 | 4.400 | 0.141 | 5.210 | 0.845 | C |
| 17 | 4970 | 0.802 | 3.50 | 6.25 | 11.52 | 0.091 | 7.30 | 154.80 | 0.124 | 6.400 | 0.151 | 5.464 | 1.171 | C |
| 18 | 4490 | 0.826 | 3.50 | 6.25 | 11.52 | 0.091 | 7.30 | 154.80 | 0.124 | 5.100 | 0.156 | 5.233 | 0.974 | C |
| **Nawy, E.W. and Neuwerth, A.M., 1977** | | | | | | | | | | | | | | |
| 7 | 4700 | 0.815 | 5.00 | 10.85 | 11.06 | 0.980 | 3.80 | 105.00 | 0.098 | 32.025 | 0.222 | 32.228 | 0.994 | N |
| 8 | 4300 | 0.835 | 5.00 | 10.85 | 11.06 | 0.980 | 3.80 | 105.00 | 0.098 | 28.700 | 0.228 | 30.908 | 0.929 | N |
| 9 | 4300 | 0.835 | 5.00 | 10.75 | 11.16 | 1.176 | 3.80 | 105.00 | 0.098 | 34.825 | 0.248 | 32.662 | 1.066 | N |
| 10 | 5100 | 0.795 | 5.00 | 10.75 | 11.16 | 1.176 | 3.80 | 105.00 | 0.098 | 33.950 | 0.235 | 35.389 | 0.959 | N |

*Continued.*

| Beam | $f'_c$ psi | $\beta_1$ | b (in) | d (in) | l/d | $A_f$ | $E_f$ ($\times 10^6$ psi) | $f_f$ ksi | $(c/d)bal$ | Moment Calculations for Comp. Failure (expt.) k-ft | c/d | $M_{th}$ lb-ft | $M_{expt}/M_{th}$ | Actual Failure |
|---|---|---|---|---|---|---|---|---|---|---|---|---|---|---|
| 11 | 5700 | 0.765 | 5.00 | 10.80 | 11.11 | 1.372 | 3.80 | 105.00 | 0.098 | 37.100 | 0.243 | 39.732 | 0.934 | N |
| 12 | 4400 | 0.830 | 5.00 | 10.80 | 11.11 | 1.372 | 3.80 | 105.00 | 0.098 | 34.388 | 0.262 | 35.270 | 0.975 | N |
| **Sonobe, Y. et al., 1997** | | | | | | | | | | | | | | |
| AK-0.81-300 | 4270 | 0.837 | 7.87 | 9.65 | 9.79 | 0.615 | 9.40 | 193.64 | 0.127 | 45.544 | 0.239 | 39.917 | 1.141 | C* |
| AK-1.82-300 | 4270 | 0.837 | 7.87 | 9.65 | 9.79 | 1.382 | 9.40 | 193.64 | 0.127 | 57.255 | 0.335 | 53.438 | 1.071 | C* |
| AK-3.23-300 | 4270 | 0.837 | 7.87 | 9.65 | 9.79 | 2.453 | 9.40 | 193.64 | 0.127 | 74.822 | 0.418 | 63.938 | 1.170 | C* |
| AK-0.81-780 | 11000 | 0.650 | 7.87 | 9.65 | 9.79 | 0.615 | 9.40 | 193.64 | 0.127 | 60.833 | 0.176 | 61.576 | 0.988 | CR |
| CK-0.81-300 | 4270 | 0.837 | 7.87 | 9.65 | 9.79 | 0.615 | 14.14 | 155.20 | 0.215 | 53.676 | 0.285 | 46.479 | 1.155 | CR* |
| GR-1.26-780 | 11000 | 0.650 | 7.87 | 9.65 | 9.79 | 0.957 | 4.33 | 78.30 | 0.142 | 58.231 | 0.151 | 53.344 | 1.092 | CR |
| **Al-Salloum et al.,1996** | | | | | | | | | | | | | | |
| group 2 | 4540 | 0.823 | 7.87 | 6.20 | 17.14 | 1.758 | 5.17 | 101.62 | 0.132 | 25.348 | 0.341 | 23.453 | 1.081 | C* |
| group 3 | 4540 | 0.823 | 7.87 | 8.30 | 12.81 | 0.786 | 6.29 | 128.60 | 0.128 | 35.949 | 0.234 | 30.350 | 1.184 | C* |
| **Al-Salloum et al.,1997** | | | | | | | | | | | | | | |
| comp-00 | 5132 | 0.793 | 7.87 | 7.50 | 13.12 | 0.786 | 6.29 | 128.44 | 0.128 | 30.311 | 0.236 | 27.259 | 1.112 | C* |
| comp-25 | 5132 | 0.793 | 7.87 | 7.50 | 13.12 | 0.786 | 6.29 | 128.44 | 0.128 | 28.615 | 0.236 | 27.259 | 1.050 | C* |
| comp-50 | 5291 | 0.785 | 7.87 | 7.50 | 13.12 | 0.786 | 6.29 | 128.44 | 0.128 | 28.827 | 0.233 | 27.631 | 1.043 | C* |
| comp-75 | 5291 | 0.785 | 7.87 | 7.50 | 13.12 | 0.786 | 6.29 | 128.44 | 0.128 | 36.034 | 0.233 | 27.631 | 1.304 | C |
| **Zhao et al., 1997** | | | | | | | | | | | | | | |
| GB1 | 4350 | 0.833 | 6.00 | 8.66 | 10.45 | 0.662 | 6.53 | 145.14 | 0.119 | 27.562 | 0.247 | 25.590 | 1.077 | C* |

GR = Fiber wound Glass bars, AK-Sand braided Aramid bars, CK-Sand braided carbon bars, CR-Fiber wound Carbon bars

comp-75, has shown high moment because of 75% comp. reinforcement

| | | | | | | | | | | | | | | |
|---|---|---|---|---|---|---|---|---|---|---|---|---|---|---|
| GB5 | 4530 | 0.824 | 6.00 | 8.66 | 10.45 | 0.662 | 6.53 | 145.14 | 0.119 | 29.719 | 0.244 | 26.096 | 1.139 | C* |
| GB9 | 5775 | 0.761 | 6.00 | 8.66 | 10.45 | 0.662 | 6.53 | 145.14 | 0.119 | 29.292 | 0.227 | 29.086 | 1.007 | C* |
| GB10 | 5775 | 0.761 | 6.00 | 8.66 | 10.45 | 0.662 | 6.53 | 145.14 | 0.119 | 29.131 | 0.227 | 29.086 | 1.002 | C* |
| **Brown and Barthalomew, 1996** | | | | | | | | | | | | | | |
| D-1 | 5080 | 0.796 | 4.00 | 4.50 | 14.67 | 0.221 | 6.00 | 80.00 | 0.184 | 4.860 | 0.223 | 4.723 | 1.029 | C* |
| D-2 | 5080 | 0.796 | 4.00 | 4.00 | 16.50 | 0.221 | 6.00 | 80.00 | 0.184 | 4.330 | 0.235 | 3.908 | 1.108 | C* |
| **Brown and Barthalomew, 1993** | | | | | | | | | | | | | | |
| 1 | 5200 | 0.790 | 6.00 | 4.81 | 5.39 | 0.110 | 6.50 | 130.00 | 0.130 | 5.193 | 0.136 | 5.196 | 1.000 | C* |
| 2 | 5200 | 0.790 | 6.00 | 4.81 | 5.39 | 0.110 | 6.50 | 130.00 | 0.130 | 4.900 | 0.136 | 5.196 | 0.943 | C* |
| 4 | 5200 | 0.790 | 6.00 | 4.81 | 5.39 | 0.110 | 6.50 | 130.00 | 0.130 | 5.330 | 0.136 | 5.196 | 1.026 | C* |
| 5 | 5200 | 0.790 | 6.00 | 4.81 | 5.39 | 0.110 | 6.50 | 130.00 | 0.130 | 5.423 | 0.136 | 5.196 | 1.044 | C* |
| 6 | 5200 | 0.790 | 6.00 | 4.81 | 5.39 | 0.110 | 6.50 | 130.00 | 0.130 | 4.977 | 0.136 | 5.196 | 0.958 | C* |

*Note*: above values are taken as an average of the range stated for strength, and realistic modulus value based on the range of 6 to 7 msi stiffness for a 70% fiber fraction.

**Theriault and Benmokrane, 1998**

| | | | | | | | | | | | | | | |
|---|---|---|---|---|---|---|---|---|---|---|---|---|---|---|
| BC2NB | 7698 | 0.665 | 5.12 | 6.51 | 9.07 | 0.368 | 5.52 | 112.62 | 0.128 | 14.736 | 0.185 | 13.649 | 1.080 | C* |
| BC2HA | 8292 | 0.650 | 5.12 | 6.51 | 9.07 | 0.368 | 5.52 | 112.62 | 0.128 | 14.515 | 0.181 | 14.082 | 1.031 | C* |
| BC2HB | 8292 | 0.650 | 5.12 | 6.51 | 9.07 | 0.368 | 5.52 | 112.62 | 0.128 | 15.178 | 0.181 | 14.082 | 1.078 | C* |
| BC2VA | 14120 | 0.650 | 5.12 | 6.51 | 9.07 | 0.368 | 5.52 | 112.62 | 0.128 | 16.725 | 0.142 | 19.061 | 0.877 | C |
| BC4NB | 6698 | 0.715 | 5.12 | 5.32 | 11.10 | 0.736 | 5.52 | 112.62 | 0.128 | 15.178 | 0.281 | 12.425 | 1.222 | C* |
| BC4HA | 7814 | 0.659 | 5.12 | 5.32 | 11.10 | 0.736 | 5.52 | 112.62 | 0.128 | 15.743 | 0.273 | 13.115 | 1.200 | C* |
| BC4HB | 7814 | 0.659 | 5.12 | 5.32 | 11.10 | 0.736 | 5.52 | 112.62 | 0.128 | 15.768 | 0.273 | 13.115 | 1.202 | C* |
| BC4VA | 13555 | 0.650 | 5.12 | 5.32 | 11.10 | 0.736 | 5.52 | 112.62 | 0.128 | 20.925 | 0.216 | 18.184 | 1.151 | C |
| BC4VB | 13555 | 0.650 | 5.12 | 5.32 | 11.10 | 0.736 | 5.52 | 112.62 | 0.128 | 21.736 | 0.216 | 18.184 | 1.195 | C |

*Continued.*

| Beam | $f'_c$ psi | $\beta_1$ | $b$ (in) | $d$ (in) | $l/d$ | $A_f$ | $E_f$ ($\times 10^6$ psi) | $f_f$ ksi | $(c/d)bal$ | Moment Calculations for Comp. Failure | | | | Actual Failure |
|---|---|---|---|---|---|---|---|---|---|---|---|---|---|---|
| | | | | | | | | | | (expt.) k-ft | $c/d$ | $M_{th}$ lb-ft | $M_{expt}/M_{th}$ | |
| **Benmokrane and Masmoudi, 1996** | | | | | | | | | | | | | | |
| Series-1 | 7550 | 0.673 | 7.87 | 10.33 | | 0.455 | 5.47 | 112.62 | 0.127 | 43.410 | 0.136 | 39.111 | 1.110 | C* |
| Series-2 | 7550 | 0.673 | 7.87 | 10.33 | | 0.740 | 5.47 | 112.62 | 0.127 | 48.130 | 0.170 | 48.287 | 0.997 | C* |
| Series-3 | 6530 | 0.724 | 7.87 | 9.45 | | 1.122 | 5.47 | 112.62 | 0.127 | 54.170 | 0.219 | 47.505 | 1.140 | C* |
| Series-4 | 6530 | 0.724 | 7.87 | 9.45 | | 1.748 | 5.47 | 112.62 | 0.127 | 62.720 | 0.266 | 56.469 | 1.111 | C* |
| **Vijay and GangaRao, 1996.** | | | | | | | | | | | | | | |
| C2 | 6500 | 0.725 | 6.00 | 10.63 | 10.16 | 0.884 | 5.50 | 85.00 | 0.163 | 40.348 | 0.212 | 44.274 | 0.911 | C* |
| C4 | 6500 | 0.725 | 6.00 | 10.38 | 10.41 | 0.884 | 5.50 | 85.00 | 0.163 | 35.636 | 0.214 | 42.623 | 0.836 | C* |
| M1 | 4500 | 0.825 | 6.00 | 10.69 | 10.10 | 0.614 | 5.50 | 81.00 | 0.169 | 36.741 | 0.200 | 33.097 | 1.110 | C* |
| M2 | 4500 | 0.825 | 6.00 | 10.38 | 10.41 | 1.228 | 5.50 | 81.00 | 0.169 | 39.101 | 0.274 | 41.233 | 0.948 | C* |
| **Swamy and Aburawi, 1997** | | | | | | | | | | | | | | |
| F-1-GF | 6089 | 0.746 | 6.06 | 9.00 | 9.19 | 0.823 | 4.93 | 84.95 | 0.148 | 36.900 | 0.213 | 30.993 | 1.191 | C |
| **Cosenza et al, 1997** | | | | | | | | | | | | | | |
| | 4349 | 0.833 | 19.68 | 5.71 | 23.45 | 1.375 | 6.09 | 128.44 | 0.125 | 42.400 | 0.236 | 34.957 | 1.213 | C |
| Total Beams = 64 | | | | | | | | | | | | | | |
| Average | | | | | | | | | | | | | 1.068* | |
| Std. Dev. | | | | | | | | | | | | | 0.089* | |

* Only those results that are considered for calculation of average value and standard deviation (results published after 1993 or later with concrete strengths between 27.5 to 68.9 MPa (4 to 10 ksi), and at least two published results for the compression failure mode by the author.)

B: bending; S: shear; C: compression; T-tension; R-rupture; T/C: Both tension bar rupture and concrete crushing; N: not specified/not clear.

# Index

reinforced polymers (CFRP), 6, 103–104,
    108, 109–110
    design for flexural strengthening, 151–190
    sizing with, 56–57
Carbon particles, 12
C-bars, 88–89
Cement, 2, 20–21, 103. *See also* Concrete
C-glass, 49
Chemical aging, 83–84
Chemical vapor deposition (CVD), 54
Chlorendics, 40
Chord modulus, 26
City Hall of Gossau, St. Gall, Switzerland, 108
Civil Engineering Research Foundation (CERF),
    96
Classification
    of carbon fibers, 51
    of concrete, 21–22
    of resins, 4–5
Clay particles, 12
Clean-up guidelines, resin, 120
Coarse aggregate, 20
Column wrapping
    corrosion prevention, 109
    Hotel Nikko, Beverly Hills, California, USA,
        108
    for seismic resistance by Caltrans, 108–109
Compatibility, strain, 137
Composites
    advantages of, 17–18, 19–20
    basic anatomy of, 11–13
    cement, 2, 20–21
    classification of, 6
    constituents, 1–2
    defined, 1–2
    fibers used in, 3, 11–13, 16
    fillers used in, 12–13
    manufacturing, 63–76
    mechanical and thermal properties of, 4,
        10–11
    naturally occurring, 2
    plastic, 2
    pultrusion, 14–15
    resins used in, 4, 11–13
    wood, 2
Compression failure, 132–133, 226–228, 244,
    245f
    tension and, 147, 148f
Compression molding, 73, 74f
Computation of deflection and crack width, 151,
    166–171, 180–181, 190
Computer Numeric Control (CNC) machines, 65
Concrete
    admixtures, 21
    age-strength relationships, 23–24

aggregates in, 20
bar location and strength of, 273
bonding with FRP-ER, 104–105
cement in, 2, 20–21
cover, 273
curing of, 23–24
flexural strength of, 27–28, 125
ingredients of, 2, 20–21
modulus of elasticity of, 25–27
nominal flexural strength of singly reinforced
    beams of, 143–150
Poisson's ratio, 28–29
shear behavior of wrapped, 191–195
shear modulus of, 29
shear strength contribution in FRP-reinforced
    beams, 248–249
splitting strength of, 28
steel-reinforced, 106–107, 118
strength, 22–24
    characteristics, 24–29
stress-strain relationship, 24–25
tensile strength, 27–28
types of, 21–22
unit weight, 22
Constructed Facilities Center (CFC), 66, 73
Contamination of fillers, 55–56
Continuous mandrels, 68
Continuous winders, 68
Corrosion prevention, 109
Cotton flock, 47t
Coupling agents, 84–85
Cover, concrete, 273
Crack width
    ACI provisions for, 319–321
    deflection and, 151, 166–171, 180–181, 190,
        317–318
    flexural strength and, 322–340
Creep
    concept of, 340
    /relaxation, 89–90
    -rupture in FRP-ER, 130–131, 163, 340–341
        stress limits, 341–345
Crosslinking, 35
Cured phase of resins, 4
Cure time, polymer, 33
Curing
    concrete, 23–24
    resin, 65

# D

Deflections
    of beams reinforced with GFRP bars, 290–319
    calculation of, 288–289